T0231145

VC++ MFC Extensions by Example

John E. Swanke

CRC Press
Taylor & Francis Group
Boca Raton London New York

CRC Press is an imprint of the
Taylor & Francis Group, an **informa** business

CRC Press
Taylor & Francis Group
6000 Broken Sound Parkway NW, Suite 300
Boca Raton, FL 33487-2742

First issued in hardback 2017

© 1999 by John Swanke
CRC Press is an imprint of Taylor & Francis Group, an Informa business

No claim to original U.S. Government works

ISBN 13: 978-1-138-41240-8 (hbk)
ISBN 13: 978-0-87930-588-8 (pbk)

Visit the Taylor & Francis Web site at
http://www.taylorandfrancis.com

and the CRC Press Web site at
http://www.crcpress.com

Acquisitions Editor: Berney Williams
Editor: Liza Niav
Layout Production: Michelle o'Neal and Kris Peaslee
Cover Art Design: John Freeman and Robert Ward

To the only one for me,
my wife,
Cathy Krinitsky

Acknowledgments

I would like to thank Paul Swanke, one of my many brothers (Roy, Karl, Mark, Rob, Kurt and one sister, Jeanne), for keeping me honest. Every time I think I know all there is to know about something, he sends me an e-mail that changes everything.

I would also like to thank John Bradberry for his priceless advice. As someone relatively new to Gate's stuff, I get a perspective in which there are now only calluses. Our careers just seem to be synergistically entwined. And we don't even talk about Adams' group anymore.

I would like to thank Berney Williams, my acquisitions editor and coach at CMP Books. If you have even an inkling of writing a book, drop him a line — they don't come any better.

I would also like to thank Liza Niav for not only untying my language knots but for being able to pick the right font at the same time — the talent boggles the mind.

I would like to thank Michelle O'Neal for not only tying everything together but for doing it with style. It's such a shame I have to spend months and months writing a book just so I can work with her a few days.

And finally I would like to again thank my wife, Cathy Krinitsky, without whom this occasion would not be possible. She's the pearl in my oyster, the scent in my candle, the nugget in my truffle, the fizz in my champagne.

Table of Contents

Section I

Chapter 13   Messaging and Communication . . . 467

Chapter 14   Multitasking 521

Chapter 15   Potpourri. 547

Appendix A   Message and Override
Sequences 569

What's on the Downloadable Files? 654

The files can be downloaded from:

ftp://ftp.cmpbooks.com/pub/VC_MFC_Extensions.zip

List of Figures

Introduction

For getting a quick start on a problem, I have found nothing beats a good example—one that's general purpose enough to work in any application, yet not so general that it's filled with loose ends. The code for the example can't be buried deep inside another project with no clue as to where the example begins and hundreds of other examples end. With this book, I tried to choose good examples and presented them as a collection of time-saving code snippets. I hope this format will be both informative to an MFC novice and valuable to a proficient MFC programmer.

The subjects and examples presented in this book are actually a continuation of ideas from an earlier book, *Visual C++ MFC Programming by Example*. Although the current book is intended to stand on its own, you might find the content of the earlier book just as useful.

What's in Store

The examples in this book are grouped into chapters that cover several different aspects of an MFC application, from interface issues to application control. Similar chapters have been organized into one of the following sections.

Basics

Although I tried to make this book example-oriented, I found that knowing just a few of the basics up front can help make understanding something later much easier. You can certainly skip the first section if you like, but don't be surprised if I refer to it in later sections.

User Interface Examples

The examples in this section concentrate on the user interface of an application. Because MFC is interface intensive, the vast majority of the examples in this book will be in this section. Topics include menus, toolbars, status bars, views, dialog boxes and bars, control windows, and some basic examples of creating plain windows and drawing. This section is topped off with some example applications, including text editors and wizards.

Internal Processing Examples

The examples in this section, while still applying to most MFC applications, are more representative of the unseen, non-user interface part of MFC. Topics include messaging, communication and timers, making sounds, and binary strings.

About the Downloadable Files

Downloadable from ftp://ftp.cmpbooks.com/pub/VC_MFC_Extensions.zip is a working Visual C++ v5.0 and v6.0 project for every example in this book. If you want to find the project that corresponds to a particular book example, just locate its number among the directory names on the CD or downloadable files. Except where noted, most of the downloadable examples were created as an MDI application using all of the AppWizard defaults and a project name of "Wzd".

About the SampleWizard

Also downloadable is the SampleWizard utility, which can help you add the examples in this book directly to your applications. This utility guides you through a catalog of examples. When an individual example is selected, the instructions and code necessary for including the example in your project are listed. You will also be given the opportunity to substitute the example's project name ("Wzd") with your own.

SampleWizard can be found in the \SWD directory on the downloadable files. SampleWizard makes use of the \Wizard subdirectory found in each example's directory on the downloadable files and contains all of the particulars for that example. Simply execute SW.EXE. The rest should be intuitive. I have found it particularly useful as a user tool in the Developer Studio. Make sure to make the directory the current project directory and these examples will be copieddirectly into your project directory.

Basics

Whether you got here by reading the first book in this series or by years of experience, the purpose of this section is to review what you need to know to understand the examples in this book. Often, programming is simply a matter of trying different approaches until something works. Although this usually gets the job done (and is sometimes unavoidable in the absence of time or documentation), it isn't as rewarding or efficient as knowing what to do the first time. Knowing what to do with MFC usually involves an understanding of four basics: how the Windows API creates a window, how MFC wraps and improves on the Windows API, how MFC communicates with a window, and how MFC draws. Beyond these basics, this section also takes a look at toolbars and status bars, which are not what they appear, and how MFC communicates with things that aren't windows (e.g., serial ports and Internet sites).

The chapters in this section include the following.

Overview

Chapter 1 reviews how MFC wraps and improves on the Windows API. If you've read the first book in this series you will find this chapter to be a quick overview of the Basics there. This chapter was included so that this book could stand on its own for an advanced reader. However, if you find a discussion here too brief, please refer to the first book.

Control Bars

In Chapter 2, we take a look at the control bars supported by MFC. Standard control bars include toolbars, status bars, and rebars. MFC adds to this mix with dialog bars and docking bars. We also take a look at the mysteries of how MFC keeps your control bars from overlapping each other and the view.

Communication

Chapter 3 looks at the different ways your application can communicate with the outside world. The most basic of these, sending window messages, is discussed in Chapter 1. Chapter 3 examines the other avenues, including: LAN and Internet communication; serial and parallel ports; DDE; and window hooks and pipes.

Chapter 1

Overview

In this chapter, we will review the nature of a Windows application—how it creates its windows, how those windows talk to each other and how to draw in those windows. We will then look at how the Microsoft Foundation Classes (MFC) and the Developer Studio make creating a Windows application much easier.

The Nature of Windows

When the Windows operating system starts an application, it begins by creating a program thread, which is simply an administrative chunk of executable memory that shares execution time with other applications in the system. If this application will be interacting with a user through the screen, it is the responsibility of this program thread to create windows on the screen.

The program thread creates these windows by calling the operating system's Application Program Interface (API). The actual function name is `::CreateWindowEx()`, which requires (among other things): a screen position, a window size, and the style of window to create.

Window Classes

Since a lot of the windows created by a thread will share the same look (e.g., all of the buttons drawn by your application), these similar styles have been lumped together into structures called Window Classes. Notice that these are structures and not C++ classes. A Window Class must be specified when creating a window.

Messaging

The operating system sends a window message to a window to tell that window that it's been clicked by the user. Each window processes a window message with its very own window procedure. For example, the window procedure for a button might turn around and send a message to its application's main window to tell it to do something.

Each window's procedure is also responsible for actually drawing its window on the screen. The operating system will send the window a WM_PAINT message when it's time to be drawn.

All similar windows share the same window procedure. For example, all button controls use the same exact window procedure so that all buttons look and act the same. In this case, the window procedure is located in the operating system; the address is specified in a window's Window Class structure. All button controls are created with the same Window Class, which is itself called BUTTON.

Client and Nonclient Area

When a window procedure draws a window on the screen, it draws two parts: a client area and a nonclient area.

To draw the nonclient area, a window procedure always calls on the same operating system procedure that all windows call. This operating system procedure will then draw the familiar frame, menu bar, and caption bar that are common to most windows. The type of nonclient area drawn by the operating system procedure depends on the style of the window. For example, you don't see the frame, menubar, or caption bar around a button window, because its style specified that no nonclient area was to be drawn for it.

The client area of a window is always drawn by that window's own window procedure. The procedure can still be supplied by the operating system, as we saw with button controls, so that windows of the same type are all drawn by the same procedure. The procedure can also be supplied by you to draw graphic figures or listings.

Overlapped, Popup, and Child Windows

Beyond the Window Class, hundreds of other window styles can be specified for a window, usually to affect how the window is drawn or how it acts. Three styles create the three most basic window types: overlapped, popup, and child.

Overlapped windows have all of the characteristics of an application's main window. The nonclient area includes a resizable frame, a menu bar, a caption bar, and minimizing and maximizing buttons.

Popup windows have all of the characteristics of a message box or dialog box. The nonclient area includes a nonresizable frame and a caption bar.

Child windows have all of the characteristics of a control (e.g., a button). There is no nonclient area—the window procedure of the window is expected to draw everything.

These windows also behave differently, as is discussed in the following text.

Parent and Owner Windows

Since a user interface can be made up of dozens of windows, controlling them all from the program thread would be a nightmare. For example, if a user minimizes an application, should the program thread be responsible for minimizing every window that makes up the interface? Instead of direct control, every window an application creates can be assigned a controlling window through the `::CreateWindowEx()` call. If the controlling window is then minimized, all controlled windows are also minimized automatically by the operating system. If the controller is destroyed, so is every controlled window.

Each controlled window can also be the controller of another window, such that minimizing or destroying a window will only affect that part of the user interface. No matter what or where the window is, you can create another window inside of it.

The controlling window for a child window is called a *parent window*. A parent window will also *clip* its child windows, meaning a child window cannot draw outside of its parent. Child windows are typically controls, such as buttons, that automatically generate a message to their parent window when the user interacts with the child window. This allows several controls to be processed in one centralized location—the parent window's window procedure.

The controlling window for a popup or overlapped window is called an *owner window.* Unlike a parent window, an owner window does not clip its owned windows. When an owner window is minimized, so are its owned windows. However, when a an owner window is hidden, an owned window will still appear.

Please refer to Figure 1.1 to see what windows make up a Windows application.

Figure 1.1   The Windows That Comprise a Windows Application Interface

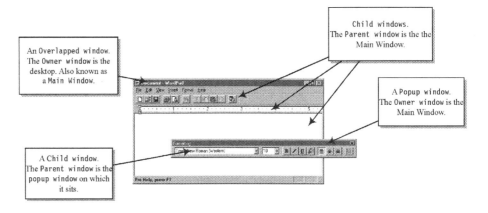

Windows Messaging

As mentioned before, each window has its own window procedure that responds to messages from the operating system or other windows. The operating system, for example, might send a message to a window that the user just clicked that window with the mouse cursor. If this was a button window labeled "Load File", its window procedure might respond by sending a message to the application's main window procedure to load a file. The main window procedure might then respond by loading the file and drawing the file contents in its client area.

Sending or Posting a Message

A message can be transmitted to a window by either sending or posting. Sent messages are processed immediately, while posted messages are placed in a queue to be handled FIFO whenever the application is idle.

There's little difference between sending a message to a window procedure and calling that window procedure directly as a routine. However, you can ask the operating system to intercept all sent messages in your application for interesting effects. You can't easily intercept a direct call to a window procedure.

Messages relating to user input (e.g., mouse and key clicks) are usually posted so that this input can be buffered by a slow system. All other messages are sent. In the previous example, the system posted the mouse click message, while the button window sent the "Load File" message to the main window.

Message Types

There are three types of messages: window, command, and control notification.

Window messages are those types of messages that are used by the operating system and other windows to control a window. They include messages like "Create", "Destroy" and "Move". The mouse click message in the previous example was a window message.

Command messages are a special type of window message that are sent from one window to another to process a request from the user. In our example, the message from the button window to the main window was a command message.

Control notifications, like command messages, are sent by a control window to a parent window when the user interacts with it. However, control notifications are not sent with the intention of performing a command for the user, but as a way to allow the parent window to change the control, such as loading it with more data to display. Our example didn't have a control notification.

Receiving Messages

A window procedure doesn't look that much different from any other function or method. Messages come in, are sorted by type, and are then processed. Parameters supplied by the calling function further distinguish the message. Command messages are further sorted by command ID, which is in wParam. The DefWindowProc() function sends any messages you don't want to process to the operating system. All messages to a window pass through here. Even messages to paint the nonclient area of a window pass through here, although these are usually passed along to DefWindowProc().

A sample window procedure for a main window follows.

```
MainWndProc( HWND hWnd,UINT message,WPARAM wParam,LPARAM lParam )
{
    switch( message )
    {
    case WM_CREATE:
         :    :    :
             break;
    case WM_PAINT:
         :    :    :
             break;
    case WM_COMMAND:
        switch (id)
        {
            case IDC_LOAD_FILE:
                 :    :    :
                 break;
        }
             break;
    default:
        return( DefWindowProc( hWnd, message, wParam, lParam ) );
    }
    return( NULL );
}
```

Subclassing a Window Procedure

As mentioned earlier, you define the address of a window procedure in the Window Class. Any windows created from that Window Class then pass all of their messages to that procedure. If you need to add your own special processing to a window when you are using a Window Class provided by the system, you will have to use subclassing.

You subclass a window by pointing it to your window procedure so that all messages can be processed by you. If you want to process only one or two messages, simply pass the rest onto the original window procedure. Note that with subclassing, you are not modifying the original Window Class. Instead, you are directly modifying the window object, which keeps a copy of the window procedure address.

In contrast, superclassing does involve modifying the original Window Class, which is then used to create windows. However, superclassing is rarely used in MFC applications because the benefits can be more easily and safely achieved with MFC.

Please refer to Figure 1.2 for an overview of subclassing.

Figure 1.2   Subclassing a Window Process

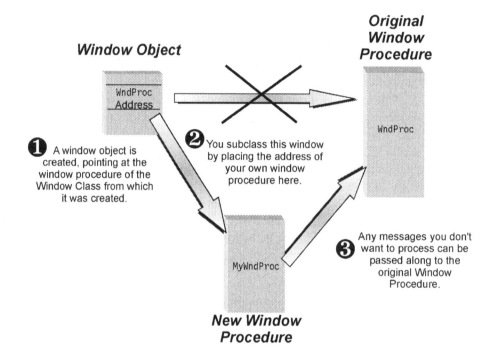

Windows Drawing

The Windows API provides several calls for drawing. There are point, arc, line, shape, and circle drawing functions, as well as functions to fill shapes and draw bitmaps. As with any graphics API, you pass the dimensions, color, width, and location of the window to the functions, which take care of the rest. To simplify graphic calls to the Windows API, several arguments have been consolidated into one reusable object called a device context.

Device Contexts

A graphic context is an object that contains several characteristics common to drawing, including the location, the line width, and the fill-pattern color. This object can be set up once and used repeatedly. You don't actually create a device context yourself. Instead, you use one of several possible API calls and the system returns one with ready-to-use values. For example, a device context created by the system for the screen will contain the current location of the drawing tool on the screen, where you can draw on the screen, what color you will be drawing, etc.

Drawing Tools

A device context does not, however, contain all of the characteristics needed for drawing. Several characteristics exist in additional graphic objects that the device context references. Each of these objects represents the drawing characteristics of a particular drawing tool (e.g., color and width for a pen or patterns for a brush). These tools include: pens, which draw a straight line; brushes, which fill enclosed areas with a pattern; fonts, which determine how your drawn text will appear; and palettes, which determine what colors you will be using. Two other drawing tools—bitmap and region—are somewhat more abstract. A bitmap tool is like a brush tool, except that it can only fill an area with a bitmap pattern. And the region tool is like a pen tool, except that it cuts instead of draws. For example, you can use the region tool to cut the word "STOP" from a bitmap image for an interesting effect.

Mapping Modes

The device context also keeps track of your mapping mode. By setting a mapping mode, you can specify your x and y calling arguments in inches or centimeters and each drawing function will automatically determine how many pixels to draw. The available mapping modes are seen in Table 1.1.

Table 1.1   Available Mapping Modes

Mode	Usage Description
MM_TEXT	This is the default mapping mode. The value in x and y is exactly equivalent to one (1) screen pixel or printer dot and a positive y goes down the screen or printed page.
MM_HIENGLISH	The value in x and y is equivalent to 1/1000 of an inch on the screen or printed page. Windows determines how many pixels are required for the current screen device to equal 1/1000 of an inch. A positive y goes up the screen or printed page.

Mode	Usage Description
MM_LOENGLISH	The value in x and y is equivalent to 1/100 of an inch on the device and y goes up.
MM_HIMETRIC	The value in x and y is equivalent to 1/100 of a millimeter on the device and y goes up.
MM_LOMETRIC	The value in x and y is equivalent to 1/10 of a millimeter on the device and y goes up.
MM_TWIPS	The value in x and y is equivalent to 1/1440 of an inch on the device and y goes up. This is usually used for text drawing—one twip is equivalent to $1/20^{th}$ of a font point.
MM_ANISOTROPIC	You determine how many pixels x and y represent by setting up a Window View and a Viewport View (described in the following section).
MM_ISOTROPIC	Same as MM_ANISOTROPIC, except that both x and y must create the same number of pixels.

Windows View vs. Viewport View

The MM_ANISOTROPIC and MM_ISOTROPIC mapping modes allow you to define your own conversion ratios for converting x and y coordinates into pixels. This is done by defining two rectangles called views. First, you define a rectangle that represents the total area in which you will be drawing (e.g., 0,0,1000,1000). Next, you define a rectangle that would represent those same exact coordinates on the screen or printer where your drawing will eventually appear (e.g., 0,0,500,500). The first rectangle is called a Window View and the second rectangle is called a Viewport View. When you define both of these rectangles in your device context and use one of these mapping modes, you can blindly draw using Window View coordinates and they will automatically be converted into Viewport View coordinates. You can also zoom in, pan out, and invert your drawing just by changing the Viewport View coordinates.

Logical Units vs. Device Units

When drawing in a mapping mode other than MM_TEXT, the coordinates you pass to a drawing function are in Logical Units. Logical Units can be in inches, centimeters, or pixels. The drawing function itself draws in device units. For example, a line drawing function might draw 254 pixels to represent a logical unit of 1 inch. The number 1 is, therefore, in logical units and the number 254 is in device units. This doesn't become an issue unless you want your user to be able to interact with the drawing using a mouse. The coordinates passed back to your application by a mouse will be in device units and, therefore, you must convert those coordinates back into logical units using yet another Windows API call.

Drawing Functions

The Windows API has a multitude of drawing functions. A sampling of these follows.

- Functions to draw a point [e.g., `SetPixel()`].
- Functions to draw lines [e.g., `LineTo()`, `Arc()`, and `Polyline()`].
- Functions to draw shapes [e.g., `Rectangle()`, `Polygon()`, and `Ellipse()`].
- Functions to fill and invert shapes [e.g., `FillRect()`, `InvertRect()`, and `FillRgn()`].
- A function to scroll the screen [`ScrollDC()`].
- Functions to draw text [e.g., `TextOut()` and `DrawText()`].
- Functions to draw bitmaps and icons [e.g., `DrawIcon()` and `BitBlt()`].

Dithered vs. Nondithered Colors

All of the drawing functions available can draw in color, from lines to shapes to text. However, unless your system has lots of video memory, chances are your colors will be dithered. A dithered color is actually a collection of other primary colors that, when displayed by several closely knit pixels, gives the appearance of the desired color.

For most purposes, dithered colors are okay. However, since dithered colors appear somewhat blurred, they usually aren't okay for a graphics application where lines can't bleed into the shapes they encompass. Your choices for a graphic application are as follows.

Get lots of video memory.   Dithering is required because although every pixel on a screen has its very own RGB color, there usually isn't enough video memory on a typical system to store a unique RGB value for every pixel. For example, at 32 bits per pixel, a 800 by 600 pixel screen would need almost 2Mb of video memory to contain all the color values.

Draw only in standard colors. A video card for Windows is guaranteed to define at least 20 standard colors because it uses those colors to create the dithering effect.

Configure your own colors. Beyond the standard colors, a video card has room to define over 200 more colors. You can define these colors in the device context palette and draw in just those colors. Most graphic applications follow this approach, which is described in more detail in Example 31 and in my previous book, *Visual C++ MFC Programming by Example*.

Device Independent Bitmaps

Each pixel, therefore, can have its very own RGB color, where each color — red, green and blue — can be in the range 0–254. A row of pixels with the same color (e.g., black) would appear as a line on the screen. A rectangle of pixels with different colors can create any picture. If you were to store the color values of each of these pixels to a file, you would have a bitmap file. The header of this file would not only indicate how many color values the file contained, but also how many values made up a row.

If each of the color values in that bitmap file contained a full RGB value, it would be a Device Independent Bitmap, because its colors are fully defined within the bitmap. If each of the color values were actually byte indexes into a color table, it would still be Device Independent if it also contained the color table. Color indexing like this is used to shrink the size of a bitmap. A series of 8-bit indexes takes up one-fourth of the space of a 32-bit RGB value.

In contrast, a Device Dependent Bitmap is made up of indexes into the color table that is defined in a system's video card at some point in time. This is the color table you configure in the device context palette to display nondithered colors. If a bitmap points into this color table, it cannot exist outside of that device.

Metafiles

You can also draw to a file, called a metafile. A metafile can be "replayed" at some future point like a graphic batch file. Metafiles are also preferable to storing a picture in a bitmap because they can be stretched with better results than a bitmap, which can get distorted.

When to Draw

This may seem like a silly point, but on a multitasking system in which your application is jostling for space on the screen with other applications, you can't draw at just any time—even though you are only drawing to your own window. Usually, you will only draw when your window receives the WM_PAINT message or WM_DRAWITEM for an owner drawn control window. The system sends this message only when your window is at least partially visible and has just been exposed because either another window closed or new information needs to be drawn.

If you draw at any other time, whatever you draw is likely to be drawn over by someone else's drawing routine or even your own routine the next time it recieves a WM_PAINT or WM_ERASEBKGND message.

The Nature of MFC

So far, we have only discussed the functionality that is available to your application from the Windows API. This API is not object-oriented. You can't, for example, create an instance of a window and call its member functions to act on that window. Neither can you derive a class from a Window Class to add your own functionality (e.g., add your own window procedure).

The Microsoft Foundation Classes provide your C++ application with a simulated object-oriented access to the Windows API. Functionally, each MFC class centers around a Windows resource object, such as a window, and the API functions that control that resource.

For example, the MFC CWnd class creates and controls a window. Whenever the operating system creates a window, it creates an administrative chunk of memory called a window object and returns a pointer to that object called a window handle. The MFC CWnd class stores this handle as a member variable, which is then used by each member function of CWnd to control that window. For example, CWnd's MoveWindow() member function calls the Windows API ::MoveWindow() to move the window belonging to that window handle. Because MFC is written in C++, you can derive your own class from CWnd to add functionality to it.

However, because this is only a simulation of object-oriented design, an MFC class has no more control over the internal workings of the operating system than any other API call. You can't, for example, modify the way a window opens if that functionality is internal to the API.

Creating and Destroying MFC Classes

Creating an MFC class is more involved than creating a plain C++ class. For MFC classes that wrap a system resource like a window, not only do you need to create an instance of your MFC class, but you also need to call a member function of that class to create the system resource. Therefore, creation of an MFC class is almost always a two-step process.

1. Create a class instance
2. Create the system resource

Note: MFC classes don't simply create the system resource in their own class constructors because the successful creation of a system resource isn't a forgone conclusion. Since a class constructor is notoriously hard to access (it doesn't even return an error status), a member function is much easier to work with.

Destroying an MFC class can be equally involved. If the class instance is destroyed first, it can simply destroy the resource in its class destructor. However, if the resource goes first, the Windows API has no way to know that there's an instance of an MFC class that must also be destroyed. Surprisingly, this is only a problem between CWnd classes and their derivations and window resources. Other types of resources can't be destroyed by the user. Ex: your user can't click a button and destroy a device context.

However, a user can close a window by clicking on its close button. When this happens, the CWnd object must also know enough to destroy itself to prevent memory leaks. Luckily, when a window resource is destroyed, it sends a windows message that a CWnd object can intercept and use to destroy itself, too.

Because an MFC class object and a system resource are two separate entities, you can also programmatically detach and reattach the two. For example, you can attach a CWnd class object to a window object that already exists by using Attach(). All MFC classes that control a system resource have both Attach() and Detach().

The Nature of the Developer Studio

To help orchestrate MFC classes into an application, the Developer Studio provides several wizards and editors.

The AppWizard or Application Wizard can be used to generate the basic class files you need for an application. The classes generated are all derived from MFC classes. After being compiled, they're linked to the MFC library to create the application.

The ClassWizard is used to create additional class files for an application or to add new member functions to an existing class. The created classes can also be derived from MFC.

The Dialog Editor is used to create dialog box templates by dragging control window icons into a blank frame. The created template is saved as a resource in an application. This resource can then be used to create a dialog box at run time. The ClassWizard can be invoked directly from the Dialog Editor to create a dialog class that can create this dialog box.

The Toolbar Editor is used to create a toolbar and bitmap resource that can be used to create an application toolbar.

The Cursor, Icon, and Bitmap Editors are simple graphic editors used to create cursors, icons, and bitmap resources that can be used by the application.

The Menu Editor is used to create menu bar and popup menu resources that can be used by the application.

The String Editor is used to create string resources so that text strings can be separated from an application and easily converted from one language to another (i.e., English to French, not C++ to Java).

The Text Editor obviously allows you to edit class files.

Windows and MFC Summary

To review the relationships, this is how Windows, MFC, and the Developer Studio all work together. The Windows operating system creates and supports your application (including window creation), MFC wraps this functionality in C++ classes, and the Developer Studio helps you create these classes.

We will now review just what classes MFC has to offer.

The Basic Classes

Most MFC classes are derived from three base classes: CObject, CCmdTarget, and CWnd. As we saw before, the CWnd class wraps the Windows API for creating and controlling a window. It also allows you to add your own message processing to a window procedure. The CCmdTarget class allows classes that don't create a window to process messages too, but only command messages, which we will discuss later. The CObject class provides a lot of basic functionality to its derived classes, from getting the size of a class object to saving that object out to a disk file.

CObject

The CObject class itself provides precious little functionality. Six other companion macros do the brunt of the work. Together, they allow classes derived from CObject to get their class's name and object size at run time, create an object of this class without knowing the class's name, and store and retrieve an instance of this class to an archive device without knowing its name.

The following macros allow an instance of a class to know its class name and object size at run time.

```
DECLARE_DYNAMIC( CYourClass )                    // in the .h file
IMPLEMENT_DYNAMIC( CYourClass, CYourBaseClass )   // in the .cpp file
```

Use CObject::GetRuntimeClass() to get the particulars of a class at run time that uses these macros.

These next macros include the functionality of the preceding macros but also allow a class instance to be created without knowing its class name:

```
DECLARE_DYNCREATE( CYourClass )                    // in the .h file
IMPLEMENT_DYNCREATE( CYourClass, CYourBaseClass )   // in the .cpp file
```

Use CObject::CreateObject() to create an instance of a class that uses these macros without knowing its class name.

These next macros include all of the preceding functionality, but also allow a class instance to be saved to disk without knowing its class name:

```
DECLARE_SERIAL ( CYourClass )     // in the .h file
IMPLEMENT_SERIAL ( CYourClass , CYourBaseClass , schema )
                              // in the .cpp file
```

CCmdTarget

Classes derived from CCmdTarget can receive and process command messages from your application's menu or toolbar(s). The CCmdTarget class is discussed in much more detail in the section "Class Messaging" on page 31.

CWnd

As discussed, the member functions of CWnd encapsulate the Windows API responsible for creating and maintaining windows. The CWnd class is derived from CCmdTarget and, therefore, can also receive and process command messages. All other MFC classes that control a window are derived from CWnd.

Note: The following letters are used in this chapter to designate from which of the preceding base classes an MFC class is derived.

- O indicates the class is derived from CObject.
- OC indicates a class is derived from CObject and CCmdTarget.
- OCW indicates a class is derived from CObject, CCmdTarget, and CWnd.

Application Classes

The AppWizard will generate several classes for your application derived from up to four of the following MFC classes.

CWinApp will become your application's **Application Class** and is responsible for initializing and running your application. This is the program thread we discussed previously.

CFrameWnd will become your application's **Frame Class** and is responsible for displaying and routing user commands and displaying the application's main window.

CDocument is your application's **Document Class** and is responsible for loading and maintaining a document. A document can be anything you want it to be from a manuscript to the settings of a network device.

CView is your application's **View Class** and is responsible for providing one or more views into the document.

Note: We will be using the term Application Class, Frame Class, etc., throughout this book to refer to your derivation of these four base classes.

Which of these base classes the AppWizard includes in your application depends on what type of application you're creating.

- A **Dialog Application** simply has a dialog box for its user interface and no frame, document, or view class. A Dialog Application uses just a derivation of the Application Class, CWinApp. The dialog box is created using MFC's CDialog class, which is discussed in the section "Dialog Classes" on page 25.

- A **Single Document Interface (SDI) Application** can load and edit one document at a time and uses all four base classes mentioned previously.

- A **Multiple Document Interface (MDI) Application** can load and edit several documents at once and uses all four base classes plus two derivations of CFrameWnd called CMDIFrameWnd and CMDIChildWnd.

Document View

Your derivations of both the CDocument and CView classes are responsible for the Document and View. MFC applications are "document-centric", meaning the application lives to load, view, edit, and save documents, whether those documents are text files, graphic shapes, or binary configuration files. The job of the Document class is to load a document from disk into its member variables. Then, one or more View classes are created to display those member variables. A document can have more than one view just by creating multiple View class objects for a Document class object. Because a Document class does not have an associated window, it is not derived from CWnd. It is, however, derived from CCmdTarget and, therefore, can process command messages.

CWinApp (OC)

The Application Class is the first object created when your application runs and the last to execute before it terminates. At startup, the Application Class is responsible for creating the rest of your application.

- For a Dialog Application, the Application Class creates a dialog box using CDialog.

- For an SDI application, the Application Class creates one or more Document Templates (see the following section) and then opens an empty document using that template.
- For an MDI application, the Application Class creates one or more Document Templates and then opens an empty document using that template, all within a Main Frame Class.

Your Application Class is derived from CWinApp and gets the name CXxx-App from the AppWizard, where Xxx is the name of your application.

Document Templates

A Document Template defines what Frame, Document, and View class to create when your application opens a document. To create a Document Template, you create an instance of either the CSingleDocTemplate class for SDI applications or CMultiDocTemplate class for MDI applications and initialize it with three class pointers.

```
pDocTemplate = new CMultiDocTemplate(
    IDR_APPTYPE,
    RUNTIME_CLASS( CAppDoc ),        // Your Document Class
    RUNTIME_CLASS( CChildFrame ),    // Your Frame Class
    RUNTIME_CLASS( CAppView )        // Your View Class
    );
```

The RUNTIME_CLASS() macro seen here returns a pointer to a class's CRuntimeClass structure, which is added to a class using the DECLARE_DYNCREATE and IMPLEMENT_DYNCREATE macros. A Document Template opens a document by creating an instance of all three of these classes using their CRuntimeClass::CreateObject() functions.

Threads

The CWinApp class is itself derived from CWinThread. The CWinThread class wraps the Windows API that creates and maintains application threads in your system. You can, in fact, multitask within your own application by creating another instance of this CWinThread class. Please see Examples 56 and 57. The CWinApp class represents the primary thread of execution in your application.

CFrameWnd (OCW)

The Frame Class is the next object created when your application runs and is responsible for displaying and directing user commands to the rest of your application.

For an SDI application, your Frame Class is derived from CFrameWnd and the AppWizard automatically assigns it the name CMainFrame.

For an MDI application, your Frame Class is derived from CMDIFrameWnd and the AppWizard again assigns it the name CMainFrame. MDI applications also have a **Child Frame Class** for each document that is opened. Each Child Frame Class is derived from CMDIChildWnd and the AppWizard automatically assigns a Child Frame Class the name CChildFrm.

There is no Frame Class for a Dialog Application. As mentioned earlier, a Dialog Application is composed of an Application Class and a Dialog Class.

CDocument (OC)

The Document Class is usually the next object created by your application, either to open a new document or to open an existing document. The Document Class is responsible for loading a document into its member variables and allowing the View Class to edit those member variables. A document can consist of anything from a graphic file to the settings of a programmable controller.

Your Document Class is derived from CDocument and the AppWizard automatically assigns it the name CXxxDoc, where Xxx is your application's name.

CView (OCW)

After an instance of the Document Class is created, an instance of the View Class is created. The View Class is responsible for depicting the contents of the Document Class. It might also allow your user to edit the document. A window splitter class, CSplitterWnd, allows your document to have more than one view at a time. These views can be created from several instances of the same View Class or from different View Classes entirely.

The AppWizard allows you to derive your View Class from one of several MFC base classes, including CTreeView, CEditView, CRichEditView, CListView, etc. Each of these base classes imparts a different set of functionality to your application. All of these classes are derived from the CView class. Whatever base class you choose, the AppWizard automatically assigns your derivation of it the name CXxxView, where Xxx is your application's name.

As mentioned previously, you can create three types of MFC applications from these four base classes: Dialog, SDI, and MDI. We will now review these application types in more detail.

A Dialog Application

A Dialog Application is made up of an Application Class, which is derived from `CWinApp`, and a dialog box, which is created using a class derived from the `CDialog` class.

An SDI Application

An SDI Application is made up of an Application Class derived from `CWinApp`, a Frame Class derived from `CFrameWnd`, a Document Class derived from `CDocument`, and one or more View Classes per document derived from one of several `CView`-derived View classes.

An MDI Application

An MDI Application is made up of an Application Class derived from `CWinApp`, a Frame Class derived from `CMDIFrameWnd`, one or more Child Frame Classes derived from `CMDIChildWnd`, a Document Class per child frame derived from `CDocument`, and one or more View Classes per document derived from `CView`.

Other User Interface Classes

Along with Frame and View Classes, MFC provides several other classes that support your user interface.

Common Control Classes wrap the common controls like buttons.

The Menu Class does for the menu what the `CWnd` class does for a window.

Dialog Classes wrap the dialog box and the Common Dialogs.

Control Bar Classes wrap the control bars (toolbars, dialog bars, status bar).

Property Classes wrap the Property Sheet and Page.

Common Control Classes (OCW)

The Common Control Classes encapsulate the functionality of the common controls (e.g., buttons, list boxes, etc.). These classes are derived from CWnd to inherit window member functions like ShowWindow() and MoveWindow(). When these classes create a window, they use one of the common control Window Classes. For example, when you use the CButton common control class to create a button, it uses the BUTTON Window Class to create the actual window.

```
Create( _T( "BUTTON" ), lpszCaption, dwStyle, rect, pParentWnd, nID );
```

Table 1.2 lists the Common Control Classes, the controls they create, and the Window Classes they use.

Table 1.2   Common Control MFC Classes and Their Windows Classes

MFC Class	Common Control	Windows Class
CAnimateCtrl(OCW)	Animation Control	SysAnimate32
CButton(OCW)	Button Control	BUTTON
CComboBox(OCW)	Combo Box Control	COMBOBOX
CEdit(OCW)	Edit Control	EDIT
CHeaderCtrl(OCW)	Header Control	SysHeader32
CListBox(OCW)	List Box Control	LISTBOX
CListCtrl(OCW)	List Control	SysListView32
CProgressCtrl(OCW)	Progress Control	msctls_progress32
CScrollBar(OCW)	Scroll Bar Control	SCROLLBAR
CSliderCtrl(OCW)	Slider Control	msctls_trackbar32
CSpinButtonCtrl(OCW)	Up/Down Button Control	msctls_updown32
CStatic(OCW)	Static Control	STATIC
CTreeCtrl(OCW)	Tree Control	SysTreeView32
CTabCtrl(OCW)	Tab Control	SysTabControl32
CDateTimeCtrl(OCW)	Date/Time Picker Control	SysDateTimePick32
CMonthCalCtrl(OCW)	Month Calendar Control	SysMonthCal32
CHotKeyCtrl(OCW)	Hot Key Control	msctls_hotkey32
CToolTipCtrl(OCW)	Tool Tip Control	tooltips_class32

Not all MFC Common Control Classes simply wrap a common control Window Class. Three MFC classes actually provide functionality not found in the common controls. Table 1.3 shows these classes, the MFC class from which they're derived, and the additional support they provide.

Table 1.3   Derived MFC Classes and Their MFC Base Class

MFC Class	MFC Class Derivation	Functionality Added
CBitmapButton	CButton	Better support for bitmaps on buttons.
CCheckListBox	CListBox	Check boxes in a list box.
CDragListBox	CListBox	User draggable items in a list box.

The Menu Class (0)

The CMenu class wraps the Windows API that creates and maintains menus. CMenu also has two member functions, Attach() and Detach(), that allow you to wrap an existing menu object the same way a CWnd object can wrap an existing window.

Dialog Classes

The CDialog class wraps the Windows API that creates dialog boxes. Dialog boxes are popup windows that, when created, can populate themselves with the control windows defined in a dialog template.

Common Dialog MFC Classes

The MFC library also has six classes that wrap the Windows APIs that creates the Common Dialogs. Common Dialogs are dialog boxes that come prepopulated with controls to prompt your user for some commonly requested information, such as filenames for loading and saving, colors, fonts, and printing parameters. They save you the job of writing these dialogs yourself, while also presenting your user with a familiar dialog box.

Table 1.4 shows the purpose of the Common Dialog, the Windows API that provides it, and the MFC Common Dialog class that wraps that.

Table 1.4   Common Dialog MFC Classes and Their Use

Common Dialog Box	Window API Call	MFC Class
Select Color	`::ChooseColor()`	`CColorDialog`
Open/Save File	`::GetOpenFileName()` `::GetSaveFileName()`	`CFileDialog`
Find or Replace Text	`::FindText()` `::ReplaceText()`	`CFindReplaceDialog`
Select a Type Font	`::ChooseFont()`	`CFontDialog`
Setup Print Page	`::PageSetupDlg()`	`CPageSetupDialog`
Print	`::PrintDlg()`	`CPrintDialog`

For an example of using the Common Dialogs, please see Examples 23 and 24.

Control Bar Classes (OCW)

The Control Bar Classes wrap the Windows API that provides your application with toolbars, status bars, dialog bars, and rebars. We cover these classes in much more depth in the next chapter.

The `CToolBar` **(OCW) and** `CToolBarCtrl` **(OCW) classes** help you to create and maintain toolbars.

The `CStatusBar` **(OCW) and** `CStatusBarCtrl` **(OCW) classes** create and maintains status bars.

The `CDialogBar` **(OCW) class** creates and maintains dialog bars.

The `CRebar` **(OCW) and** `CRebarCtrl` **(OCW) classes** create and maintain rebars.

Property Classes

The Property Classes wrap the Windows API that can provide your application with Property Pages and Property Sheets. One or more Property Pages appear in a Property Sheet to create the tabbed view familiar to Windows users. This view is typically used to select program options.

The `CPropertySheet` **(OCW) class** creates a Property Sheet. Although `CPropertySheet` is not derived from `CDialog`, it's quite similar.

The CPropertyPage (OCW/CDialog) **class** creates a Property Page. And it *is* derived from CDialog.

Drawing Classes

The CDC class wraps the device context we discussed earlier and all of the drawing functions that require a device context—which is pretty much all of them. In addition to CDC, there are four other MFC classes derived from CDC that provide additional functionality.

The CClientDC **class** is typically used to conveniently create and destroy a device context for you. CClientDC is usually created on your stack. Its constructor creates a device context for the client area of your window by calling CDC::GetDC(). Then when your routine returns, CClientDC's destructor destroys that context by calling CDC::ReleaseDC(). No muss, no fuss, no forgotten device contexts to release and cause resource leaks.

The CWindowDC **class** does for the nonclient area of your window what the CClientDC class does for your client area.

The CPaintDC **class** calls CWnd::BeginPaint() when it's constructed to get the device context. The device context in this case only allows you to draw to the area of your window's client area that has been invalidated—as opposed to drawing to the entire client area. The CPaintDC class also calls CWnd::End-Paint() when it's deconstructed.

The CMetaFileDC **class** is used to create a metafile. As we discussed earlier, a metafile is a disk file that contains all of the drawing actions and modes necessary to draw a figure. You can create a metafile by opening a metafile device context and then using your drawing tools to draw to it as if it were a screen or printer device. The generated file can then be reread to create the figure at a future point to one of the other devices.

Drawing Tools and Classes

MFC also has a class to wrap each of the drawing tool characteristics: CPen for pens, CBrush for brushes, CFont for fonts, CPalette for palettes, CBitmaps for bitmaps, and CRegion for regions. Each of these classes creates the associated graphic object, which can then be selected into the device context.

Other MFC Classes

Not all MFC classes affect the user interface. Several MFC classes also wrap the Windows APIs that control files, databases, and window sockets. Several MFC classes also maintain data collections (e.g., lists, arrays, maps, etc.).

File Classes

The CFile (0) class wraps the Windows API for creating and maintaining a flat file. Three MFC classes have been derived from CFile to provide additional functionality.

The CMemFile **class** allows you to create a file in memory instead of on disk. When you construct a CMemClass object, the file is immediately opened and you can use its member functions to read and write to it as if it were a disk file.

The CSharedFile **class** is similar to the CMemFile class except that it's allocated on the global heap, which makes it available for you to share using the clipboard and DDE.

The CStdioFile **class** allows you to read and write text strings terminated with carriage control and line feed characters.

CArchive **and Serialization**

The CArchive class uses the CFile class to save the class objects of your document to disk in a process called serialization. With serialization, the member variables in your classes and whole class objects can be stored to an archive device in sequence so that they can be restored in the exact same sequence later.

Database Classes

The MFC library has classes to support two types of databases.

The Open Database Connectivity (ODBC) classes wrap the ODBC API that most database vendors support. If your application uses MFC's ODBC classes, it can support any Database Management System (DBMS) that supports the ODBC standard.

The Data Access Objects (DAO) classes support a newer database API that has been optimized for use with the Microsoft Jet database engine. You can still access ODBC compliant database systems and other data sources through the Jet engine.

ODBC Classes

There are three main ODBC classes.

The CDatabase (0) **class** opens a DBMS database using the ODBC API. After you've constructed a CDatabase object, you can use its OpenEx() member function to establish a connection to a database. Calling the Close() member function of CDatabase closes the connection.

The CRecordset **class** is used to store and retrieve records through a database connection.

The CDBVariant **class** represents a column in a record set without concern for the data type.

DAO Classes

The DAO classes have three classes similar to the ODBC classes.

The CDaoDatabase (0) **class** opens a DAO database.

The CDaoRecordSet (0) **class** holds records.

The COleVariant **class** represents a record column.

The DAO Classes also include three more classes.

The CDaoWorkSpace (0) **class** manages a database session, allowing transactions to be Committed (stored to the database) or Rollbacked (undone).

The CDaoQueryDef (0) **class** represents a query definition.

The CDaoTableDef (0) **class** represents a table definition, including the field and index structure.

Data Collection Classes

The Data Collection Classes maintain and supports arrays, lists, and maps of data objects.

The CArray **class** and its derivatives support an array of data objects. An *array* is made up of one or more identical data objects (e.g., integers, classes, etc.) that are contiguous in memory and can therefore be accessed with a simple index. The CArray class can grow or shrink its size dynamically. There are several derivations of CArray (e.g., CByteArray, CDWordArray, etc.) that allow you to create a type-safed array. However, there's also a CArray<type,arg_type> template class with which you can make any type type-safed.

The CList **class** and its derivations support a linked-list of data objects. A *linked-list* is made up of one or more identical data objects (e.g., integers, classes, etc.) that aren't continuous in memory and are doubly-linked so that you can go forwards and backwards through the list. There are several derivations of CList (e.g., CPtrList, CObList, etc.) that allow you to create a type-safed list. However, there's also a CList<type,arg_type> template class that allows you to make CList type-safe for any type.

The CMap **class** and its derivations support a dictionary of data objects. A *data dictionary* stores one or more identical data objects (e.g., integers, classes, etc.) under a binary or text key. You can use this key to retrieve a data item. For example, since Windows doesn't keep track of which MFC CWnd objects belong to which window, your application uses a CMap object to associate window handles with their companion CWnd objects. There are several derivations of CMap (e.g., CMapWordToPtr, CObToString, etc.) that allow you to create a type-safed dictionary. However, there's also a CMap<class KEY,class ARG_KEY,class VALUE,class ARG_VALUE> template class with which you can make any type type-safed.

Communication Classes

The MFC library contains classes that allow your application to communicate over a network or over the Internet. These classes are discussed in detail in Chapter 3.

Class Messaging

Our last stop before ending this chapter is to review how MFC processes messages. As mentioned earlier, each window has a window procedure that can process any message sent to that window. We also saw that a window procedure is typically written as a case statement that sorts out the different messages and processes them individually. In practice, this case statement can get very large and very hard to maintain. The MFC solution is to redirect these messages into the member functions of your CWnd derived class, instead. You can then handle window messages from the safety and comfort of a class.

MFC even sends some messages to non-CWnd derived classes for processing in a technique called Command Routing discussed in the section "How MFC Processes a Received Message" on page 32.

How MFC Receives a Posted Message

As mentioned earlier, messages can either be sent or posted. A sent message is essentially the same as calling a window process directly as if it were any other function. A posted message, however, goes into a message queue that was set up for you by the operating system when it initially started your application. Mouse and keyboard clicks typically get posted to this queue, which your application then removes one by one, sending them onto the windows that were clicked or the window that had input focus when a key was pressed.

The Windows API provides two calls, GetMessage() and PeekMessage(), to allow your application to remove messages from this queue. The MFC class CWinThread wraps these calls into a member function called Run(). Run() is the last function called when your MFC application first executes. Run() then sits there peeking in the message queue, waiting for the user to press or click something. Run() also takes this time to perform some background maintenance of your MFC classes, as well as provide you with an opportunity to do your own maintenance.

You can take over any aspect of this mechanism for some powerful results. Please see Examples 47 and 48.

Once a posted message has been removed from the queue and sent to a window, its processing is identical to a sent message.

How MFC Processes a Received Message

The basic window procedure for all MFC controlled windows is the static function AfxWndProc(). Whenever you decide to subclass a window using MFC, the address of this function is stuck in the appropriate window object by MFC. Then when a message comes in, it is the job of AfxWndProc() to call one of your member functions to process the messages you care about and pass the rest on to the original window procedure.

But beyond this simple book-keeping process, MFC has built two additional features into its messaging that AfxWndProc() must contend with: Command Routing and Message Reflection.

Command Routing   Almost all Command Messages generated by the menu and the toolbar wind up being sent to the Main Frame Class's window for processing. Since a lot of these messages could probably be better processed in another application class, such as the View or Document Class, AfxWndProc() will automatically pass these messages around your application in a process called Command Routing. This process also allows MFC classes that don't control a window (e.g., the Document Class) to process a Command Message.

Message Reflection   As mentioned earlier, Control Notifications are sent to a control's parent window so that the parent window can potentially update the control with data, such as add more items to a list box control. Since putting control-specific code into the parent window class goes against object-oriented programming, the AfxWndProc() of the parent window class will automatically bounce these messages back to the control class in a process called Message Reflection to allow it to be able to update its own look.

One last consideration for AfxWndProc() is size. Using standard C++ techniques the easiest way to direct incoming messages to a member function would be to have an overridable dummy function in the base class for each possible message. If you derive from this class and override this function, voila, you can process this message yourself. Unfortunately, there are so many window messages the Utable in the base class would be prohibitively large. So there are no dummy functions in the base class. Instead, each class you derive from the base class has an embedded address map of your member functions and the messages they process. AfxWndProc() accesses this *Message Map* directly and if there isn't an entry for the message that just came in, the message is quickly shuttled on to the original window procedure. We even drop down into assembler for extra speed.

AfxWndProc() accomplishes all of this functionality using just six helper functions, which are called in the following order.

1. Messages to all windows created by MFC are processed by AfxWndProc().

2. AfxWndProc() receives the message, finds the CWnd object the message belongs to, and then calls AfxCallWndProc().

3. AfxCallWndProc() saves the message (message identifier and parameters) for future reference and then calls WindowProc(). This saved message is used in the event you don't process the message.

4. WindowProc() sends messages on to OnWndMsg().

5. WindowProc() sends any messages not processed by OnWndMsg() to DefWindowProc().

6. OnWndMsg() sorts the message by type. For WM_COMMAND messages, OnWndMsg() calls OnCommand(). For WM_NOTIFY messages, OnWndMsg() calls OnNotify(). Everything left over is a window message and OnWndMsg() searches your class's Message Map for a handler that will process it. If it can't find a handler, it returns the message to WindowProc(), which sends it on to DefWindowProc().

7. OnCommand() checks to see if this is a Control Notification (lParam is not NULL). If it is, OnCommand() tries to reflect the message back to the control making the notification. If it isn't a control notification—or if the control rejected the reflected message—OnCommand() calls OnCmdMsg(). OnNotify() also tries to reflect the message back to the control and, if unsuccessful, calls the same OnCmdMsg() function.

8. OnCmdMsg(), depending on what class received the message, will potentially route Command Messages and Control Notifications in a process known as Command Routing. For example, if the class that owns the window is a Frame Class, Command and Notification messages are also routed to the View and Document Classes looking for a Message Handler for the window.

For an overview of this process, please see Figure 1.3.

Figure 1.3 Overview of MFC Message Processing

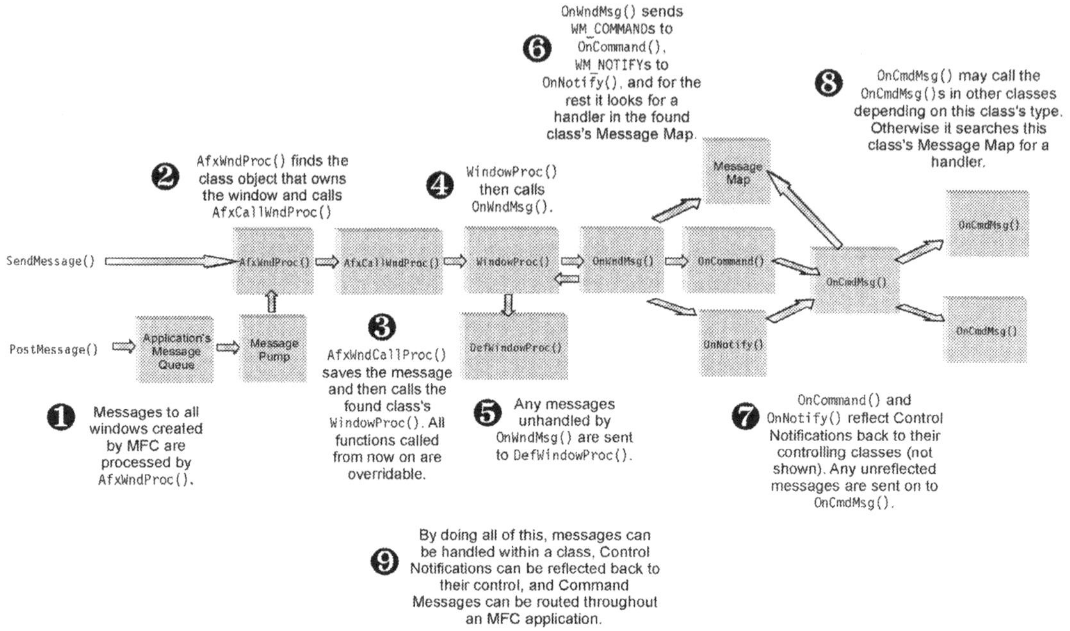

⑥ OnWndMsg() sends WM_COMMANDs to OnCommand(), WM_NOTIFYs to OnNotify(), and for the rest it looks for a handler in the found class's Message Map.

⑧ OnCmdMsg() may call the OnCmdMsg()s in other classes depending on this class's type. Otherwise it searches this class's Message Map for a handler.

② AfxWndProc() finds the class object that owns the window and calls AfxCallWndProc()

④ WindowProc() then calls OnWndMsg().

SendMessage() → AfxWndProc() → AfxCallWndProc() → WindowProc() → OnWndMsg() → OnCommand()

Message Map

OnCmdMsg()

OnCmdMsg()

OnNotify()

OnCmdMsg()

PostMessage() → Application's Message Queue → Message Pump

DefWindowProc()

① Messages to all windows created by MFC are processed by AfxWndProc().

③ AfxWndCallProc() saves the message and then calls the found class's WindowProc(). All functions called from now on are overridable.

⑤ Any messages unhandled by OnWndMsg() are sent to DefWindowProc().

⑦ OnCommand() and OnNotify() reflect Control Notifications back to their controlling classes (not shown). Any unreflected messages are sent on to OnCmdMsg().

⑨ By doing all of this, messages can be handled within a class, Control Notifications can be reflected back to their control, and Command Messages can be routed throughout an MFC application.

OnCmdMsg() **and UI Objects**

Not only is OnCmdMsg() a helper function for processing Command Messages, it's also used to automatically enable, disable, and even checkmark your application's menu items, status bar panes and toolbar buttons. In other words, since OnCmdMsg() already has all of the infrastructure necessary to pass a command message from a menu item to your class members, why not turn it around and let your class members change the appearance of that menu item?

Whenever a menu drops down, your Main Frame Class cycles through all of the command IDs in that menu by calling OnCmdMsg() for each. But instead of it looking for a member function to handle the command, OnCmdMsg() is looking for a member function that you might have added to checkmark that menu item, change its text, or disable it.

Whenever your application is idly waiting for a new posted message (which is how we started this discussion of MFC messaging), it also loops through all of your toolbar buttons and status bar panes with OnCmdMsg(), again looking for your own user interface handler to change their appearance.

Summary

In this chapter, we reviewed the nature of a Windows application, including its windows, its messaging, and its drawing. We also reviewed the nature of MFC—how it wraps the Windows API to give you a fairly good (but not perfect) C++ interface to creating a Windows application. We also reviewed some of the more important MFC classes, and how MFC directs window messages to the member functions of your classes.

Yes, we covered a lot in this chapter, but hopefully it was a review of what you already knew. If you would like to learn more, please refer to my previous book, *Visual C++ MFC Programming by Example*.

In the next chapter, we discuss just one topic, Control Bars, in much more depth.

Chapter 2

Control Bars

In Chapter 1, we saw that common control windows were special child windows provided by the Windows API to create buttons, list boxes, edit boxes, etc. Although a toolbar may look like a collection of these control windows, it is, in fact, just one long control window that paints buttons on itself. A status bar is also just a long, skinny control window painting panes within its border. Both toolbars and status bars represent a special category of common control window called a common control bar.

In this chapter, we will look at how to create common control bars using just the Windows API. Then, we will look at creating them using several MFC classes that provide a great deal more functionality than is available through the API. Finally, we will look at the styles available for the buttons and panes drawn by these bars.

Common Control Bars

Currently, there are three common control bars: the toolbar, the status bar, and the rebar. Toolbars and status bars are available in all versions of Windows. Rebars, on the other hand, are only available with Windows 98 or later and systems with Internet Explorer 4.0 installed. We will discuss rebar control bars later in this chapter.

Toolbars are the familiar windows that contain multiple buttons along the top and sides of your application's main window. The status bar is usually located at the bottom of your main window, indicating the keyboard states and displaying help messages.

At first glance, you might think that the buttons in a toolbar are a row of child windows created using the common control class, Button. A status bar, you might think, is populated with Static controls. In fact, these two control bars paint and control these images themselves. Please see Figure 2.1.

Figure 2.1   The Toolbar and Status Bar Control Windows

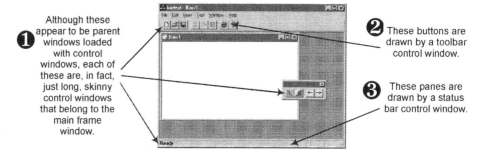

Although these appear to be parent windows loaded with control windows, each of these are, in fact, just long, skinny control windows that belong to the main frame window.

❷ These buttons are drawn by a toolbar control window.

❸ These panes are drawn by a status bar control window.

So, why go to all the bother of recreating this functionality rather than using the existing common controls? Control bars can save a lot of overhead by doing it themselves. More than fifty percent of the time it takes to create any window is spent in the background allocating and initializing resources. A toolbar with ten or more buttons can paint much quicker in one paint handler than in ten.

Creating Control Bars with the API

We will take a quick look now at what goes into creating and initializing control bars with just the API. Since a lot of this work is done for us by MFC, we won't be getting too bogged down in details here.

The Windows API call to create a toolbar or status bar is identical to creating any common control. To create a toolbar, you would use

```
HWND hWnd = ::CreateWindowEx( ...,"ToolbarWindow32",... );
```

A toolbar control must also be initialized with a bitmap object that defines its set of button faces. To define this bitmap object, you send this control a TB_ADDBITMAP window message, as seen here.

```
::SendMessage( hWnd, TB_ADDBITMAP, nNumButtons, &tbab );
```

```
::SendMessage( hWnd, TB_ADDBITMAP, nNumButtons, &tbab );
```

The `tbab` parameter used here is a `TBADDBITMAP` structure that contains the handle of the bitmap object. The `nNumButtons` parameter tells the toolbar how many button faces this bitmap represents.

Once the bitmap has been defined, you can start telling the toolbar control which buttons to paint and where to paint them with the `TB_ADDBUTTONS` message, as seen here.

```
::SendMessage( hWnd, TB_ADDBUTTONS, nNumButtons, tbbuttons );
```

This time, the `tbbuttons` parameter represents an array of `TBBUTTON` structures that defines each button. This structure tells which button face to paint for this button and which command message ID to generate when the button is clicked.

To create a status bar, you would use

```
HWND hWnd = ::CreateWindowEx( ...,"msctls_statusbar32",... );
```

A status bar is made up of panes that usually display text. You create these panes with one `SB_SETPARTS` window message.

```
::SendMessage( hWnd, SB_SETPARTS, nParts, Widths );
```

The `nParts` parameter tells the control how many panes to create and the `Widths` parameter is an integer array that defines the size of each pane in pixels.

You use the `SB_SETTEXT` message to put text in a pane, where the `nPane` parameter tells the status bar control into which pane to put this text.

```
::SendMessage( hWnd, SB_SETTEXT, nPane, "pane text" );
```

Control Bar Styles

As with any common control, toolbars and status bars can be created with the standard set of Window Styles (`WS_CHILD`, `WS_VISIBLE`, etc.). However, toolbars and status bars can also be created with their very own set of styles. Toolbar styles have the prefix `TBSTYLE_` and status bars styles have the prefix `SBARS_` (although there's currently only one status bar style). A sampling of both of these styles may be found in Table 2.1.

Table 2.1   Some Toolbar and Status Bar Styles

Style	Usage Description
TBSTYLE_FLAT	The buttons in a toolbar are drawn flat, meaning the borders around a button are not drawn unless the mouse cursor is dragged over the button.
TBSTYLE_LIST	The button text is displayed to the left of the button.
TBSTYLE_TOOLTIPS	The control generates a TTN_NEEDTEXT notification if the user allows the mouse cursor to hover over a button in the control. It's your responsibility to provide the text for this tooltip.
TBSTYLE_WRAPABLE	The control positions its buttons on multiple rows to fit the control bar into the current size of the control.
SBARS_SIZEGRIP	The bar displays the familiar gripper bar in the lower corner of the control. Since the control is normally positioned at the bottom of your application window, the bar affects the whole window. This is currently the only status bar style.

In addition to these control styles, both of these control bars also share a common set of styles called the Common Control Styles. These styles have the prefix CCS_ and were intended to include all of the styles that were common to all control windows. However, because of the divergent nature of control windows, these styles are mostly used by control bars. The Common Control Styles may be found in Table 2.2. Please see Figure 2.2 for examples of how these styles affect a control bar.

Table 2.2   Common Control Styles

Style	Usage Description
CCS_NORESIZE	The size of the created control bar will be the size specified in the ::CreateWindow() API call. This style overrides the next four styles, which cause the specified size to be ignored.

Style	Usage Description
CCS_TOP CCS_BOTTOM	The control bar is initially aligned and stretched along the top or bottom of its parent window frame. The height is set to a system standard. The size specified in the ::CreateWindow() API call is ignored.
CCS_LEFT CCS_RIGHT	The same as the previous description, except the control bar is aligned with the left or right side of the parent frame and the width is set to a system standard.
CCS_NODIVIDER	A common control bar will automatically draw a line along its top—typically used to separate a toolbar from a menu—unless this style is set.
CCS_ADJUSTABLE	The user is allowed to dynamically configure his or her toolbars, although not in the nice way that the Developer Studio provides. Please see Example 9 for the Developer Studio method.
CCS_NOMOVEY CCS_NOMOVEX CCS_NOPARENTALIGN	These styles either have no apparent effect or their affect duplicates the behavior of one of the styles listed previously.

Figure 2.2   Example Toolbar Styles

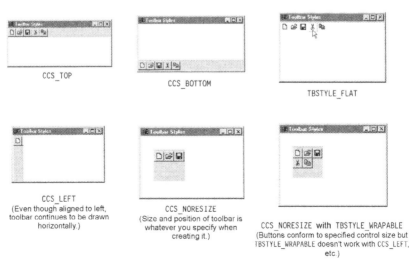

CCS_TOP

CCS_BOTTOM

TBSTYLE_FLAT

CCS_LEFT
(Even though aligned to left, toolbar continues to be drawn horizontally.)

CCS_NORESIZE
(Size and position of toolbar is whatever you specify when creating it.)

CCS_NORESIZE with TBSTYLE_WRAPABLE
(Buttons conform to specified control size but TBSTYLE_WRAPABLE doesn't work with CCS_LEFT, etc.)

As seen in these tables and figures, the variety of window styles available for control bars is rather lacking, even with the common control styles. A control bar will automatically draw a line at the top of the control—but not the bottom, left, or right. There isn't a style to make a common control bar vertical. (You can align a bar with the left or right border of the main window, but the buttons will disappear out of the control). Moreover, the control bar you create will not automatically share the main window's client area with any other control. In fact, you must manually do your own resizing and moving to keep your view and control bars visible. Another problem with common control bars is that the user can't move them. The toolbars in most applications today can be moved to any side of your application's main window or floated in their very own special window frame. However, the control bars provided by the Windows API do not support any of this functionality. To remedy this lack of functionality, MFC provides several classes that wrap and enhance your application's control bars.

Creating Control Bars with MFC

Two classes provided by MFC, `CToolBarCtrl` and `CStatusBarCtrl`, simply wrap the Windows API to create a toolbar and a status bar (the same way the `CWnd` class wraps the Windows API to create a generic window, as we saw in Chapter 1).

`CToolBarCtrl` and `CStatusBarCtrl`

To create a toolbar using MFC, you can use

```
CToolBarCtrl tb;
tb.Create ( DWORD dwStyle, const RECT& rect, CWnd* pParentWnd,
    UINT nID );
```

To add a bitmap to this toolbar, you can use

```
tb.AddBitmap( int nNumButtons, CBitmap* pBitmap );
```

Finally, to define actual toolbar buttons, you can use

```
tb.AddButtons( int nNumButtons, LPTBBUTTON lpButtons );
```

where the `lpButtons` argument is the same as the `TBBUTTONS` array we used previously to define buttons using the API.

To create a status bar, you can use

```
CStatusBarCtrl sb;
sb.Create( DWORD dwStyle, const RECT& rect, CWnd* pParentWnd,
    UINT nID );
```

To add panes and text, use

```
sb.SetParts( int nParts, int* pWidths );
sb.SetText( LPCTSTR lpszText, int nPane, int nType );
```

where both pWidths and nPane arguments were used previously in the API version.

As you can see, the immediate benefit of these two classes is that they make talking to the API easier for a C++ program. However, because they simply wrap the API and add no functionality, the toolbars and status bars they create have no more functionality than their API versions. They still can't be moved or docked by the user or oriented vertically. They can be used to more easily set up and control the buttons and panes within your control bars. However, to get any of the functionality that you've come to expect from your control bars, you need to use two other MFC classes: CToolBar or CStatusBar.

CToolBar **and** CStatusBar

To create a toolbar using CToolBar, you would again use the familiar Create() member function as seen here.

```
CToolBar tb;
tb.Create( CWnd* pParentWnd, DWORD dwStyle, UINT nID );
```

As you can see, creating a toolbar is already a little easier without the rect argument to worry about. The biggest savings in effort, however, comes when loading the bitmap and button definitions.

```
tb.LoadToolBar( UINT id );
```

Rather than define the bitmap and each button programmatically, you use the Toolbar Editor to create a resource that you refererence here. The LoadToolBar() function then takes care of translating your resource into API calls to set up the toolbar.

The CStatusBar class is equally as easy to use.

```
CStatusBar sb;
sb.Create( CWnd* pParentWnd, DWORD dwStyle, UINT nID );
sb.SetIndicators( const UINT* lpIDArray, int nIDCount );
```

Rather than define the text and number of panes for your status bar, you use the String Editor to create a group of strings that are defined in the IDArray array, which you pass to SetIndicators().

Not only is loading and configuring your bars easier with these classes, you also get the following enhanced functionality.

- With the CToolBar class, the colors in the bitmap supplied for a toolbar are automatically adjusted to reflect the current button colors. Light gray is the standard Windows button face color, but if your user decides to go with a new color scheme for their desktop (e.g., light mauve) your toolbar buttons would look out of place if they were still light gray. Therefore, CToolBar converts them to the current button color. All of these color substitutions can be found in Table 2.3.

Table 2.3 CToolBar **Bitmap Color Translations**

Toolbar Bitmap Color	System Color
black	COLOR_BTNTEXT
dark gray	COLOR_BTNSHADOW
light gray	COLOR_BTNFACE
white	COLOR_BTNHIGHLIGHT

- Other member functions are provided in CToolBar and CStatusBar that build on the functionality of the Windows API. For example, CToolBar::SetButtonText() allows you to directly set the text of a toolbar button without worrying about all the API calls needed to make this change

- The automatic ability for toolbar buttons and status bar panels to be enabled, disabled, or put in a checked state are provided by these classes. In other words, unless your toolbar is created using CToolbar, your OnCmdUI() message handlers won't enable or disable your toolbar buttons.

Using the principle of C++ inheritance, you might think that CToolBar and CStatusBar would be derived from CToolBarCtrl and CStatusBarCtrl so that you can get the functionality of both classes from just one class. Instead, Microsoft decided to derive CToolBar and CStatusBar from a new base class called CControlBar. Does that mean Microsoft had to reproduce all of CToolBarCtrl's functionality in CToolBar? No. Because both CToolBar and CToolBarCtrl talk to a control bar by sending messages to its window handle, you can actually have an instance of both CToolBar and CToolBarCtrl controlling the same toolbar at the same time through that one common window handle. In fact, a member function of CToolBar called GetToolBarCtrl() will automatically create an instance of the CToolBarCtrl class that points to the same toolbar. A similar function is provided for status bars called `GetStatusBarCtrl()`. To see how this all works together, please see Figure 2.3.

Figure 2.3   **Control Bar Class Hierarchy**

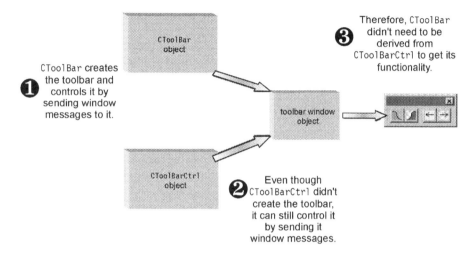

CControlBar

Deriving both `CToolBar` and `CStatusBar` from the same `CControlBar` class allowed Microsoft to factor the following functionality into `CControlBar`.

- If a status bar or toolbar doesn't process any of the following window messages, `CControlBar` offers them back to the owner of the bar for processing.

WM_NOTIFY	WM_DELETEITEM
WM_COMMAND	WM_COMPAREITEM
WM_DRAWITEM	WM_VKEYTOITEM
WM_MEASUREITEM	WM_CHARTOITEM

Normally, windows messages are discarded if not processed by the window for which they were intended. This functionality allows toolbars, for instance, to pass button commands back to the frame window so that they can be processed using the same command handlers that process the menu.

- As mentioned previously, the control bars provided by the Windows API are only able to draw a border along their tops. The `CControlBar` class can draw a border on any side of its toolbar or status bar.

- As mentioned previously, the control bars provided by the Windows API can't orient their toolbars vertically. They can be aligned to the left or right side of a window, but their buttons would disappear out of the control horizontally. Even with the TBSTYLE_WRAPABLE style, a bar won't reliably appear as just a single column of buttons. The CControlBar class can realign its toolbar buttons into a single column.

Note: Status bars cannot be oriented vertically, even with CControlBar. However, a vertical status bar would be not very aesthetically pleasing, so not much is lost by this limitation.

- As mentioned previously, you have to manually make sure your Windows API control bars aren't overlapping each other or the view window. The CControlBar class, along with a little help from other two classes, will automatically make sure your control bars aren't stepping on each other. A discussion of those two other classes and the method used to prevent overlapping control bars can be found in the section "How Views and Bars Share the Client Area" on page 55.
- The CControlBar class also supports tool tip and fly-by help for the buttons and panes in your toolbars and status bars. *Tool tip help* (a.k.a. *bubble help*) causes a small white help window to appear over a button or pane when the mouse cursor is held over it. *Fly-by help* appears in the status bar when the mouse cursor is scrolling through menu items or over a toolbar button.

There are several new control bar styles created for use with the CControlBar class, which can be found in Table 2.4. With the exception of CCS_ADJUSTABLE, you should use these styles with CToolBar and CStatusBar control bars instead of the CCS_ styles listed previously. You will continue to use the TBSTYLE_ and SBARS_ styles.

Table 2.4   MFC Control Bar Styles

Style	Usage Description
CBRS_ALIGN_TOP CBRS_ALIGN_BOTTOM	A control bar is aligned and stretched along the top or bottom of parent window frame. If the bar is dockable, it is allowed to dock to the top or bottom of the parent frame.
CBRS_ALIGN_LEFT CBRS_ALIGN_RIGHT	Same as the previous description, except that it aligns or docks to the left or right side of parent window frame.
CBRS_ORIENT_HORZ	A docking control bar can be docked to the top or bottom of the parent window frame.
CBRS_ORIENT_VERT	Same as the previous description, except the docking control bar can be docked to the left or right side of the window frame.
CBRS_ALIGN_ANY CBRS_ORIENT_ANY	A docking control bar can be aligned or docked to any side of the parent window frame.
CBRS_TOOLTIPS	Tool tips help (a.k.a. bubble help) appears when the user holds the mouse cursor over a toolbar button or status bar panel.
CBRS_FLYBY	Fly-by help appears in the status bar when the user holds the mouse cursor over a toolbar button or status bar pane.
CBRS_BORDER_TOP CBRS_BORDER_LEFT CBRS_BORDER_RIGHT CBRS_BORDER_BOTTOM CBRS_BORDER_ANY CBRS_BORDER_3D CBRS_HIDE_INPLACE CBRS_NOALIGN CBRS_FLOATING	MFC's control bar classes use these styles internally.

The CControlBar class alone will automatically take care of keeping your toolbars and status bars from overlapping each other and the view. Unfortunately, its method is very regimented. No two bars can share the same row or column. Moreover, your user will still be unable to move them or float them. This added functionality requires two more MFC classes, CDockBar and CDockContext.

Docking Bars

To allow your user to move their toolbars around using the mouse, an MFC application creates four unique control bars, called docking bars, along each side of its main window. You may not have always noticed these docking bars, but they paint that empty space you see surrounding a movable toolbar. More often than not, docking bars are collapsed into invisibility along the sides of your application's main window waiting for the next time your user needs to dock a bar to that side. Once a toolbar is docked to a docking bar, they expand to surround the toolbar.

The CFrameWnd class provides a member function called EnableDocking() that can optionally create up to four of these bars around the outside of your window frame. CFrameWnd's EnableDocking() function creates docking bars using the CDockBar class. CDockBar is itself derived from CControlBar, just like CToolBar and CStatusBar, to make use of its shared functionality, including the ability of CControlBar to draw a border around its bar.

Please see Figure 2.4 for look at docking bars.

Note: Status bars, even when derived and controlled by CControlBar, do not support docking.

Figure 2.4   Example Views of Docking Bars

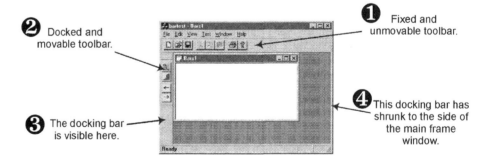

② Docked and movable toolbar.

① Fixed and unmovable toolbar.

④ This docking bar has shrunk to the side of the main frame window.

③ The docking bar is visible here.

Turning Docking On

Docking is not automatically enabled by your MFC application. To enable docking, you must either select "Docking toolbars" when using the

AppWizard to create your project or you can add the same three lines it adds to your CMainFrame class's OnCreate() function, as seen here.

CFrameWnd::EnableDocking() creates the docking bars.

CControlBar::EnableDocking() adds the ability to the control bar to move and dock.

CFrameWnd::DockControlBar() actually docks the control bar to the docking bar.

CFrameWnd::EnableDocking()

To enable your main frame or child frame for docking, you would use

```
EnableDocking( DWORD dwStyle      // which side to create
                                  //     a docking bar on:
//      BRS_ALIGN_TOP             // for the top
//      CBRS_ALIGN_BOTTOM         // for the bottom
//      CBRS_ALIGN_LEFT           // for the left side
//      CBRS_ALIGN_RIGHT          // for the right side
//      CBRS_ALIGN_ANY            // for all four sides
     );
```

You can also OR these styles for additional effect. For example, using CBRS_ALIGN_LEFT with CBRS_ALIGN_TOP alone would prevent your user from docking to either the bottom or right side of their frame window because there wouldn't be a docking bar there to accept it.

The CFrameWnd::EnableDocking() function creates up to four docking bars at a time. Each docking bar is created using the CDockBar class. The CDockBar class is just like any other MFC control class. To create the control, you first create an instance of the class and then call its Create() member function to create the window. To create a docking bar window, CDockBar uses the AfxWndControlBar Window Class.

The CFrameWnd class stores a pointer to each instance of CDockBar it creates in a pointer list called m_listControlBars. This is an undocumented public member variable of type CPtrList.

If you would like to use your own derivation of CDockBar for your docking bars, you will have to write your own version of CFrameWnd::EnableDocking(), since the name CDockBar is hard coded into EnableDocking(). Please see the section "Playing with Your Bars" on page 69 for more on this subject.

CControlBar::EnableDocking()

To add the functionality necessary to allow a control bar to dock, you should call

```
m_wndToolBar.EnableDocking( dwStyle    // which side can this bar
                                       //    be docked to:
//     CBRS_ALIGN_LEFT                 // for left side only
//     CBRS_ALIGN_RIGHT                // for right side only
//     CBRS_ALIGN_TOP                  // for top only
//     CBRS_ALIGN_BOTTOM               // for bottom only
//     CBRS_ALIGN_ANY                  // for any side
    );
```

Again, these styles may be ORed for interesting affects. A style of CBRS_ALIGN_LEFT with CBRS_ALIGN_TOP, for instance, would tell this toolbar that it can only dock to the top or left side of the main window—even though there might be docking bars on all four sides.

The docking functionality that EnableDocking() adds to this control bar is contained in an instance of another MFC class, CDockContext. The functions in CDockContext take over whenever the user either double-clicks or drags a toolbar. When the user releases a dragged toolbar, CDockContext determines to which docking bar and at which spot a toolbar should dock.

CFrameWnd::DockControlBar()

To actually dock a control bar to a docking bar, you call the following.

```
DockControlBar(
    ( CWnd * )&m_wndToolBar    // a pointer to the CControlBar object
    LPRECT lpRect              // location to dock the bar (optional)
    );
```

Here CFrameWnd's DockControlBar() determines which docking bar to use and docks the control bar there. The method DockControlBar() uses to choose a docking bar depends on proximity based on the lpRect argument. It also obviously depends on how you enabled the control bar. For example, if you had specified CBRS_ALIGN_LEFT alone when enabling the control bar, CFrameWnd's DockControlBar() function would only dock the bar to left side of the application window. A style of CBRS_ALIGN_ANY with no location specified causes CFrameWnd's DockControlBar() to automatically select the first existing docking bar, which is typically the top docking bar.

CFrameWnd's `DockControlBar()` docks your bar by calling `CDockBar`'s `Dock-ControlBar()` function. Although both `CFrameWnd`'s and `CDockBar`'s `DockControlBar()` functions have the same name, they do entirely different things. `CDockBar`'s version accepts a bar for docking by storing a pointer to your bar in an internal array. `CDockBar`'s version also does some other chores, which are discussed in the following section.

CDockBar::DockControlBar()

The call to `CDockBar`'s `DockControlBar()` function is as follows.

```
DockControlBar( CWnd *pBar,LPRECT lpRect );
```

You never really need to call this function yourself—`CFrameWnd`'s `DockControlBar()` does that for you. The control bar pointer and location rectangle with which you call `CFrameWnd`'s `DockControlBar()` is passed to `CDockBar`'s `DockControlBar()`. `CDockBar`'s `DockControlBar()` then stores the control bar pointer and location rectangle in a member array in `CDockBar` called `m_arrBars`. This array is used whenever the docking bar is displaying its docked bars. A null pointer in this array tells `CDockBar` to start a new row. If you don't specify an exact position when calling `DockControlBar()`, `CDockBar` simply adds the new position to the end of this array and creates a new row (adds a null at the end). If you do specify a position, `DockControlBar()` identifies which row to stick the new control bar on and inserts it there.

CDockContext::DockControlBar()

The `CDockContext` class also calls `DockControlBar()` when it has to dock a bar. `CDockContext`'s `DockControlBar()` first determines which docking bar to dock to (not unlike `CFrameWnd`'s `DockControlBar()` function) and then calls that bar's `CDockBar::DockControlBar()` function, passing the user's current mouse position.

Automatic Sizing and Moving

Not only can the user move toolbars within a docking bar, but the docking bar itself will automatically move its bars if one or more disappear off the side of the window frame when the user is resizing their application. Partially obscured bars are left alone, but if a bar becomes entirely invisible, it's moved to a new row (or column, if the bar is vertically aligned). As you can see, docking a control bar to a particular location can be very transitory. A docking bar area can be in constant flux.

Much more on this subject can be found in the section "How Views and Bars Share the Client Area" on page 55.

Please see Figure 2.5 for an overview of docking control bars.

Figure 2.5   Docking Bar Overview

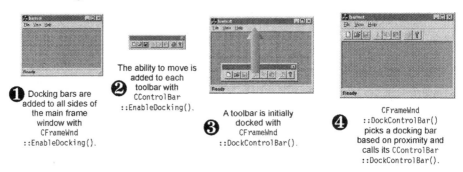

Summary of Docking Toolbars

Docking control bars, therefore, took two MFC classes and four functions which are as follows:

1. The CDockBar class creates up to four docking bars along the sides of your application's main window that can control and contain other control bars. If a docking bar contains no other control bars, it shrinks to nothing and seemingly disappears from your application, waiting for a new bar to be dropped on it.

2. The CDockContext class processes your user's mouse clicks and drags to move your control bars around a docking bar and is created for your control bar by CControlBar::EnableDocking().

3. CFrameWnd's EnableDocking() function creates instances of CDockBar for each side of its window.

4. CControlBar's EnableDocking() function creates an instance of CDockContext to allow a bar to process user mouse input to move and dock the bars.

5. CFrameWnd's DockControlBar() function determines which docking bar to dock a toolbar to.

6. CDockBar's DockControlBar() function is used by a docking bar to take control of a control bar.

So that's how your user can move and dock their toolbars. But wait, there's more (although not much more).

I

2

If you're familiar with MFC toolbars, you're also probably aware that you can drag a toolbar away from the side of the main window and that it will suddenly be surrounded by its own frame—in other words, it will float. For this functionality, MFC supplies one last class called CDockMiniFrameWnd.

Floating Bars

The CDockMiniFrameWnd class is itself eventually derived from CFrameWnd. (Please see "The CMiniFrameWnd Class" on page 53.) Essentially, it's a simplified frame window that contains just one docking bar. There's no need for four docking bars this time because this frame window won't have a view. When the CDockContext class realizes that the user has dragged their toolbar into open space, it spontaneously creates an instance of CDockMini-FrameWnd and docks the toolbar to a docking bar, which was automatically created inside. Please see Figure 2.6 for an example floating toolbar.

Figure 2.6   A Floating Toolbar

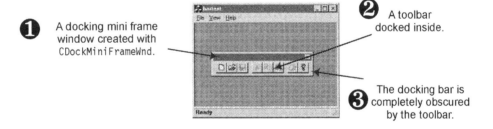

❶ A docking mini frame window created with CDockMiniFrameWnd.

❷ A toolbar docked inside.

❸ The docking bar is completely obscured by the toolbar.

The CMiniFrameWnd **Class**

The CDockMiniFrameWnd class is actually derived from CMiniFrameWnd, which is itself derived from CFrameWnd. CMiniFrameWnd, therefore, has all the functionality of CFrameWnd, but is more easily tailored to a custom application than either the main frame or child frame can be. It can have its own menu and toolbar and, in effect, be somewhat of a mini SDI application right in the middle of your current application. However, you can't use the App-Wizard to create this mini SDI application and mixing view classes and document classes within the same project might become somewhat of a logistical nightmare. As a result, you might be better served to simply spawn and create a separate SDI application than to create one within your application. CMiniFrameWnd's best job in life may be simply to support CDockMiniFrameWnd and floating toolbars.

Floating toolbars have two features that docked toolbars don't. First, the user is able to change the size of the toolbar just by dragging a corner of the mini frame window. And a user can also dock one or more additional toolbars in the same mini frame window. Please see Figure 2.7 for an example of these features.

Figure 2.7   Resized and Multi Floating Toolbars

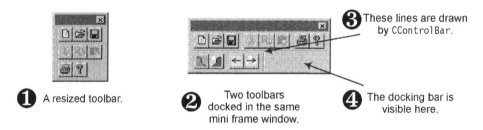

1 A resized toolbar.

2 Two toolbars docked in the same mini frame window.

3 These lines are drawn by CControlBar.

4 The docking bar is visible here.

Note: A resizeable toolbar cannot also be docked with another toolbar in a mini frame window.

Turning this new floating bar functionality on and off required two additional control bar styles that can be found in Table 2.5.

Table 2.5   MFC Floating Control Bar Styles

Style	Usage Description
CBRS_SIZE_FIXED	Prevents the user or system from resizing the control bar.
CBRS_SIZE_DYNAMIC	Allows the user or system to resize the control bar, although the user can only resize a floating toolbar.
CBRS_FLOAT_MULTI	Allows multiple control bars to be docked within the same mini frame window.

You can programmatically float a toolbar by calling CFrameWnd's FloatControlBar() function.

Summary of MFC's Advanced Control Bar Classes

So there you have it. The CToolBar, CStatusBar, and CControlBar classes allowed your control bars to share the client area with the view, along with a lot of other functionality. The CDockBar and CDockContext classes allowed your user to move the toolbars. The CDockMiniFrameWnd class allowed your user to float their toolbars. But I still haven't explained how all of the screen components share the screen until now.

How Views and Bars Share the Client Area

The client area of your application can have two types of child windows vying for space: view windows and control bar windows. In the case of an MDI application, the view window is the `MDIClient` area. In the case of a SDI application, the view window is created by the `CView` class. To determine how much screen real estate each window gets is based on a simple round robin approach. It starts by calling the main window's `GetClientRect()` function to find out how much total space is available. This space is then passed around as a rectangle argument to every child window in the main window, each taking out a chunk along the side and positioning itself there before passing a slightly smaller rectangle onto the next window, until all windows have had a chance to find a spot. Whatever space is leftover is then given to the view window.

Not all child windows participate in this process—only those created by classes derived from `CControlBar`. That means bars created with classes like `CToolBarCtrl` or created directly through the API don't get a spot. Since the view window isn't created by a `CControlBar` class, it doesn't stake out a spot, either, but it does get whatever's leftover. However, sometimes there's nothing left and the view disappears.

This whole process starts when something changes in your application's main window—either a toolbar is added or deleted, moved or resized—or when the size of the main window changes. With a floating toolbar, the user may have changed its shape or added a new bar. Any of these actions usually results in `CFrameWnd`'s `RecalcLayout()` being called. For example, the `OnSize()` message handler in `CFrameWnd` calls this function directly. You can also programmatically call this function, as seen in the next section.

CFrameWnd::RecalcLayout()

The syntax of this function is

```
CFrameWnd::RecalcLayout( BOOL bNotify );
```

where the `bNotify` argument should be set to `TRUE` if your frame is encompassing an OLE frame and you would like that frame to recalculate its layout, too.

This function does precious little other than to call a base class function, CWnd::RepositionBars(), that does the actual round robin work. RecalcLayout() does identify the view window for RepositionBars(). If there is no view window, as is the case with floating toolbars, RecalcLayout() uses RepositionBars() to close the frame around any bars it contains.

CWnd::RepositionBars()

The syntax of this function is

```
CWnd::RepositionBars(
    UINT nIDFirst, UINT nIDLast,    // child windows to poll
    UINT nIDLeftOver,               // id of leftover window
    UINT nFlags,                    // can be set to simply inquire
                                    //     how much space the
                                    //     control bars need

    LPRECT lpRectParam,             // can return inquired size
    LPCRECT lpRectClient,           // substitute client area
    BOOL bStretch                   // cause bars to stretch to fit
                                    //     client area

    );
```

RepositionBars() works by first determining the client area of its window—in other words, how much space in which to place the view and bars. To get this size, it simply calls CWnd::GetClientRect(). RepositionBars() then sends a WM_SIZEPARENT window message (which was created especially for MFC) to any child windows it contains that fall within the nIDFirst and nIDLast arguments but isn't the leftover window, nIDLeftOver. Only two arguments are passed to these windows: the client area rectangle and whether or not each control bar should be stretched to fill the client area. The OnSizeParent() message handlers of each of these windows is then responsible for determining how much of the client area it wants and subtracting that space from the client area rectangle so that that space is unavailable to the next window.

When all of the child windows—control bars, in this case—have had an opportunity to stake out an area of the client area, RepositionBars() uses the remaining space to size and position the leftover window, which in this case is the view window.

Although there are several types of bars, the WM_SIZEPARENT window message is handled by just one message handler in CControlBar. That's why only bars created with CControlBar are eligible to stake out a spot. The method used by `CControlBar`'s `OnSizeParent()` message handler to determine how much space to stake out is discussed in the next section.

CControlBar::OnSizeParent()

The syntax of this message handler is:

```
LRESULT CControlBar::OnSizeParent( WPARAM, LPARAM lParam )
```

```
LRESULT CControlBar::OnSizeParent( WPARAM, LPARAM lParam )
```

where the `lParam` contains the client area rectangle.

The `OnSizeParent()` message handler receives the rectangle being passed around to each window, determines how much space its own control bar needs, subtracts that space from the rectangle, and then passes it on to the next window. To find out how much space its own control bar needs, `OnSizeParent()` calls `CalcDynamicLayout()`.

The `CalcDynamicLayout()` function is itself overlaid by every control bar class derived from `CControlBar` (e.g., `CToolBar`, `CStatusBar`, etc.) to determine the size required by that type of bar. In the case of dockable toolbars, `CalcDynamicLayout()` might even change the shape of the bar before it even figures out its size.

The size that `CalcDynamicLayout()` returns to `OnSizeParent()` is more of a request than a requirement. For example, a stretched hortizontal toolbar will request a width of 32,767 pixels. Obviously, not too many monitors can handle that resolution. In fact, no monitors can handle it. It's true. `CalcDynamicLayout()` is really asking `OnSizeParent()` to make its width the maximum size it can using the currently available space—or, in other words, to stretch the bar across the client area of your main window.

`OnSizeParent()` is, therefore, the final judge of just how much of the client area its bar will take up, always trying to fit its bar so that it takes up the smallest amount of space along the side of the window frame. Once it has made that determination, the bar is physically moved to that location and its size is debited from the client area rectangle as mentioned previously.

The method used by `CalcDynamicLayout()` to calculate the size of its bar and even potentially reshape it is discussed in the next section.

CalcDynamicLayout() **and** CalcFixedLayout()

Two functions help calculate the size of a bar for `OnSizeParent()`.

CalcDynamicLayout() is used to potentially layout and calculate the size of a toolbar with the CBRS_SIZE_DYNAMIC style.

CalcFixedLayout() calculates all other bar sizes.

OnSizeParent() doesn't call CalcFixedLayout() directly. Instead, it calls CalcDynamicLayout(), which determines if this is a dynamic toolbar; if not, it calls CalcFixedLayout().

The syntax for CalcDynamicLayout() is

```
virtual CSize CCalcDynamicLayout( int nLength, DWORD dwMode );
```

```
virtual CSize CCalcDynamicLayout( int nLength, DWORD dwMode );
```

where the nLength argument is to make a request for a specific control bar length and dbMode is a combination of the Layout Modes found in Table 2.6. Both of these arguments are only applicable to toolbars with the CBRS_DYNAMIC_SIZE style.

Note: If you call CalcDynamicLayout() with a valid length argument, the bar must be docked and neither the LM_MRUWIDTH or LM_COMMIT layout modes can be used.

Table 2.6   CalcDynamicLayout() **Layout Modes**

Mode	Usage Description
LM_LENGTHY	Set if the length argument indicates a height instead of a width.
LM_COMMIT	Tells the function to save whatever size it calculates in a "Most Recently Used" variable that can be retrieved using the next mode. It also locks in a toolbar orientation and shape—please see the section "CToolBar::CalcFixedLayout() and CToolBar::CalcDynamicLayout()" on page 59.
LM_MRUWIDTH	Tells the function to return the last size it calculated when the LM_COMMIT mode was used.

Mode	Usage Description
LM_STRETCH	Tells the function to return a size for a stretched bar. If LM_HORZ is used, the bar is stretched horizontally; otherwise, it is stretched vertically. Bars are stretched to fill up the empty space between the bar and the side of the main window. Docked and floating bars aren't stretched because the docking bar they sit on fills up any empty space.
LM_HORZ	Indicates that the bar is horizontally or vertically oriented.
LM_HORZDOCK LM_VERTDOCK	Tells the function to return a size for a docked toolbar. Docked toolbars have only one row or column of buttons.

The syntax for CalcFixedLayout() is

```
CSize CalcFixedLayout( BOOL bStretch, BOOL bHorz );
```

```
CSize CalcFixedLayout( BOOL bStretch, BOOL bHorz );
```

where the bStretch argument is set if you would like a stretched return size rather than the control bar's real size. The bHorz argument applies only to toolbars that can be stretched vertically.

The default action for CalcDynamicLayout() in CControlBar is simply to call CalcFixedLayout(). The default action for CalcFixedLayout() in CControlBar is simply to return a size of 32767,0 for a horizontally stretched bar, 0,32767 for a vertically stretched bar, or 0,0 for an unstretched bar. Obviously, then, an underived CControlBar would always be invisible.

For each type of control bar (e.g., toolbar, docking bar, etc.), however, the derived versions of CalcFixedLayout() and CalcDynamicLayout() are much more involved, as we will see in the next section.

CToolBar::CalcFixedLayout() and CToolBar::CalcDynamicLayout()

CToolbar's CalcFixedLayout() and CalcDynamicLayout() both call the same helper function, CalcLayout(), to determine the toolbar size. CalcLayout() starts by asking the toolbar how many buttons it has and what their configurations are. If the toolbar has the CBRS_SIZE_FIXED style, CalcLayout() simply returns the current size of the toolbar based on button styles and whether this bar has more than one row.

If, however, the toolbar has the CBRS_SIZE_DYNAMIC style, CalcLayout() may potentially change the shape of the toolbar before calculating its size.

The method it uses to lay out the toolbar depends on the following layout modes in the following order.

- If the toolbar is floating or the LM_MRUWIDTH mode is used, the toolbar will revert to the last most recently used size and return that size. Typically, this mode is used to restore the size of a floating toolbar if it's been docked.

- If the toolbar is docked vertically or horizontally, CalcDynamicLayout() will reshape the toolbar into a single row or column of buttons and request the maximum width or height available in the client area.

- If the length is not −1, the toolbar will become that length if the bar is docked and the LM_COMMIT mode is not used. Length refers to width if the LM_LENGTHY layout mode is set.

To commit this layout, you must also use the LM_COMMIT layout mode. Without LM_COMMIT, the size returned by CalcDynamicLayout() will be valid for the requested layout, but the toolbar itself will remain whatever shape it was before the operation. LM_COMMIT also causes CalcDynamicLayout() to save whatever size it calculates here for the next time the LM_MRUWIDTH mode is used.

Laying Out the Toolbar

To change the shape of a toolbar (make it vertical or give it multiple rows to create a particular size), CalcDynamicLayout() applies the TBSTATE_WRAP button state to the toolbar button just before the new row is to start. As you can see, this is a button state, not a toolbar style. There is no reliable way to tell a toolbar to wrap at a particular button. You could give a toolbar the TBSTYLE_WRAPABLE style so that it automatically wraps into rows if the width of the bar is too narrow. However, you don't get the same degree of control that using TBSTATE_WRAP directly affords you. In fact, to make a toolbar vertical, the TBSTATE_WRAP state is simply applied to every single button in the toolbar. Please see Figure 2.8 for how the TBSTATE_WRAP state is used to create multiple rows.

Figure 2.8   Wrapping Toolbar Rows

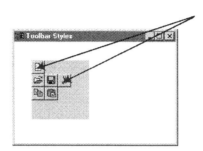

The TBSTATE_WRAP toolbar state was applied to these two buttons in this single toolbar control window.

Note: Why is CalcDynamicLayout() fooling around with the shape of its toolbar right in the middle of another operation to get its size? Why isn't the toolbar already the correct shape? Because there is no better time to reshape a toolbar than at the same time all of its neighbors are in flux, too. Just changing the shape of one bar would cause holes to appear underneath its previous location, and overlaps to occur in its new location. Therefore, whenever a toolbar needs to be reshaped, the system simply changes its styles and then calls RecalcLayout() to actually change the shape.

CStatusBar::CalcFixedLayout()

Since a status bar can never be resized or reoriented, its CalcDynamicLayout() function simply calls CalcFixedLayout(). The status bar version of CalcFixedLayout() then simply returns the height required by the status bar based on its border size and the size of the font being used. The width is always returned as 32,767 pixels wide to cause the status bar to stretch to both sides of the window frame.

CDockBar::CalcFixedLayout()

Again, since a docking bar can never be resized or reoriented, its CalcDynamicLayout() function simply calls CalcFixedLayout(). The docking bar version of CalcFixedLayout() determines how much space it needs to enclose all of its docked control bars. But before it makes this calculation, CalcFixedLayout() first has to do a little layout work of its own.

CalcFixedLayout() loops through all of the bars docked to it (remember, m_arrBars contains this list) and calls each of their CalcDynamicLayout() functions. This is the exact same layout function we reviewed previously for CToolBar. The return requested size is then used to find a spot in the docking bar to put that bar. If a bar goes off the left or bottom of the parent frame, CalcFixedLayout() moves it back. If a bar is so far out of the frame it's invisible, CalcFixedLayout() creates a new row, or column if it's vertical, and moves the bar there.

Once CalcFixedLayout() has done this for each of its docked bars, it calculates how much room it needs in the parent frame and returns that value to OnSizeParent(), which deducts it as usual from the total space available.

Note: The bars located in a docking bar never receive a WM_SIZEPARENT message themselves. Since they are child windows of the docking bar, the RepositionBars() function discussed previously doesn't "see" them, but sees only their parent, the docking bar. The docking bar's CalcFixedLayout(), therefore, performs somewhat of a pseudo, surrogate OnSizeParent() function for these control bars, making sure they don't overlap one another.

Summary of Sharing Client Area

Here is a brief review of how the control bars in your application's main window share that space with each other.

1. CFrameWnd's RecalcLayout() function is called when something in the main window changes.

2. RecalcLayout() then calls CWnd's RepositionBars() function, telling it to allow all the bars in the window to stake out a space and then give the leftover space to the view window.

3. RepositionBars() loops through all of the child windows in the window and sends them a WM_SIZEPARENT message. It passes a slowly shrinking rectangle to each of these bars that represents what's still available in the frame.

4. CControlBar's OnSizeParent() message handler processes the WM_SIZEPARENT message for all bar types by first asking the bar how much space it wants and then subtracting that from what's available.

5. `OnSizeParent()` finds out how much its control bar wants by calling `CalcDynamicLayout()`, which is overlaid by each bar type to determine the required size based on that type.

5a. A docking bar does a little layout work of its own on its docked bars before returning a requested size.

And that's it. Please see Figure 2.9 for an overview of this process.

Figure 2.9   Summary of Sharing Client Area

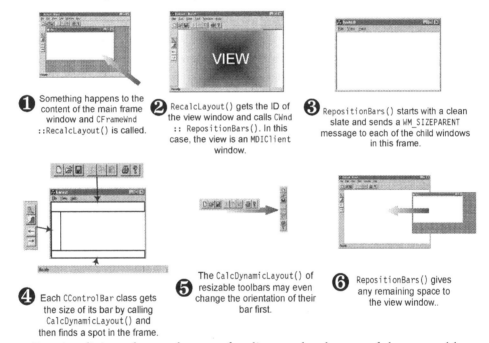

➊ Something happens to the content of the main frame window and `CFrameWnd::RecalcLayout()` is called.

➋ `RecalcLayout()` gets the ID of the view window and calls `CWnd::RepositionBars()`. In this case, the view is an `MDIClient` window.

➌ `RepositionBars()` starts with a clean slate and sends a `WM_SIZEPARENT` message to each of the child windows in this frame.

➍ Each `CControlBar` class gets the size of its bar by calling `CalcDynamicLayout()` and then finds a spot in the frame.

➎ The `CalcDynamicLayout()` of resizable toolbars may even change the orientation of their bar first.

➏ `RepositionBars()` gives any remaining space to the view window..

For simplicity sake, we have so far discussed only two of the control bars available to your MFC application. In addition to toolbars and status bars, there are dialog bars and rebars.

Dialog Bars

A dialog bar is a cross between a toolbar and a modeless dialog box. Just as the `CToolBar` class creates and controls a toolbar child window, the `CDialogBar` class creates and controls a child modeless dialog box. The `CDialogBar` class is itself derived from `CControlBar` to again make use of the common functionality there, such as drawing border lines, docking, and floating.

A dialog bar is made from a dialog template that should have the child style and no border. For ergonomic reasons, it should have the same general dimensions of a toolbar or status bar (i.e. it should be long and narrow or tall and skinny). Fat dialog bars tend to obscure your application's view.

Although dialog bars can be floated and docked, they aren't automatically resizable by your user. You can, however, override its `CalcFixedLayout()` function and programmatically control its size from there.

CDialogBar::CalcFixedLayout()

Just as with `CToolBar` and `CDockBar`, `CDialogBar` also has its own derivation of the `CalcFixedLayout()` function to stake out a section of the main window. In this case, this function simply returns some permutation of `m_sizeDefault`, which is the original size defined in the dialog template from which it was created. For example, if the dialog bar is to dock and stretch to fill the top of the application's window, `CalcFixedLayout()` returns `32767,m_sizeDefault.cy`.

If you want to control this functionality, simply override `CalcFixedLayout()` in your derivation of `CDialogBar` and supply your own values.

Rebars

The last control bar available to your application is called a rebar, although it's also been called a Cool Bar. The term rebar may itself may be a shortened version of Resize Bar.

A rebar has a Windows API equivalent and can be created as seen in the following code fragment, although it's only available in the latest version of Windows 98 or by installing Internet Explorer v4.01 or above. Again, we won't get bogged down in details.

```
HWND hWnd = ::CreateWindowEx( ...,"ReBarWindow32",... );
```

```
HWND hWnd = ::CreateWindowEx( ...,"ReBarWindow32",... );
```

In fact, a rebar seems to be a Windows API attempt to provide some of the functionality previously available only in an MFC docking bar so that users of non-MFC applications can also move their toolbars around. However, when it comes to the ability to move toolbars, a rebar far outperforms a simple docking bar. Not only can toolbars be moved, they can also be resized. Not just toolbars are supported, but any type of control window, including edit boxes, buttons, combo boxes, list boxes, and animation controls.

Each control window is contained in what's called a band—one control per band, but there can be unlimited bands per rebar. A band adds a gripper bar, your own title, and draws its background with a bitmap provided by you. Please see Figure 2.10 for the layout of a rebar.

Figure 2.10   Rebar Control Window

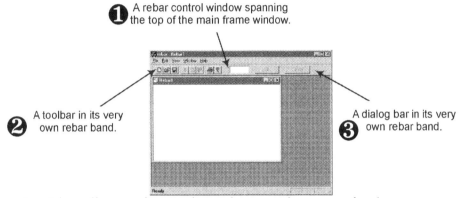

As with toolbars and status bars, there are three sets of styles you can use to affect the look and behavior of a rebar: the standard window styles (WS_VISIBLE, etc.), the common control styles (CCS_VERT, etc.) and a control-specific set of styles, which has an RBS_ prefix for rebars. The more interesting of the rebar-specific styles may be found in Table 2.7.

Table 2.7   Rebar Control Bar Styles

Style	Usage Description
RBS_BANDBORDERS	The rebar control draws lines between adjacent bands.
RBS_VERTICALGRIPPER	The gripper bar will be displayed vertically in a vertical rebar (Win98 and IE4.01 only).
RBS_AUTOSIZE	The rebar will change the layout of its bands when the size of the control bar changes (Win98 and IE4.01 only).

As is common with human behavior, a rebar doesn't get along with its docking bar ancestor. You can create more than one rebar in your application and even orient it vertically, but you can't make them dockable. You can, of course, still create docking bars and dock other toolbars in your application.

To add a new control window to a rebar, you can use

```
::SendMessage( hWnd, RB_INSERTBAND, uIndex, (LPARAM)prbbi );
```

where uIndex is the point at which you would like to insert the band and prbbi refers to a REBARBANDINFO structure that defines the child window to put in the band.

MFC has only recently provided support for rebars in version 6.0, but the support is very similar to toolbar and status bar support, with a class that simply wraps the API and a class that provides higher level functionality.

CReBar and CRebarCtrl

The CReBarCtrl class wraps the Windows API for creating a rebar the same way CToolBarCtrl wraps a toolbar creation. To create a rebar and add a new band using the CReBarCtrl class, you can use:

```
CReBarCtrl rb;
rb.Create( DWORD dwStyle, const RECT& rect, CWnd* pParentWnd,
    UINT nID );
rb.InsertBand( UINT uIndex, REBARBANDINFO* prbbi );
```

where the uIndex and prbbi arguments are the same as when inserting a band with the API.

Of course, the more advanced features have again been saved for the class that doesn't just wrap the API. In this case, CReBar builds on the functionality found in the other MFC classes: CToolBar, CStatusBar, and CDialog-Bar. You can add bars created with these classes directly to a rebar created with CReBar, as seen here.

```
CReBar rb;
rb.Create( CWnd* pParentWnd, DWORD dwCtrlStyle, DWORD dwStyle,
    UINT nID );
CToolBar tb;
tb.Create( CWnd *pParentWnd );
tb.LoadToolBar( UINT nID );
rb.AddBar( &tb, LPCTSTR lpszText, CBitmap* pbmp, DWORD dwStyle );
```

The lpszText and pbmp arguments of AddBar() are the band text and bitmap, which are optional.

If you want to include another type of window in a band, like an animation control, just put that single control in a dialog template and create a dialog bar. Then use CReBar::AddBar() to add to that bar.

This ability to directly load another CControlBar makes CReBar particularly useful. However, this class has no other added functionality over CReBarCtrl.

CReBar::CalcFixedLayout()

Just as CToolBar and CDialogBar, CRebar has its own derivation of CalcFixed-Layout(), which returns a request for size to OnSizeParent(). In this case however, CReBar's CalcFixedLayout() simply interrogates all of its bands for their size while keeping a running total and passes that total back to the top. Since rebars themselves can't be docked or floated, the user certainly can't resize the bar itself, so there's no CalcDynamicLayout().

Command Bars

One last control bar bears discussion, mostly because it's used in several Microsoft applications, including the Developer Studio. A Command Bar is nothing more than an ordinary toolbar with flat buttons, a style you can create yourself by using the TBSTYLE_FLAT toolbar style. A Command Bar also has a gripper bar drawn at the front to help the user grab the bar. Otherwise, finding a toolbar border to drag from around flat buttons would be difficult. Please see Example 8 for one way of imitating the Command Bar look.

The menu in the Developer Studio is also a Command Bar. In other words, the menu in the Developer Studio is actually a flat toolbar with button text, but no button icons. When one of these buttons is pushed, a popup menu is created directly below that button, giving the illusion of a standard menu drop down.

Using toolbars this way doesn't add any extra functionality to your application—in fact, it obviously takes extra work to accomplish—but it does make it look different.

Control Bar Widget Styles

Each type of control bar has a unique set of styles for giving its controls a distinguished look.

Toolbar Button Styles

Some of the more interesting toolbar button styles include the following.

TBSTYLE_SEP is typically used to create a gap between toolbar buttons to create groups of buttons. The index that normally points to the button face in the bitmap becomes the size of the gap in pixels. As example of this can be found in my previous book, *Visual C++ MFC Programming by Example.*

TBSTYLE_AUTOSIZE is available to your application only with Windows 98 or if Internet Explorer 4.01 is installed. This style causes each button size to be based on the text within and is used frequently with toolbars that are used to mimic menu bars such as in Command Bars.

TBSTYLE_DROPDOWN causes a toolbar to generate a TBN_DROPDOWN control notification rather than a command message for the button. This has the same limited use seen with TBSTYLE_AUTOSIZE and is also used to implement a menu toolbar.

TBSTYLE_EX_DRAWDDARROWS is used in conjunction with TBSTYLE_DROPDOWN to create a button with a drop down arrow embedded in it. It must be applied, however, using the TB_SETEXTENDEDSTYLE window message.

You can also apply several states to your toolbar buttons to change their appearance. However, other than the TBSTATE_WRAP state we discussed previously, most other states are already handled for you by MFC via the OnCmdUI message handler. This includes disabling or checking a button.

Status Bar Panel Styles

Some of the more interesting status bar pane styles include the following.

SBPS_NOBORDERS causes the pane not to appear recessed into the bar. This style is automatically applied to the first pane of an MFC application's status bar where the help messages appear. If you would like this pane to be recessed too, you will need to apply the next two styles to pane zero (0).

SBPS_NORMAL causes the pane to appear recessed and unstretched.

SBPS_STRETCH causes this pane to expand to fill up any empty area of a stretched status bar. This style can be applied to only one pane, usually the first, to accommodate the help message that appears there.

SBPS_POPOUT causes the pane to appear bubbled out. You can get this same effect by using the SetCheck(TRUE) member function of CCmdUI as part of updating the user interface.

SBPS_OWNERDRAW allows you to draw the individual pane by deriving your own class from CStatusBar and overriding the DrawItem() member function. You can use this style to display icons or use color in your status panes.

Rebar Band Styles

Some of the more interesting rebar band styles include the following.

RBBS_BREAK causes this bar to start a new row or column.

RBBS_CHILDEDGE causes the band to draw an edge at the top and bottom of its child window.

RBBS_FIXEDSIZE prevents the band from being resized—no gripper bar is displayed.

RBBS_NOVERT hides this band if the rebar control has been vertically oriented.

RBBS_GRIPPERALWAYS **and** RBBS_NOGRIPPER cause the gripper bar to always be displayed or to never be displayed, respectively. These two styles are available only on Window 98 and Internet Explorer 4.01 systems.

Playing with Your Bars

Although MFC's control bar classes have alleviated a lot of the tedium and headache associated with using control bars, they have been implemented in a way that makes modifying their default behavior sometimes impossible to do without depending on undocumented features. In some cases, it may even require rewriting MFC functionality, although not much.

Here, then, are some ways to modify and enhance your control bars.

Overriding `CControlBar::CalcDynamicLayout()`

This is perhaps the most common way to enhance your control bars. Overriding `CalcDynamicLayout()` allows you to determine the size of your control bar. Since MFC doesn't currently support the ability for the user to resize dialog bars, you can add this functionality yourself by first providing some mechanism for your user to request a new size, then reflect that new size in an overload of the `CalcDynamicLayout()` function.

There also isn't any automatic way to define a toolbar with multiple columns like the three column toolbar that appears with Dialog Editor. Instead, you need to override `CalcDynamicLayout()` in your own derivation of the toolbar class where you would apply the `TBSTATE_WRAP` state to the appropriate buttons and then return the appropriate size. You can even get fancy by reorienting the toolbar based on the docking bar to which the toolbar is docked.

Add a `WM_SIZEPARENT` Message Handler

As mentioned earlier, all child windows in your application's main window receive the `WM_SIZEPARENT` window message. However, only the `CControlBar` class currently has a message handler for `WM_SIZEPARENT`, which allows `CControlBar` windows to stake out a spot in the client area. Therefore, to add a noncontrol bar window to this process simply means adding a `WM_SIZEPARENT` message handler to your window's class. However, since the other types of windows you might want to participate in this process can be stuck in a dialog bar where you get a `WM_SIZEPARENT` message handler for free, you probably would never need to create this handler yourself.

Overriding `CMainFrame::RecalcLayout()`

To take control of just how your control bars are laid out in your application's main window, you can override `CMainFrame::RecalcLayout()` or `CChildFrm::RecalcLayout()`. By performing your own manual layout, you can:

- decide which bar gets to stake out a spot first and
- always put a particular bar in a particular location.

Unfortunately, MFC has made it impossible to modify this process without rewriting two of their functions: `CFrameWnd::RecalcLayout()` and `CWnd::RepositionBars()`. If you're squeamish about rewriting MFC functionality because you must incorporate Microsoft's future enhancements to these functions, I would make the case that these are small functions whose content is unlikely to change.

If the benefits to changing these functions is enough to overcome any reluctance, you can start your version of `RecalcLayout()` and `Reposition-Bars()` by copying the source from the `\src` directory of your MFC distribution. These functions are contained in `wincore.cpp` and `winfrm.cpp`. Since their content is straightforward, I'll leave their modification up to you.

Deriving from `CDockBar`

To modify and enhance your docking bars, you will again need to rewrite an MFC function. Once again, the function is small and unlikely to change in future versions of MFC. Docking bars are created using the `CDockBar` class, normally inside of `CFrameWnd`'s `EnableDocking()` function. Unfortunately, when `EnableDocking()` creates this class, it uses the hard coded name, `CDockBar`, making it impossible to substitute your own derivation of `CDockBar`. Therefore, if you really want to use your own version of `CDockBar`, you will need to reproduce the functionality of `EnableDocking()` in your own function inside of `CMainFrame`. Once more, you can start with the original code in MFC's `\src` directory in `winfrm.cpp`.

What do you get by adding your own version of `CDockBar`? Since a docking bar is just like any other window, you can take over how it's drawn. For example, you can fill this normally drab area with your company logo. You can also take over the method by which the bars in this docking bar are positioned. For an example of this, please refer to my artical "Making MFC Docking Bars Cool" in the 5/99 issue of WDJ (*Windows Developer's Journal*) availble at web address: `www.wdj.com`

Control Bar Examples

The examples in this book that relate to control bars include the following.
- Example 1   Putting a Static Logo in the Toolbar
- Example 2   Putting an Animated Logo in the Toolbarr
- Example 8   Adjusting the Appearance of Your Command Bar
- Example 9   Creating Programmable Toolbars

Summary

We saw in this chapter that toolbars and status bars are, in fact, just long, skinny control windows that painted buttons and panes on themselves. We saw how to create these control bars using the Windows API and the MFC classes that made this job much easier. We saw how bars were docked and how they shared the client area of your main window with each other and the view window. Finally, we looked at a few possible ways to modify and enhance your control bars.

Next, we turn our attention to different ways your MFC application can communicate with its environment.

Chapter 3

Communication

In Chapter 1, we touched briefly on how to communicate with your application's windows using `SendMessage()` and `PostMessage()`. But what if you wanted to communicate with another application entirely or even an application on another system?

In this chapter, we will review six methods available to your application for communicating with the outside world, including applications on other systems, on the Internet, or over a serial line. We will see how this communication link will allow us to control another application or even use its functionality. We will also discover a couple of ways to share large amounts of data with another application.

Interprocess Communication

Communication between applications, on the same system, or over a network is called Interprocess Communication (IPC). Your MFC application has the following six avenues available for communicating with another application.

Windows messaging allows you to communicate with the windows of any other application. This is the same windows messaging we used to communicate with our own application's windows.

Dynamic Data Exchange (DDE) allows you to pass large amounts of data between applications by maintaining it in globally allocated memory. You could do this yourself by manually putting data in a globally allocated chunk of memory and then passing that pointer using windows messaging. However, DDE provides a standard that allows any application that conforms to it to play.

Message Pipes can be used to set up a permanent communication channel between applications through which data can be read and written as if your applications were accessing a flat file. You give up the speed of DDE data transfers, but Message Pipes allow you to seamlessly send data to applications on other systems.

Window Sockets have all the functionality of a Message Pipe, but follow a communication standard that allows you to communicate with applications on non-Windows systems, such as UNIX. You should, in fact, use Window Sockets in favor of Message Pipes and DDE in any new development.

Serial/Parallel communication allows your application to talk to an application or device through a serial or parallel port.

Internet communication allows your application to upload or download a file from an Internet address.

Communicating Strategy

Although each of these communication methods is invoked with a different Windows API or MFC class, the procedure for using them is almost identical.

1. Use the Windows API or a MFC class to open a communication link with the other application.
2. Read and write to that other application. With some methods, that can mean sending and receiving messages. With other methods, the process is not that much different from reading and writing to a flat file.
3. Then close the link.

Synchronous vs. Asynchronous Communication

Each of these communication methods allows you to communicate synchronously or asynchronously. *Synchronous communication* causes your application to pause until finished reading or writing to the other application. *Asynchronous communication* allows your application to continue while the system finishes the read or write operation for you and then informs you when it's done with an event flag or by calling a function you specify.

Although asynchronous communication sounds preferable, especially in an application that will be talking to several other applications at once, it isn't very object-oriented. You need to either step out of an object to provide a static callback function or you need to worry about processing an event flag. Setting up an asynchronous link is also more complicated and, therefore, more prone to bugs.

The solution is to use synchronous communication, but to put each read or write operation in its very own thread so that it doesn't stop your application from communicating with other applications. When the operation is finished, the thread can automatically store any read data back into your application. The thread can also set an event flag, but only in the more hospitable environment of MFC classes. Examples 51 and 52 show how to do this.

Note: Your application's ability to create its own thread has only just been available with Windows 95 and Windows NT. If you will be developing for a Windows 3.1 platform, you will need to use asynchronous communication. With asynchronous communication, the operating system itself is putting the read or write operation in its own special thread.

Note: In Window Socket jargon, an application that communicates synchronously is considered to be "blocking".

We will now present each communication method, including how to open and close a communication link and how to read and write over it.

Windows Messaging

The same messaging system that allows your application to control its windows and process command messages from the user can also be used to control and process commands from other applications. What makes this all possible is that each window object is globally allocated such that any application can send a message to any other application's windows. The trick is knowing which window to talk.

Opening and Closing

To open a communication link with another application, simply determine the appropriate window handle to which you should send messages. However, you don't know this handle at compile or link time because each handle is created when the window is created and can be different each time. You can't ask the other window for its handle because you need its handle to make the request. There are three other techniques for getting that window handle.

1. When a parent application creates a child application, it passes the handle of one of its windows as a creation argument. The child application can then send that handle the window handle of one of its own windows and communication can proceed. Typically, the handles that these applications exchange are to plain (but always hidden) windows created simply to communicate through. Closing communication is then simply a matter of closing these windows.

2. If the target application is already running, you can instead look for the appropriate window using the Windows API call `FindWindow()`. This function will return a handle to any window that matches a particular name or uses a particular Window Class. To help `FindWindow()` find the right window, you can, therefore, create a plain messaging window as discussed previously using a highly unique caption or Windows Class name.

3. You can also use the Windows API call `BroadcastSystemMessage()`, which will send a message to every application currently running. You can use `BroadcastSystemMessage()` to broadcast a special message that contains your application's own window handle. Applications that process this message can then reply with their own window handles, thus completing the circuit. To see this last method in action, please see Example 49.

Reading and Writing

To now communicate with the other application is simply a matter of plugging this window handle into ::SendMessage() or ::PostMessage(). You can also wrap the window handle in a CWnd class using Attach(), and then use CWnd's other methods to send and post messages. Sending a message provides your application with synchronous communication; posting provides asynchronous communication.

Although you send the standard window messages (e.g., WM_CLOSE, WM_MOVE, etc.) to the target window, you will typically want to create your own window message to perform some custom task.

You could create your window message ID by simply defining it in the range above WM_USER, as seen here.

```
#define WM_MYMESSAGE WM_USER+1
```

An arguably more robust method of creating a new message ID is to register your message with the system by using

```
#define IDString "MyMessage"
MsgID = ::RegisterWindowMessage( IDString );
```

where IDString is a unique text string of which every application is aware, and MsgID is the created message ID. Therefore, rather than assigning a numeric value for your message ID that might get mismatched among applications, you can use a descriptive text string, such as OpenFile, and ::RegisterWindowMessage() creates a message ID for you. If one application registers OpenFile first, subsequent applications's calls to ::RegisterWindowMessage() using OpenFile returns that same message ID so that there's never a danger of mismatch.

To see how these functions are used, please refer to Example 49.

Review

Sending messages to another application using windows messaging is then simply a matter of getting a window handle that belongs to the other application and then sending it a message. For an overview of this process, please see Figure 3.1.

Figure 3.1   Windows Messaging

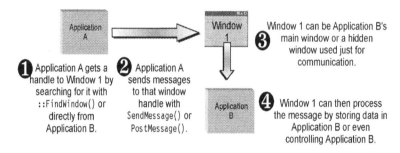

① Application A gets a handle to Window 1 by searching for it with `::FindWindow()` or directly from Application B.

② Application A sends messages to that window handle with `SendMessage()` or `PostMessage()`.

③ Window 1 can be Application B's main window or a hidden window used just for communication.

④ Window 1 can then process the message by storing data in Application B or even controlling Application B.

Dynamic Data Exchange (DDE)

`SendMessage()` and `PostMessage()` allow you to send two integer values at a time to the other application. To send one thousand bytes would, therefore, take 250 messages. To send more at a time, you can stuff your data in globally allocated memory and pass its handle as one of these parameters. This approach, however, will only allow you to exchange data with other applications that you design. To communicate data with applications you don't design, you should follow the DDE standard, which has been implemented in the Dynamic Data Exchange Library (DDEML).

Client/Server

With DDE we are also introduced to the concept of a Client Application and a Server Application. In the case of DDE, almost all data resides in the Server Application, where it is accessed by the Client Application. Communication is then a matter of the Server putting out an "Open for Business" sign and a Client reading or writing data to the Server.

Client/Server is discussed in more detail in the section "Client/Server" on page 100.

Opening and Closing

To open a communication link, the Server Application starts by advertising itself with

```
// initialize DDE services
UINT DdeInitialize(
    LPDWORD pidInst,            // a pointer to the instance
    PFNCALLBACK pfnCallback,    // a callback function (see below)
```

```
    DWORD afCmd,                    // command and filter flags
    DWORD ulRes                     // reserved
    );
// create a string handle to the name of the service we are creating
HSZ DdeCreateStringHandle(
    DWORD idInst,       // returned by DdeInitialize() above
    LPTSTR psz,         // pointer to service name string
    int iCodePage       // CP_WINANSI or CP_WINUNICODE for Unicode
    );
// register the service
HDDEDATA DdeNameService(
    DWORD idInst,       // returned by DdeInitialize() above
    HSZ hsz1,           // string handle to service name
    0L,                 // reserved
    UINT afCmd          // service name flags
    );
```

A Client Application can then connect to the Server with

```
UINT DdeInitialize( ... );      // as seen above
HCONV DdeConnect(
    DWORD idInst,       // returned by DdeInitialize() above
    HSZ hszService,     // handle to service name string
    HSZ hszTopic,       // handle to topic name string (optional)
    PCONVCONTEXT pCC    // pointer to structure with context data
    );
```

To handle this connection in the Server Application, the Server's callback routine must handle an XTYP_CONNECT message.

```
HDDEDATA CALLBACK DdeCallback( type ... )
{
    switch( type )
    case XTYP_CONNECT:
        return TRUE;    // return FALSE to reject connection
}
```

The Client can then close this link with

```
BOOL DdeDisconnect(
    HCONV hConv      // handle returned by DdeConnect() above
    );
```

Reading and Writing

To request data from the Server, the Client can then use

```
HDDEDATA DdeClientTransaction(
    NULL, 0,
    HCONV hConv,            // handle returned by DdeConnect() above
    HSZ hszItem,            // string handle of name of data to read
                            //     and return
    UINT wFmt,              // format of data based on clipboard formats
    XTYP_REQUEST,           // <<<<<<<<<<<<<<<transaction type
    DWORD dwTimeout,        // should request time-out
    LPDWORD pdwResult       // pointer to result
    );
```

The data that's returned is actually a handle to a DDE data object, which can either be accessed with

```
LPBYTE DdeAccessData(
    HDDEDATA hData,         // handle returned by DdeClientTransaction()
    LPDWORD pcbDataSize     // pointer to data length
    );
```

or copied automatically into a local buffer with

```
DWORD DdeGetData(
    HDDEDATA hData,     // handle returned by DdeClientTransaction()
    LPBYTE pDst,        // pointer to destination buffer
    DWORD cbMax,        // amount of data to copy
    DWORD cbOff         // offset to beginning of data in hData
    );
```

The Server processes requests for data in its callback routine, as seen here. A data handle to requested data is created using DdeCreateDataHandle().

```
case XTYP_ADVREQ:
{
    HDDEDATA hData =
        DdeCreateDataHandle(
            DWORD idInst,      // returned by server's DdeInitialize()
            LPBYTE pSrc,       // pointer to source buffer
            DWORD cb,          // length of source buffer
            DWORD cbOff,       // offset from beginning of source buffer
            HSZ hszItem,       // handle to item name string
            UINT wFmt,         // format of data in clipboard formats
            UINT afCmd         // creation flags
            );
            return hData;
}
```

For the Client to write data to the Server, it can either put the data into another data handle or pass the data directly with

```
HDDEDATA DdeClientTransaction(
    LPBYTE pData,          // pointer to data to pass to server
                           //     or data handle
    DWORD cbData,          // length of data
                           //     or 0ffffffffh if pData is a data handle
    HCONV hConv,           // handle returned by DdeConnect() above
    HSZ hszItem,           // string handle of name of data to read
                           //     and return
    UINT wFmt,             // data format based on clipboard formats
    XTYP_POKE,             // <<<<<<<<<<<<<transaction type
    DWORD dwTimeout,       // should request time-out
    LPDWORD pdwResult      // pointer to result
    );
```

The Server must then accept this data in its callback routine with

```
case XTYP_POKE:
    DdeGetData();     // as before
```

Other DDE Functions

The DDEML provides a lot more functionality than what's presented here. For example, you can set up an advise loop that allows both Server and Client to be immediately advised if the data they share is modified.

The DDEML also supports a standard that allows a Client to ask the Server to perform a command. Which command to execute is expressed in a text string, such as `Open file.dat`, where `Open` is the command which the Server parses and executes if it's in its repertoire of commands.

MFC Support

Your SDI and MDI applications automatically support three DDE command strings.

`[open( "file.dat" )]` tells your application to open the specified file.

`[print( "file.dat" )]` tells your application to print the specified file.

`[printto( "file.dat","printer","driver","port" )]` tells your application to print the specified file to the specified device.

Your MFC application also has built-in support for the command line flag `/dde`, which simply hides your application initially so that your application operates in the background when printing files.

Other than this limited support, MFC does not support DDE. Why not? For one reason—applications have moved away from running on single systems to having multiple parts that can run on multiple systems. The shared global memory of one system is not accessible outside of that system. In other words, DDE is out of fashion.

Window messaging also can't make the leap between machines. In fact, Message Pipes were created to make that leap.

Review

DDE allows larger amounts of data to be shared among applications by orchestrating the use of global memory. This is also its downfall, since global memory can't be shared over a network. For an overview of this process, please see Figure 3.2.

Figure 3.2   **Dynamic Data Exchange**

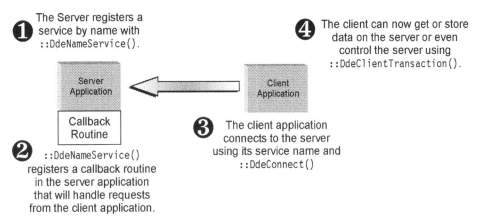

❶ The Server registers a service by name with ::DdeNameService().

❹ The client can now get or store data on the server or even control the server using ::DdeClientTransaction().

❷ ::DdeNameService() registers a callback routine in the server application that will handle requests from the client application.

❸ The client application connects to the server using its service name and ::DdeConnect()

Message Pipes

To allow continuous communication between two applications, the Windows API provides *Message Pipes*, which are a system of buffers and handles that even allow your application to talk to an application on another system. There are two types of message pipes: anonymous and named. *Anonymous pipes* are usually used to communicate between a parent "shell" application and its child processes. *Named pipes* can talk to any application, including those on another system. We will be reviewing only named pipes in this chapter.

> **Note:** Message Pipes are actually outdated just like DDE. We present them here for the sake of completeness. For new development, you should consider using Window Sockets instead, which are presented in the section "Window Sockets" on page 86.

Opening and Closing

For your application to create a named pipe to which other applications can connect, you would use

```
HANDLE CreateNamedPipe(
    LPCTSTR lpName,          // name of pipe in the form:
                            // where you provide
                            // servername and pipename
```

```
    DWORD dwOpenMode,              // open mode:
                                  // PIPE_ACCESS_DUPLEX means
                                  //     pipe is read/write
                                  // PIPE_ACCESS_INBOUND means read only
                                  // PIPE_ACCESS_OUTBOUND means write only
    DWORD dwPipeMode,             //     pipe-specific modes:
                                  // PIPE_TYPE_BYTE means data is
                                  //     written as a stream of
                                  //     potentially unrelated bytes.
                                  // PIPE_TYPE_MESSAGE means each time
                                  //     a block of data is written, it's
                                  //     treated like a message packet.
    DWORD nMaxInstances,          // max number of instances of this pipe
    DWORD nOutBufferSize,         // output buffer size, in bytes
    DWORD nInBufferSize,          // input buffer size, in bytes
    DWORD nDefaultTimeOut,        // time-out time, in milliseconds
    LPSECURITY_ATTRIBUTES lpSecurityAttributes
                                  // pointer to security attributes
    );
```

If another application wanted to connect to this pipe, it would call

```
HANDLE hPipe = CreateFile(
    LPCTSTR lpFileName,           // pipe name created above
    DWORD dwDesiredAccess,        // GENERIC_READ and/or
                                  //     GENERIC_WRITE
    DWORD dwShareMode,            // FILE_SHARE_READ and/or
                                  //     FILE_SHARE_WRITE
    LPSECURITY_ATTRIBUTES lpSecurityAttributes,
                                  // security
    DWORD dwCreationDisposition,  // always OPEN_EXISTING
    DWORD dwFlagsAndAttributes,   // file attributes:
                                  // FILE_FLAG_OVERLAPPED makes this
                                  //     an asynchronous connection
    HANDLE hTemplateFile          // copy attributes from this file
    );
```

To close a pipe, you can call

```
CloseHandle(
    HANDLE hPipe
    );
```

Reading and Writing

Once a pipe connection has been established, you can read and write to it using the following two functions.

```
BOOL ReadFile(
    HANDLE hPipe,                        // pipe handle created above
    LPVOID lpBuffer,                     // buffer that receives data
    DWORD nNumberOfBytesToRead,          // number of bytes to read
    LPDWORD lpNumberOfBytesRead,         // number of bytes actually read
    LPOVERLAPPED lpOverlapped            // used with asynchronous
                                         //    communication

);


BOOL WriteFile(
    HANDLE hPipe,                        // pipe handle
    LPCVOID lpBuffer,                    // data to write
    DWORD nNumberOfBytesToWrite,         // number of bytes to write
    LPDWORD lpNumberOfBytesWritten,      // number of bytes actually
                                         //    written
    LPOVERLAPPED lpOverlapped            // used with asynchronous
                                         //    communication
);
```

Although the Message Pipe API contains a lot of additional functionality to support asynchronous communication, you need not be concerned with it as long as you use `ReadFile()` and `WriteFile()` from within their own threads, as discussed previously. Synchronous communication is always much simpler and, therefore, less prone to bugs.

Review

Message Pipes allow you to set up a permanent connection between applications, which allows you to communicate as if you were reading and writing from a flat disk file. Please see Figure 3.3.

Figure 3.3 Message Pipes

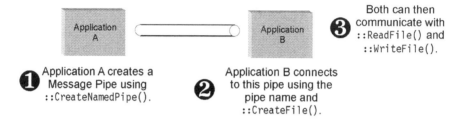

Although Message Pipes represent a quantum leap over simple messaging, you can probably forget all about them. MFC doesn't provide any classes for them and you won't find any examples in this book. Message Pipes were presented for background only because the Windows API provides a much more flexible solution to communication through Window Sockets.

Window Sockets

Window Sockets are essentially Message Pipes that conform to the UNIX sockets implementation of the Berkeley Software Distribution (v4.3) and, therefore, allow your application to talk to an application on any system that supports this standard (including, of course, the many flavors of UNIX).

Even though Window Sockets are typically used to communicate over a network, you can still use them to communicate between applications on the same system. Therefore, you have lots of flexibility when configuring your applications. You can install them all on the same system, but still have the option of putting another application on another system.

Your application has three levels of Window Socket support to choose from, including direct Windows API access, an MFC class that simply wraps that API called CAsyncSocket, and an upper level MFC class called CSocket. For the sake of simplicity, you should only use CSocket. Since CSocket is derived from CAsyncSocket, you still have access to all of the lower class's functionality, and CSocket provides some functionality that makes communicating with a non-Windows system easier.

Communication with Window Sockets is accomplished with three sockets. In addition to the socket that each application creates to talk through, a third socket is created to "listen" for new connections. In other words, a Server application creates this third socket to look for new connections from another application.

Opening and Closing

To allow your application to listen for a connection request, you would use

```
CListenSocket listenSock;
listenSock.Create(
    UINT nPort          // between 1025 and 0xffffffff set by you to
                        //     identify this listener to your other apps
    );
listenSock.Listen()     // start listening
```

The CListenSocket class used here is your own derivation of the CSocket class. We don't use CSocket directly because we need to override one of its member functions, as we will soon see.

If another application wanted to connect to this socket, it would call

```
CSocket sock;
sock.Create();                // take the defaults
sock.Connect(
    LPCTSTR lpszHostAddress,   // system address of application
                              //     with listening socket specified as:
                              //     "" or "128.23.1.22" or
                              //     "localhost" to talk to an
                              //     application server on
                              //     the same system
    UINT nHostPort            // the port number specified when creating
                              //     the listening socket
    );
```

In this example, we created the sock object on the stack for clarity. Normally, you would either make this a member variable of a class or allocate it in the heap. Once the sock object is destroyed, the connection is closed.

To complete this connection, we need to override a member function of our CListenSock class called OnAccept(). This function is called whenever CListenSocket senses that another application is knocking on the door. In

OnAccept(), we create the third and final socket, which talks directly to the other application like so.

```
CListenSocket::OnAccept()
{
    CSocket sock;
    listenSock.Accept( sock );
}
```

Again, make sure in real life to create the sock where it won't get destroyed. To close a connection on purpose, you can call

```
sock.Close();
```

Reading and Writing

Once a Window Socket connection has been established, you can read and write to it using the following two functions.

```
int numBytesReceived =          // number of bytes received
    Receive(
        void* lpBuf,            // buffer to contain data
        int nBufLen,           // length of buffer
        int nFlags = 0         // if set to MSG_PEEK, will cause
                               //     Receive() to leave received
                               //     data in socket queue, but data
                               //     is still copied to lpBuf

    );

int numBytesSent =             // actual number of bytes sent
    Send(
        const void* lpBuf,     // bytes to send
        int nBufLen,           // number of bytes to send
        int nFlags = 0
    );
```

Receive() and Send() are both synchronous functions. Therefore, you should execute them in their own threads, as seen in Example 51. Another hazard in using Receive() is that if you request more bytes than will be sent, it will never process the message. To remedy this, you should construct your messages so that they have a fixed size header that contains a total message size variable. `Receive()` can then be set up to receive the header size and then use that size variable to receive any more data.

To see a real example of this, please refer to Example 51.

Serializing Over a Window Socket

When using `CSocket`'s `Send()` and `Receive()` member functions (actually inherited from `CAsyncSocket`), you are responsible for stuffing your message into any data structures and for straightening out byte orders and string formats in communications from non-Windows MFC systems such as the Mac. However, if you use another MFC class, `CSocketFile`, this tedium will be taken care of for you—but only if both applications are created with MFC. `CSocketFile` accomplishes this using the same technique you use when saving a document: Serialization. For more on Serialization please refer to your MFC documentation or my previous book, *Visual C++ MFC Programming by Example*.

You can read from an application using serialization with

```
CSocketFile sockFile( &sock );     // where sock is your socket object
CArchive ar( &sockFile,CArchive::load );


ar >> data;                        // and any other data
```

You can write to an application using

```
CSocketFile sockFile( &sock );
CArchive ar( &sockFile,CArchive::store );


ar << data;
```

Streams vs. Datagrams

A Window Socket can be in one of two modes: stream or datagram. All of the examples presented in this book use the stream mode. Only a handful of applications use datagrams, which require less overhead but can be very unreliable. What kind of application can use something that's unreliable? One example is an application that tries to synchronize the clocks of all the systems on a network. Since this type of application is continually sending out messages, perhaps once an hour, it doesn't matter that one or two messages get lost.

Review

Window Sockets allow two or more applications to communicate, even on non-Windows platforms. A special "listening" socket is created first by an application to listen for a request to talk from another application. This listening application then creates another socket to talk through. Please see Figure 3.4.

Figure 3.4   Window Sockets

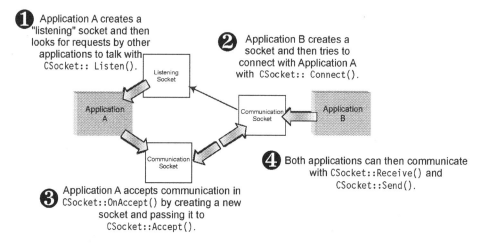

① Application A creates a "listening" socket and then looks for requests by other applications to talk with CSocket:: Listen().

② Application B creates a socket and then tries to connect with Application A with CSocket:: Connect().

③ Application A accepts communication in CSocket::OnAccept() by creating a new socket and passing it to CSocket::Accept().

④ Both applications can then communicate with CSocket::Receive() and CSocket::Send().

Serial/Parallel Communication

Serial and parallel communication are usually reserved for talking to an embedded device, such as a network node, a printer, or a telephone switch, through the ports on the back of your system. Communicating with these devices, however, isn't that much different from accessing a flat file on a disk. In fact, you use the same MFC CFile class. Again, the only difference has to do with the nature of communication: performing synchronous reads and writes in a thread and not trying to read more bytes than are available. Please see Example 52.

Opening and Closing

To open a serial or parallel port for communication you would use

```
CFile file;
CFileException e;
file.Open(
    portName,      // examples "COM1", "COM2", "LPT2"
    CFile::modeReadWrite,
    &e
    );
```

Reading and Writing

To read or write from this port you would use

```
UINT nBytes =          // actual number of bytes read
    Read(
        void* lpBuf,    // buffer to store bytes
        UINT nCount     // number of bytes to read
        );

file.Write(
    void* lpBuf,        // buffer to write
    UINT nCount         // number of bytes to write
    );
```

Configuring the Port

Although all parallel ports are created equal, you must set up a serial port to match the device to which it's talking. That means you need the same baud rate, parity, stop bits, etc. Although you can set these values using the operating system, you can also programmatically set them using the Windows API with SetCommState(). Typically, you will start by getting the current configuration.

```
DCB dcb;
::GetCommState( file.m_hFile, &dcb );
dcb.BaudRate = 1200, ...;
```

```
dcb.ByteSize = 7 or 8;
dcb.StopBits = 0,1,2 = 1, 1.5, 2;
dcb.Parity = 0-4 = no,odd,even,mark,space;
::SetCommState( file.m_hFile, &dcb );
```

Both serial and parallel communications can timeout, which can be undesirable—especially if you perform a synchronized read within a thread that should continue to read until a message comes in. To turn off this timeout, you can use ::SetCommTimeout(), like so.

```
COMMTIMEOUTS cto;
::GetCommTimeouts( file.m_hFile, &cto );
cto.ReadIntervalTimeout = 0;
cto.WriteTotalTimeoutMultiplier = 0;
cto.WriteTotalTimeoutConstant = 0;
::SetCommTimeouts( file.m_hFile, &sto );
```

Review

Serial and parallel communication, then, is achieved with the same API syntax as accessing a flat file—the systems virtual drivers take care of the specifics. Only serial communication involves extra consideration to match its characteristics (e.g., baud rate, etc.) with the device on the other side. Please see Figure 3.5.

Figure 3.5   Serial and Parallel Communication

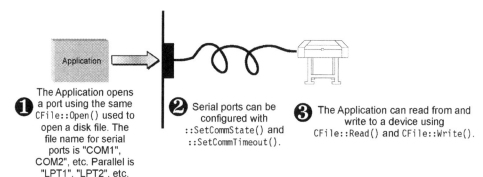

❶ The Application opens a port using the same CFile::Open() used to open a disk file. The file name for serial ports is "COM1", COM2", etc. Parallel is "LPT1", "LPT2", etc.

❷ Serial ports can be configured with ::SetCommState() and ::SetCommTimeout().

❸ The Application can read from and write to a device using CFile::Read() and CFile::Write().

Internet Communication

Several MFC classes encapsulate the Windows API's Internet Extensions (WinInet), allowing you C++ access to the Internet using one of four protocols: File Transfer Protocol (FTP), Hypertext Transfer Protocol (HTTP), gopher, or file. You connect to an Internet website using MFC's CInternetSession class. You can then access a file directly with one of four MFC classes: CStdioFile, CHttpFile, CGopherFile, or CInternetFile. You can also control a website with one of three connection classes: CFtpConnection, CHttpConnection, or CGopherConnection. Both the file classes and connection classes are created by the CInternetSession class.

Opening and Closing Files

To open an Internet file, you first open an Internet session with

```
CInternetSession();    // there are several arguments but you
                       //      will typically use the defaults
```

You can then open a particular type of file by using the OpenURL() member function of CInternetSession.

```
CStdioFile*pFile =        // see Table 3.1 for returned object type
    session.OpenURL(
        LPCTSTR pstrURL    // see Table 3.1 for sample values
        );
// There are several other arguments, however you will typically
//     use the default values.
```

The type of URL that you open determines the class object that CInternetSession creates for you, based on Table 3.1

Table 3.1   OpenURL( ) **Argument Specification**

URL Type	Internet Class Pointer Returned
file://	CStdioFile*
http://	CHttpFile*
gopher://	CGopherFile*
ftp://	CInternetFile*

File objects created this way can only be read. To write files and control the website, refer to the section "Opening and Closing Connections" on page 94.

Reading Files

Reading from any of these file types is similar to reading from a flat file. In fact, CHttpFile, CGopherFile, and CInternetFile are derived from CStdioFile, which allows you to read a file by number of bytes or by string.

```
UINT pFile->Read( void* lpBuf, UINT nCount );


pFile->ReadString( CString str );
```

CHttpFile also allows you to access an HTTP file by object and verb.

Opening and Closing Connections

The CInternetSession class also has three member functions that allow you to open a connection to a website. With a connection, not only can you read and write files, but you can also control the site. For example, an FTP connection allows you to programmatically execute FTP commands on an FTP site.

FTP Connection

To open an FTP connection, you would use

```
CFtpConnection* ftp = GetFtpConnection();
CInternetFile* pFile =
    ftp.OpenFile(
        LPCTSTR pstrFileName,
        DWORD dwAccess = GENERIC_READ,                    // and/or
                                                          // GENERIC_WRITE

        DWORD dwFlags = FTP_TRANSFER_TYPE_BINARY,         // use default
        DWORD dwContext = 1                               // use default
        );
```

To then open a file on that site, you can use

```
CFtpFile *ftpFile = ftp.OpenFile(
    // same as CStdioFile open
    );
```

With this file object, you can read and write.

Gopher Connection

To open a gopher connection, you would use

```
CGopherConnection* gopher = GetGopherConnection();
```

To then open a file on that site, you can use

```
CGopherFile *gopherFile = gopher.OpenFile(
    // same as CStdioFile Open()
    );
```

HTTP Connection

To open an HTTP connection, you would use

```
CHttpConnection* http = GetHttpConnection();
```

To then open a file on that site, you can use

```
CHttpFile* http.OpenRequest(
    // please refer to your MFC documentation for these arguments
    );
```

Other Internet Classes

Three other MFC Internet client classes of interest are `CFtpFileFind`, `CGopherFileFind`, and `CGopherLocator`. `CFtpFileFind` and `CGopherFileFind` are derived from the `CFindFile` class and can locate files over the Internet. The `CGopherLocator` class gets a gopher "locator" from a gopher server and makes the locator available to `CGopherFileFind`.

Communication Summary

The communication method you choose depends on your application. For simple, limited communication, you should use windows messaging. For extended or sophisticated communication, you should use Window Sockets. If you will be communicating with an embedded device, such as a printer, you should use serial/parallel communication. You really have no choice for Internet communication.

You shouldn't use DDE or Message Pipes at all in new development. Window Sockets are a much more flexible solution to interprocess communication and have class support in the foundation classes.

Although DDE represented a faster way to communicate data than Window Sockets, there really is no alternative to sharing data over a network other than to send one byte through the wire at a time.

Sharing Data

So far, we have covered active communication between applications in which both applications had to participate. Your application also has two ways to communicate passively by allowing other applications to access its data.

Shared Memory File uses the same globally allocated memory that DDE uses and, therefore, can't be accessed from another machine.

File Mapping allows your application to share a file, and even memory, with another application—even over the network to another Windows machine.

Note: File Mapping is still not an option if you want to share data with a non-Windows platform (e.g., UNIX). In this case, you are still stuck with Window Sockets. But don't worry, File Mapping doesn't afford your application any faster data access then Window Sockets—it still shares its data one byte at a time over the network.

Note: Your application can also share data through the clipboard. However, the clipboard is typically reserved for user interaction.

Shared Memory File

If you don't need to follow the DDE standard and you don't need to share your data across the network, you can use your own globally allocated memory to share data with other applications. In fact, MFC's CSharedFile class makes creating and accessing global memory just as easy as creating a flat file.

Creating and Destroying

Just creating and destroying an instance of the CSharedFile class will create and destroy your global memory.

```
CSharedFile file;
```

Reading and Writing

Reading and writing to this file is identical to reading and writing bytes from a flat file. Therefore, for a complete transfer of data scenario, you would write with

```
CSharedFile file;
file.Write( buffer, nBytes );
// extract global memory handle from CFile object to send another
application
HGLOBAL hgbl = file.Detach();
SendMessage( hWnd, WM_MYMESSAGE, ( WPARAM ) hgbl,0 );
```

You would read in the receiving program with

```
HRESULT OnMyMessage( WPARAM wParam, LPARAM lParam )
{
    CSharedFile file;
    // encapsulate global memory handle in this CFile object
    file.SetHandle( ( HGLOBAL ) wParam );
    file.Read( buffer,nBytes );
}
```

Review

Data can be shared through global memory using the CSharedFile class using the same function syntax as you would use accessing a flat file. Please see Figure 3.6.

Figure 3.6   Shared Memory Files

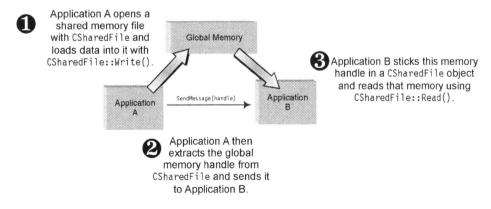

① Application A opens a shared memory file with CSharedFile and loads data into it with CSharedFile::Write().

Global Memory

③ Application B sticks this memory handle in a CSharedFile object and reads that memory using CSharedFile::Read().

Application A   SendMessage(handle) →   Application B

② Application A then extracts the global memory handle from CSharedFile and sends it to Application B.

File Mapping

If you need to share data among applications that can potentially be on other Windows platforms, you can use File Mapping. Unlike globally allocated memory, which can't be accessed by more than one application at once, File Mapping allows two or more applications to simultaneously share memory. For example, you can have an array called Bob[100] in two applications, such that storing 34 in Bob[23] in Application A causes 34 to appear in Bob[23] in Application B.

This sharing actually takes place within a file—thus, the name File Mapping. When sharing data on just one system, you will typically use the default file, which is actually the swap file used to provide your system with virtual memory. If you will be sharing over a network, you will need to open the file yourself and provide its handle.

Opening and Closing

To open a segment of the swap file for shared memory, you would use

```
m_hMap = ::CreateFileMapping(
    ( HANDLE ) Oxffffffff,      // or can be an open file handle
    0,                          // security
    PAGE_READWRITE,             // or PAGE_READONLY or PAGE_WRITECOPY
    0,                          // size -- high order
                                //      (required if no file handle)
    0x1000,                     // size -- low order
                                //      (required if no file handle)
    MAP_ID                      // unique id--required if no file handle
    );
```

This returned map handle can be turned into a memory pointer with

```
m_pSharedData = ::MapViewOfFile( m_hMap,
    FILE_MAP_WRITE,     // or FILE_MAP_READ, FILE_MAP_COPY
                        //      (FILE_MAP_WRITE is read/write)
    0,                  // offset --high order
    0,                  // offset -- low order
    0                   // number of bytes (zero maps entire file)
    );
// When using swap file, offset must be zero(0)
```

To close this shared memory, you would use

```
::UnmapViewOfFile(m_pSharedData);
::CloseHandle(m_hMap);
```

Reading and Writing

If, as mentioned previously, we were sharing an array called Bob, we could simply use the following in any application

```
int *Bob = ( int * ) pSharedData;
Bob[23] = 34;
```

Data Synchronization

All access to mapped memory on the same system is synchronized. In other words, two applications can't access the same memory location at the same time. This prevents, for instance, one application from reading from an area of memory at the same time another application is writing to it.

Mapping a file over a network, however, can run into trouble. If the file through which you are sharing is on another system, your application will actually be writing to a network buffer that, over time, will be sent to the other system. In the meantime, another application on another system could also be writing to that file. In this case, you will need to manually make sure your applications aren't clobbering each other.

Review

Memory sharing can be achieved through a file. There is no other way for two applications to share the same address space. Please see Figure 3.7.

Figure 3.7   File Mapping

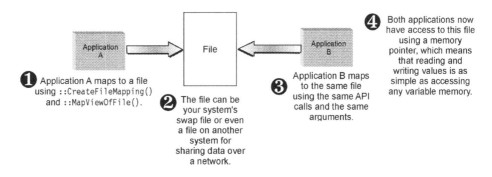

Client/Server

Not only can you send a message to another application or share data with it, but you can even access its functionality by calling its member functions indirectly. An application that shares its functionality this way is called a Server—it's there to serve you. Applications that access this functionality are called Clients—they're the customers.

Allowing one application to share its functionality with others can be accomplished with any of the communication methods presented previously with essentially the same steps.

1. Create a new command message to represent each function you want to access (e.g., IDC_CALL_HOME for the Home() method).

2. Send this message to the application that's sharing its functionality (the Server) and pass along in the same message any arguments required by the desired function.

```
wParam = x;
1Param = y;
SendMessage( hWnd,IDC_CALL_HOME,wParam,1Param );
```

3. The Server will then convert your message and arguments into an actual function call to the desired function.

```
case IDC_CALL_HOME:
    x = wParam;
    y = 1Param;
    res = Home( x,y );
    :    :    :
```

4. Any values returned by the function are returned to your application, either with this message

```
return res;
```

or with a new message

```
wParam = y;
SendMessage( IDC_CALL_HOME_REPLY,wParam,1Param );
```

5. The Client can then respond to a reply message by sticking the following values into local memory.

```
case IDC_CALL_HOME_REPLY:
    y = wParam;
```

Please see Figure 3.8 for an overview of this technique.

Figure 3.8 Indirectly Calling an External Function

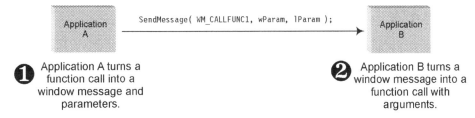

① Application A turns a function call into a window message and parameters.

② Application B turns a window message into a function call with arguments.

Passing Calling Arguments

The method you use to pass your arguments will depend on the method you're using to communicate. As seen previously, Windows Messaging allows you to pass two arguments with the message. To pass additional arguments, you can stick them in a structure and pass that structure using DDE, File Mapping, or even some globally shared memory.

You can't just simply pass a pointer to this structure to the other application because the structure exists in a different address space than the other application (a pointer in one application wouldn't point to the same data in another application). This also means you can't stick pointers in your message structure—the contents of entire arrays and argument data must be contained in this structure for it to be accessible to the other application.

When the Server Application is on another system, you can't even use DDE or global memory to pass your arguments—there's no way for the other system to access it. You can still use File Mapping to contain the arguments or you can pass them in the message itself. If you pass the arguments in the communication, any return values must also be communicated back to the Client Application. That means if the Server can potentially modify an array, the entire array must be sent back to the Client. File Mapping accomplishes the same thing, except in the background. Either way, getting large or numerous arguments back and forth can be slow—so you should plan accordingly.

Remote Procedure Calling

Arguments are most commonly passed between systems in the communication itself. In fact, there's a Windows API that will handle a lot of the tedium of accessing a Server Application over a network called Remote Procedure Call (RPC) and it sends the arguments in the message. RPC is used extensively by Microsoft's Component Object Model (COM) components, but it falls outside of the scope of this book.

Summary

In this chapter, we discovered six methods available to your application to communicate with its environment.

- **Windows messaging** for simple interprocess communication
- **Dynamic Data Exchange** for sending large amounts of data
- **Message Pipes** for communication over a network
- **Window Sockets** for communication over a network to non-Windows platforms
- **Internet Classes** for communicating over the Internet
- **Serial/Parallel Communication** for talking to devices connected to your system's serial or parallel port

We also found that Window Sockets supersede DDE and Message Pipes for the ability to distribute your application over a network.

For sharing data between applications, we reviewed two methods.

- **Share Memory Files** are created by the MFC `CShareFile` class to wrap globally allocated memory for easy C++ access.
- **File Mapping** allows sharing data, even across a network—but not to non-Windows platforms.

This concludes the pure text portion of this book. In the following sections, we will look at MFC and Visual C++ from the other angle. Actual features you might want to add to your own application will be presented, along with a step-by-step guide to implementing them. And yes, there will be more notes explaining how these features are implemented so that they might be broadened and enhanced.

User Interface Examples

The examples in this section concentrate on the user interface aspect of the applications you can create with the help of the Developer Studio, the Microsoft Foundation Classes (MFC), and Visual C++. As you might expect from a user interface development tool, the vast majority of the examples in this book will be in this section. The topics covered by chapters in this section include the following.

Applications and Environments

Examples in Chapter 4 cover the different aspects of how your application interacts with its environment. This includes everything from how it runs to how it appears to the user. The examples range from finding new places to put a logo to preventing two instances of the same application from running at the same time.

Menu, Toolbars, and Status Bars

Chapter 5 covers the next area of concern—your application's menus and controls bars, which are perhaps the primary way your user has to interact with your application. The AppWizard will automatically add a generic menu, toolbar, and status bar to your application, but these menus and bars pale in comparison to the menu and bars that even your own Developer Studio use. With a little manual effort, however, you can add this same look to your own application.

Views

If you choose to create either an SDI or MDI application, your application's view will be your user's next primary mode of interacting with your application and, in particular, the document that your application is editing. All of the examples in Chapter 6 relate to the view, from creating a view out of a Property Sheet to printing a view or dragging files into your view.

Dialog Boxes and Bars

Dialog boxes and bars allow you to prompt your user with several types of controls, such as buttons and list boxes. MFC and the Developer Studio will automatically create dialog boxes and classes for you. In Chapter 7, we look at ways to manually modify your dialog boxes to be more useful to your user.

Control Windows

Dialog Boxes are populated with buttons and edit boxes that are collectively known as Control Windows, which are child windows supplied by the operating system. Not only can you fill your dialog boxes with them, but you can put them in views, in bars, or anywhere there's a window. You can even draw them yourself. Examples given in Chapter 8.

Drawing

Bitmaps and icons allow you to add color and style to your application. All Windows interfaces are essentially alike, so logos and splash screens are really your only way to distinguish the look of your application from someone else's. Drawing is obviously also important for creating your own controls and displaying figures in a CAD application. Examples given in Chapter 9.

Help

On-line help can reduce the learning curve required for your application. Rather than try to identify the appropriate command in a manual, your user can directly interrogate the control itself. The examples in Chapter 10 demonstrate how to implement three types of on-line help.

Plain Windows

The views, dialog boxes and bars, control windows, and toolbars that make up your MFC application's interface are, in fact, all based on the lowly window. So why bother with this type of window when the others are so much more complete? Because by programming at this level, you can achieve effects that you can't achieve with a more highly evolved window. Examples are given in Chapter 11.

Specialized Applications

We round off this section with a brief review of some specific MFC applications. The examples in Chapter 12 include a couple of simple text editors, as well as two database editors. One example provides all you need to create your own Explorer-type application. Lastly, Chapter 12 includes creating a simple Wizard.

II

Chapter 4

Applications and Environments

In this chapter, we will look at different aspects of how an application interacts with its environment. This includes everything from how it runs to how it appears to the user. The examples here range from finding new places to put a logo to preventing two instances of the same application from running at the same time.

Example 1   Putting a Static Logo in the Toolbar   We will place a picture of your program's affiliation in the toolbar where it can always be seen and cherished.

Example 2   Putting an Animated Logo in the Toolbar   We will animate the last example.

Example 3   Starting Only One Instance   We will prevent more than one instance of your application from executing at one time.

Example 4   Creating a Dialog-MDI Hybrid Application   We will cross two standard application types.

Example 5   Putting Icons in the System Tray   We will put your own icon in the system tray. (The system tray is that collection of icons on the desktop next to the time display.)

Example 6   Putting a Logo in the Main Window's Title Bar   We will use the caption bar of our application to display a logo or other bitmap.

Example 1   Putting a Static Logo in the Toolbar

Objective

You would like to display a logo in the right corner of your application's toolbar, as seen in Figure 4.1.

Figure 4.1   Putting a Static Logo in the Toolbar

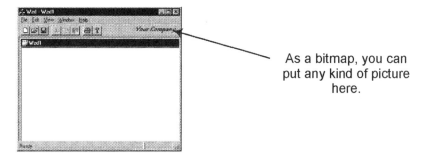

As a bitmap, you can put any kind of picture here.

Strategy

A toolbar is a single control window that paints all of the buttons in its client area. When the user clicks on a button, the toolbar determines which image they're clicking and triggers the appropriate command. We will simply be adding a child window to our application's toolbar away from these buttons. This child window will be a plain window that will display a bitmap and constantly move as the toolbar changes size.

Example 1   Putting a Static Logo in the Toolbar   **109**

Steps

Create a Plain Window Class That Will Draw the Logo

1. Use the ClassWizard to create a generic window class derived from "generic CWnd".

2. Use the bitmap class provided on the CD with this example to embed a bitmap variable in this window class, as seen here.

```
CWzdBitmap m_bitmap;
```

Note: This bitmap class was created in the previous book, *Visual C++ MFC Programming by Example*, not only to load a bitmap resource, but also to load the bitmap's palette so that it can be used to accurately recreate the colors in the bitmap.

---

3. Add a CreateBX() member function to this class, which will load our bitmap logo and also create the actual window in the toolbar.

```
void CWzdLogo::CreateBX( CWnd *pWnd, UINT nBitmapID, UINT nChildID )
{
    m_bitmap.LoadBitmapEx( nBitmapID,TRUE );
    CRect rect( 0,0,0,0 );    // will be resizing constantly anyways
    Create( NULL,_T( "" ),WS_CHILD|WS_VISIBLE,rect,pWnd,nChildID );
}
```

4. Add another member function that will allow our toolbar to move this window.

```
void CWzdLogo::MoveLogo( int nWidth, int nHeight )
{
    MoveWindow( nWidth-m_bitmap.m_Width,0,m_bitmap.m_Width,nHeight );
}
```

5. Finish this class by using the ClassWizard to add a WM_PAINT message handler that will draw our bitmap to the window.

```
void CWzdLogo::OnPaint()
{

    CPaintDC dc( this );      // device context for painting

    // get bitmap colors
    CPalette *pOldPal =
        dc.SelectPalette( m_bitmap.GetPalette(),FALSE );
    dc.RealizePalette();

    // get device context to select bitmap into
    CDC dcComp;
    dcComp.CreateCompatibleDC( &dc );
    dcComp.SelectObject( &m_bitmap );

    // draw bitmap
    dc.BitBlt( 0,0,m_bitmap.m_Width, m_bitmap.m_Height, &dcComp,
        0,0,SRCCOPY );
    // reselect old palette
    dc.SelectPalette( pOldPal,FALSE );
}
```

6. To see this plain window class in its entirety, please refer to the Logo Class for Example 1 (page 112).

Create a Custom Toolbar Class That Will Contain the Plain Window Class

1. Use the ClassWizard to create a new class from CToolBarCtrl. Edit the resulting .h and .cpp files to change all references from CToolBarCtrl to CToolBar. (The ClassWizard doesn't allow you to derive a class from CToolBar.)

2. Embed our new plain window class in our new toolbar class, as seen here.

```
CWzdLogo m_Logo;
```

Example 1   Putting a Static Logo in the Toolbar   **111**

3. Use the `ClassWizard` to add a `WM_CREATE` message handler to the toolbar class. Use this handler to call the `CreateBX()` function of the plain window class.

```
int CWzdToolbar::OnCreate( LPCREATESTRUCT lpCreateStruct )
{
    if ( CToolBar::OnCreate( lpCreateStruct ) == -1 )
        return -1;

    m_Logo.CreateBX( this,IDB_LOGO,-1 );

    return 0;
}
```

4. Use the `ClassWizard` again to add a `WM_SIZE` message handler where we will call the `MoveLogo()` function of our plain window class.

```
void CWzdToolbar::OnSize( UINT nType, int cx, int cy )
{
    CToolBar::OnSize( nType, cx, cy );

    m_Logo.MoveLogo( cx,cy );
}
```

5. To see this toolbar class in its entirety, please refer to the Toolbar Class for Example 1 (page 115).

Use the New Toolbar Class in Your Application

1. Now substitute the new toolbar class for `CToolBar` in your `MainFrm.h` file.

```
// change CToolBar to CWzdToolBar
CWzdToolbar    m_wndToolBar;
```

2. Unfortunately, this example won't work for a dockable toolbar because a dockable toolbar will only be as long as the last button it contains. Our example depends on the toolbar having plenty of extra space at the end to put our plain window. Therefore, we must make this toolbar undockable

by commenting out or deleting the following lines in `CMainFrame::OnCreate()`.

```
int CMainFrame::OnCreate( LPCREATESTRUCT lpCreateStruct )
{
        :    :    :

    // comment out or delete these lines
//  m_wndToolBar.EnableDocking( CBRS_ALIGN_ANY );
//  EnableDocking( CBRS_ALIGN_ANY );
//  DockControlBar( &m_wndToolBar );

    return 0;
}
```

Notes

- Actually, you can use this example to stick a logo potentially anywhere there's empty space in a parent window. Just embed this window in the parent window's class and use the ClassWizard to add a `WM_CREATE` message handler that will create the logo window.
- If you would like to put a logo in a dockable toolbar, please refer to Example 2. There you will find that by overloading `CalcDynamicLayout()` in your `CToolBar` class, you can inflate the size of your toolbar to include your logo. Make sure, however, to provide a vertical and horizontal version of this logo.

CD Note

- When executing the project on the CD, you will notice that the toolbar will have "Your Company" written on the right hand side.

Logo Class for Example 1

```
#if !defined WZDLOGO_H
#define WZDLOGO_H

// WzdLogo.h : header file
//
#include "WzdBitmap.h"
```

Example 1   Putting a Static Logo in the Toolbar   **113**

```
/////////////////////////////////////////////////////////////////////////
// CWzdLogo window

class CWzdLogo : public CWnd
{
// Construction
public:
    CWzdLogo();

// Attributes
public:

// Operations
public:
    void CreateBX( CWnd *pWnd, UINT nBitmapID, UINT nChildID );
    void MoveLogo( int nWidth, int nHeight );

// Overrides
    // ClassWizard generated virtual function overrides
    // {{AFX_VIRTUAL( CWzdLogo )
    // }}AFX_VIRTUAL

// Implementation
public:
    virtual ~CWzdLogo();

    // Generated message map functions
protected:
    // {{AFX_MSG(CWzdLogo)
    afx_msg void OnPaint();
    // }}AFX_MSG
    DECLARE_MESSAGE_MAP()
private:
    CWzdBitmap m_bitmap;
};


/////////////////////////////////////////////////////////////////////////

#endif
```

II   4

```cpp
// WzdLogo.cpp : implementation file
//

#include "stdafx.h"
#include "WzdLogo.h"

#ifdef _DEBUG
#define new DEBUG_NEW
#undef THIS_FILE
static char THIS_FILE[] = __FILE__;
#endif

/////////////////////////////////////////////////////////////////////////////
// CWzdLogo

CWzdLogo::CWzdLogo()
{
}

CWzdLogo::~CWzdLogo()
{
}


BEGIN_MESSAGE_MAP( CWzdLogo, CWnd )
    // {{AFX_MSG_MAP( CWzdLogo )
    ON_WM_PAINT()
    // }}AFX_MSG_MAP
END_MESSAGE_MAP()


/////////////////////////////////////////////////////////////////////////////
// CWzdLogo message handlers

void CWzdLogo::OnPaint()
{
    CPaintDC dc( this );    // device context for painting

    // get bitmap colors
    CPalette *pOldPal = dc.SelectPalette( m_bitmap.GetPalette(),FALSE );
```

Example 1   Putting a Static Logo in the Toolbar   **115**

```
    dc.RealizePalette();

    // get device context to select bitmap into
    CDC dcComp;
    dcComp.CreateCompatibleDC( &dc );
    dcComp.SelectObject( &m_bitmap );

    // draw bitmap
    dc.BitBlt( 0,0,m_bitmap.m_Width,m_bitmap.m_Height, &dcComp, 0,0,SRCCOPY );

    // reselect old palette
    dc.SelectPalette(pOldPal,FALSE);
}

void CWzdLogo::CreateBX( CWnd *pWnd, UINT nBitmapID, UINT nChildID )
{
    m_bitmap.LoadBitmapEx( nBitmapID,TRUE );
    CRect rect( 0,0,0,0 );
    Create( NULL,_T( "" ),WS_CHILD|WS_VISIBLE,rect,pWnd,nChildID );
}


void CWzdLogo::MoveLogo( int nWidth, int nHeight )
{
    MoveWindow( nWidth-m_bitmap.m_Width,0,m_bitmap.m_Width,nHeight );
}
```

Toolbar Class for Example 1

```
#if !defined WZDTOOLBAR_H
#define WZDTOOLBAR_H

// WzdToolbar.h : header file
//
#include "WzdLogo.h"
/////////////////////////////////////////////////////////////////////////////
// CWzdToolbar window

class CWzdToolbar : public CToolBar
{
```

```cpp
// Construction
public:
    CWzdToolbar();

// Attributes
public:

// Operations
public:

// Overrides
    // ClassWizard generated virtual function overrides
    // {{AFX_VIRTUAL(CWzdToolbar)
    // }}AFX_VIRTUAL

// Implementation
public:
    virtual ~CWzdToolbar();

    // Generated message map functions
protected:
    // {{AFX_MSG(CWzdToolbar)
    afx_msg void OnSize( UINT nType, int cx, int cy );
    afx_msg int OnCreate( LPCREATESTRUCT lpCreateStruct );
    // }}AFX_MSG

    DECLARE_MESSAGE_MAP()
private:
    CWzdLogo m_Logo;
};

//////////////////////////////////////////////////////////////////////

#endif
// WzdToolbar.cpp : implementation file
//

#include "stdafx.h"
#include "wzd.h"
#include "WzdToolbar.h"
```

Example 1   Putting a Static Logo in the Toolbar   **117**

```
#ifdef _DEBUG
#define new DEBUG_NEW
#undef THIS_FILE
static char THIS_FILE[] = __FILE__;
#endif


//////////////////////////////////////////////////////////////////////////////
// CWzdToolbar

CWzdToolbar::CWzdToolbar()
{
}


CWzdToolbar::~CWzdToolbar()
{
}



BEGIN_MESSAGE_MAP( CWzdToolbar, CToolBar )
    // {{AFX_MSG_MAP(CWzdToolbar)
    ON_WM_SIZE()
    ON_WM_CREATE()
    // }}AFX_MSG_MAP
END_MESSAGE_MAP()


//////////////////////////////////////////////////////////////////////////////
// CWzdToolbar message handlers

int CWzdToolbar::OnCreate( LPCREATESTRUCT lpCreateStruct )
{
    if ( CToolBar::OnCreate(lpCreateStruct) == -1 )
        return -1;

    m_Logo.CreateBX( this,IDB_LOGO,-1 );

    return 0;
}

void CWzdToolbar::OnSize( UINT nType, int cx, int cy )
```

II   4

```
{
    CToolBar::OnSize( nType, cx, cy );

    m_Logo.MoveLogo( cx,cy );
}
```

Example 2   Putting an Animated Logo in the Toolbar

Objective

You would like to play an AVI file (an animated bitmap) in the right corner of your application's toolbar, as seen in Figure 4.2.

Figure 4.2   Putting an Animated Logo in the Toolbar

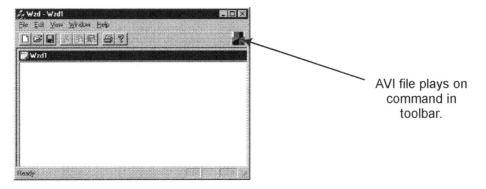

AVI file plays on command in toolbar.

Strategy

We will create our own toolbar class and put an animation control in its far corner. We will be deriving our toolbar class from MFC's CToolBar class and creating the animation control window with MFC's CAnitmateCtrl class.

Example 2   Putting an Animated Logo in the Toolbar   **119**

Steps

Add an AVI File to Your Application's Resources

1. Click on the Developer Studio's "Insert" and "Resources" menu commands to bring up the "Insert Resources" dialog box. Click on "Import" and locate the desired AVI file. When prompted for resource type, enter "AVI". This command will then copy that AVI file into your project's \res file and add an AVI folder to your project's resources. Right-click on the resource ID to give it a more appropriate name. In this example, we give it the name IDR_MFC2.

Create a Custom Toolbar Class

1. Use the ClassWizard to create a new class derived from CToolBarCtrl. Use the text editor to change all references to CToolBarCtrl to CToolBar. (Note: The ClassWizard doesn't currently allow you to derive a class from CToolBar).

2. Embed a CAnimateCtrl class variable in this new toolbar class.

```
.h:
CAnimateCtrl m_AnimateCtrl;
```

3. Use the ClassWizard to add a WM_CREATE message handler to your new toolbar class. Use this handler to create the animation control window we embedded in the last step. We will also load the AVI resource we created previously and initially play it three times.

```
int CWzdToolbar::OnCreate( LPCREATESTRUCT lpCreateStruct )
{
    if ( CToolBar::OnCreate(lpCreateStruct) == -1 )
        return -1;

    m_AnimateCtrl.Create( WS_CHILD| WS_VISIBLE| ACS_CENTER,
        CRect( 0,0,0,0 ), this, IDC_ANIMATE_CTRL );
    m_AnimateCtrl.Open( IDR_MFC2 );
```

II   4

```
    m_AnimateCtrl.Play( 0,-1,3 );      // play three times initially

    return 0;
}
```

Note that we are creating the animation control window initially with no size (CRect(0,0,0,0)). At this point, size doesn't matter because we'll be changing the size immediately.

4. Use the ClassWizard to add a WM_SIZE message handler to the toolbar class. There we will determine the current size of the toolbar and move the animation control to the far left.

```
void CWzdToolbar::OnSize( UINT nType, int cx, int cy )
{
    CToolBar::OnSize( nType, cx, cy );

    CRect rect;
    GetWindowRect( &rect );
    ScreenToClient( &rect );
    rect.left = rect.right-32;
    m_AnimateCtrl.MoveWindow( rect );
}
```

5. We will also add two helper functions to our toolbar class that will allow another class to play this AVI file.

```
void CWzdToolbar::PlayLogo()
{
    m_AnimateCtrl.Play( 0,-1,-1 );
}

void CWzdToolbar::StopLogo()
{
    m_AnimateCtrl.Stop();
}
```

6. To see the toolbar class in its entirety, please refer to the Toolbar Class for Example 2 (page 122).

Example 2   Putting an Animated Logo in the Toolbar   **121**

Use the New Toolbar Class in `CMainFrame`

1. Substitute our new toolbar class for the class currently used by `CMain-Frame`.

```
// use CWzdToolbar for toolbar in Mainfrm.h
CWzdToolbar    m_wndToolBar;
```

We must also disable this toolbar's ability to float and dock by commenting out the following lines in `MainFrm.cpp`. Otherwise, the toolbar will resize itself to include only its buttons.

```
// disable toolbar docking in Mainfrm.cpp
    // TODO: Delete these three lines if you don't want the toolbar to
    //    be dockable
//  m_wndToolBar.EnableDocking(CBRS_ALIGN_ANY);
//  EnableDocking(CBRS_ALIGN_ANY);
//  DockControlBar(&m_wndToolBar);
```

Notes

- Your animation control window won't fully extend to the actual border of the toolbar control because the toolbar control itself draws a blank border around its controls. If you would like to extend to the borders, please refer to my article "Making MFC Docking Bars Cool" in WDJ (*Windows Developer's Journal*) 5/99 at website www.wdj.com.
- An AVI file is created from two or more bitmap files that have been converted into the frames of an AVI file. Utilities to make this conversion are available as shareware packages over the Internet. They can be had for as little as $40.

CD Notes

- When executing the project on the CD, you will notice an AVI file playing in the right hand side of the toolbar.

Toolbar Class for Example 2

```cpp
#if !defined WZDTOOLBAR_H
#define WZDTOOLBAR_H


// WzdToolbar.h : header file
//
/////////////////////////////////////////////////////////////////////////////
// CWzdToolbar window

class CWzdToolbar : public CToolBar
{
// Construction
public:
    CWzdToolbar();

// Attributes
public:

// Operations
public:
    void PlayLogo();
    void StopLogo();

// Overrides
    // ClassWizard generated virtual function overrides
    // {{AFX_VIRTUAL( CWzdToolbar )
    // }}AFX_VIRTUAL

// Implementation
public:
    virtual ~CWzdToolbar();

    // Generated message map functions
protected:
    // {{AFX_MSG( CWzdToolbar )
    afx_msg void OnSize( UINT nType, int cx, int cy );
    afx_msg int OnCreate( LPCREATESTRUCT lpCreateStruct );
    // }}AFX_MSG
```

Example 2   Putting an Animated Logo in the Toolbar   **123**

```
    DECLARE_MESSAGE_MAP()
private:
    CAnimateCtrl m_AnimateCtrl;
};

/////////////////////////////////////////////////////////////////////////////

#endif
// WzdToolbar.cpp : implementation file
//

#include "stdafx.h"
#include "wzd.h"
#include "WzdToolbar.h"

#ifdef _DEBUG
#define new DEBUG_NEW
#undef THIS_FILE
static char THIS_FILE[] = __FILE__;
#endif

/////////////////////////////////////////////////////////////////////////////
// CWzdToolbar

CWzdToolbar::CWzdToolbar()
{
}

CWzdToolbar::~CWzdToolbar()
{
}


BEGIN_MESSAGE_MAP( CWzdToolbar, CToolBar )
    // {{AFX_MSG_MAP(CWzdToolbar)
    ON_WM_SIZE()
    ON_WM_CREATE()
    // }}AFX_MSG_MAP
END_MESSAGE_MAP()
```

II 4

```
/////////////////////////////////////////////////////////////////////////////
// CWzdToolbar message handlers

int CWzdToolbar::OnCreate( LPCREATESTRUCT lpCreateStruct )
{
    if ( CToolBar::OnCreate( lpCreateStruct ) == -1 )
        return -1;

    m_AnimateCtrl.Create(
        WS_CHILD|WS_VISIBLE|WS_DLGFRAME|WS_EX_CLIENTEDGE|ACS_CENTER,
        CRect( 0,0,0,0 ), this, IDC_ANIMATE_CTRL );
    m_AnimateCtrl.Open( IDR_MFC2 );
    m_AnimateCtrl.Play( 0,-1,3 );

    return 0;
}

void CWzdToolbar::OnSize( UINT nType, int cx, int cy )
{
    CToolBar::OnSize( nType, cx, cy );

    CRect rect;
    GetWindowRect( &rect );
    ScreenToClient( &rect );
    rect.left = rect.right-32;
    m_AnimateCtrl.MoveWindow( rect );
}

void CWzdToolbar::PlayLogo()
{
    m_AnimateCtrl.Play(0,-1,-1);
}

void CWzdToolbar::StopLogo()
{
    m_AnimateCtrl.Stop();
}
```

Example 3   Starting Only One Instance   **125**

Example 3   Starting Only One Instance

Objective

You would like to prevent more than one instance of your application from running on a system at any one time.

Strategy

This functionality is not currently supported by MFC or the Windows API, so we are left to use a kludge. The approach recommended by Microsoft is for your application to create something unique on the system that future versions could check against before running. For this example, we will create a uniquely named mutex resource. Mutexes are normally used to help synchronize two or more threads that are using the same data area. We use the MFC class that wraps the mutex API in Example 59. However, since we need to set the name of the mutex in this example, we will be using the API directly.

Steps

Set Up Your Application

1. Define a unique name using #define in your Application Class. One way to ensure that your name is absolutely unique all over the world is to use the GUID generator provided originally for giving COM interfaces a unique ID. You can find the GUID generator in your VC++'s \BIN directory as a program called GUIDGEN.EXE. An example definition of a unique name using GUIDGEN.EXE is as follows.

```
#define UNIQUE_NAME "{F5EFF561-ECB3-11d1-A18D-DCB3C85EBD34}"
```

Try to Create the Mutex in InitInstance()

1. Create a mutex using the previously defined name, UNIQUE_NAME, at the start of the InitInstance() function of your Application Class. Save the handle—you'll need it later to close this mutex. If another instance of your application already exists, the ::CreateMutex() function will return a handle to the mutex created by that other instance instead of creating a

new one. Call `GetLastError()` to determine if this is the handle to an existing mutex and, therefore, that another instance of your application is already running. If so, `GetLastError()` will return an error of `ERROR_ALREADY_EXISTS`. You can then exit `InitInstance()`, returning a value of `FALSE` to stop this application from running. The following shows how this might be accomplished.

```
BOOL CWzdApp::InitInstance()
{

    m_hOneInstance = ::CreateMutex( NULL,FALSE,UNIQUE_NAME );
    if ( GetLastError() == ERROR_ALREADY_EXISTS )
    {

        AfxMessageBox( "Application already running!" );
        return FALSE;

    }


        :     :     :

}
```

Close the Mutex

1. Use the `ClassWizard` to override your Application Class's `ExitInstance()` function and close this mutex's handle there.

```
int CWzdApp::ExitInstance()
{

    CloseHandle( m_hOneInstance );
    return CWinApp::ExitInstance();

}
```

Notes

- Rather than simply displaying an error that an instance of your application is already running, you might instead broadcast a message to the other instance to tell it to bring its window to the front. See Example 49 for how to broadcast a message to other applications.

- Closing the handle is done only for the sake of completeness. When your application terminates, Windows cleans up any mutexes it has created. This, incidentally, makes this approach more robust—if your application abnormally terminates and, therefore, doesn't have a chance to call

Example 4   Creating a Dialog-MDI Hybrid Application   **127**

CloseHandle(), it doesn't leave a mutex lying around to prevent your application from running again.

- Another method to determine if another instance of your application is already running would be to use CWnd::FindWindow(). FindWindow() looks for all top-level windows with a particular window title and window class name. You can either create a unique but hidden plain window (see Example 38) for which future instances of your application can look, or you can look for your application's main window. If your main window won't have a unique or constant caption, you can give it a unique window class name to look for instead. You can create this unique class name in the PreCreateWindow() function of CMainFrame. The drawback to this approach is that there's a time lag between when your instance is created and when your main window is created that a second instance could slip through. Microsoft doesn't recommend this approach.

CD Notes

- Execute the project on the CD twice. Attempting to execute it a second time will cause a message to appear that the application is already running.

Example 4   Creating a Dialog-MDI Hybrid Application

Objective

You would like to create an application that can be either Dialog or MDI, depending on a command line flag.

Strategy

We will create an MDI application using the AppWizard. This application will then check for a /d flag in its InitInstance() function and, if set, we will create a modal dialog box instead of opening an MDI frame window. Once this dialog is closed, we will leave InitInstance() with a value of FALSE, which will cause our application to terminate.

Steps

Set Up Your Application

1. Use the AppWizard to create an MDI application.
2. Use the Dialog Editor and ClassWizard to create a dialog template and class.
3. Use the ClassWizard to create your own version of the `CCommandLineInfo` class. Modify this class to check for a /d switch.

Modify `InitInstance()` to Create One of Two Application Types

1. Substitute the new `CCommandLineInfo` class for the original in your Application Class.

```
CWzdCommandLineInfo cmdInfo;
ParseCommandLine( cmdInfo );
```

2. If the /d appears on the command line, use your new dialog class to create a modal dialog box. When that box closes, return from `InitInstance()` with a value of `FALSE`, causing your application to terminate. Please refer to the following for what this would look like.

```
// if user started with /d flag, start dialog app instead
if ( cmdInfo.m_bDialogFg )
{
    CWzdDialog dlg;
    dlg.DoModal();
    return FALSE;
}
```

Notes

- Another permutation on this idea would be to create an MDI or SDI application based on a command line option. You would start by creating an MDI application with the AppWizard. Then, based on whether a certain flag was set, you would either define an SDI or MDI Document Template. For particulars, create an SDI application using the AppWizard and integrate its `InitInstance()` function with your MDI application's `InitInstance()`.

Example 4   Creating a Dialog-MDI Hybrid Application   **129**

CD Notes

- Executing the project on the CD without a /d switch will cause it to appear as an MDI application. Using the /d switch will cause it to appear as a Dialog application.

Command Line Info Class for Example 4

```
#if !defined WZDCOMMANDLINEINFO_H
#define WZDCOMMANDLINEINFO_H


// WzdCommandLineInfo.h : header file
//


/////////////////////////////////////////////////////////////////////////////
// CWzdCommandLineInfo window

class CWzdCommandLineInfo : public CCommandLineInfo
{
// Construction
public:
    CWzdCommandLineInfo();

// Attributes
public:
    BOOL m_bDialogFg;

// Operations
public:
    void ParseParam( const TCHAR* pszParam,BOOL bFlag,BOOL bLast );

// Overrides

// Implementation
public:
    virtual ~CWzdCommandLineInfo();

};


/////////////////////////////////////////////////////////////////////////////
```

II   4

```
#endif

// WzdCommandLineInfo.cpp : implementation file
//

#include "stdafx.h"
#include "wzd.h"
#include "WzdCommandLineInfo.h"

#ifdef _DEBUG
#define new DEBUG_NEW
#undef THIS_FILE
static char THIS_FILE[] = __FILE__;
#endif

/////////////////////////////////////////////////////////////////////////////
// CWzdCommandLineInfo

CWzdCommandLineInfo::CWzdCommandLineInfo()
{
    m_bDialogFg = FALSE;
}


CWzdCommandLineInfo::~CWzdCommandLineInfo()
{
}


/////////////////////////////////////////////////////////////////////////////

void CWzdCommandLineInfo::ParseParam( const TCHAR* pszParam,BOOL bFlag,
    BOOL bLast )
{
    CString sArg( pszParam );
    if ( bFlag )
    {
        m_bDialogFg = !sArg.CompareNoCase( "d" );
```

Example 5   Putting Icons in the System Tray   **131**

```
    )


    CCommandLineInfo::ParseParam( pszParam,bFlag,bLast );

)
```

Example 5   Putting Icons in the System Tray

Objective

II **4**

You would like to put an icon in the system tray. (The system tray is that collection of icons found in the lower-right side of the desktop, as seen in Figure 4.3.)

Figure 4.3   Icons in the System Tray

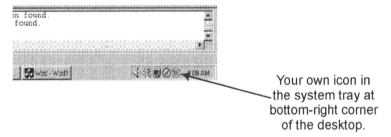

Your own icon in
the system tray at
bottom-right corner
of the desktop.

Strategy

To add an icon to the system tray, we will use the ::Shell_NotifyIcon() Window API call. To receive messages back from this icon when the user clicks on it, we will create our own windows message and manually add our own message handler.

Steps

Set Up Your Application

The system tray icon reports back to your application by sending windows messages to one of your application's windows. In an MDI or SDI application, this is typically the main window. Therefore, you would add your system tray logic to your CMainFrame class. In a Dialog Application, use your CDialog class.

1. Define a new windows message in your `CMainFrame` or `CDialog` class's include file, as seen here. This message will be sent to this class's window when the user clicks on the icon.

```
// define new window message to indicate user has clicked on icon
//      in system tray
#define WM_SYSTEMTRAY WM_USER+1
```

Create the Icon in the System Tray

1. Add the following code to your `CMainframe` or `CDialog` derived class to create the icon. The `m_hWnd` variable shown here is the window handle belonging to this class. Use the ID Editor to add `ID_SYSTEMTRAY` to your IDs. This is used to differentiate icons if your application displays more than one. See the next step for adding the `WM_SYSTEMTRAY` message, which will be sent back to your application if the user clicks on the icon.

```
// put icon in system tray
NOTIFYICONDATA nid;
nid.cbSize = sizeof( NOTIFYICONDATA );
nid.hWnd = m_hWnd;              // handle of window that will receive
                               //      messages from icon
nid.uID = ID_SYSTEMTRAY;       // id for this icon
nid.uFlags = NIF_MESSAGE|NIF_ICON|NIF_TIP;
                               // the next three parameters are valid
nid.uCallbackMessage = WM_SYSTEMTRAY;
                               // message that icon sends when clicked
nid.hIcon = AfxGetApp()->LoadIcon( IDI_SYSTEMTRAY_ICON );
                               // icon
strcpy( nid.szTip,"System Tray Tip" );
                               // bubble help message for icon
::Shell_NotifyIcon( NIM_ADD,&nid );
```

Receive Messages from the Icon

Manually add a message handler for this new message to your `CMainFrame` or `CDialog` derived class. Since you are adding this handler manually, make sure

Example 5   Putting Icons in the System Tray   **133**

to put the message map macro outside of the {{}} brackets so that the Class-Wizard can't delete them.

```
BEGIN_MESSAGE_MAP( CMainFrame, CMDIFrameWnd )
    // {{AFX_MSG_MAP( CMainFrame )
    :    :    :
    // }}AFX_MSG_MAP
    ON_MESSAGE( WM_SYSTEMTRAY,OnSystemTray )
```

2. Process the message from the icon, as seen here. The `lParam` argument contains the actual mouse message.

```
// handle system tray message
LRESULT CMainFrame::OnSystemTray( WPARAM wParam,LPARAM lParam)
{
    // wParam = the nid.uID defined above
    //     (useful if you have more then one icon in tray)
    // lParam = mouse message
    if ( wParam == ID_SYSTEMTRAY )
    {
        switch( lParam )
        {
        case WM_LBUTTONDOWN:
            break;

        case WM_RBUTTONDOWN:
            break;

        case WM_LBUTTONDBLCLK:
            break;

        }
    }
    return 1;
}
```

Remove the Icon from the System Tray

1. Before your application terminates, make sure to remove your icon from the system tray with the following code. Otherwise, the icon will remain

until your user reboots the machine. An ideal place to remove the icon would be in the WM_CLOSE message handler of the window that created the icon.

```
// delete icon from system tray
NOTIFYICONDATA nid;
nid.cbSize = sizeof( NOTIFYICONDATA );
nid.hWnd = m_hWnd;
nid.uID = ID_SYSTEMTRAY;
nid.uFlags = 0;
::Shell_NotifyIcon( NIM_DELETE,&nid );
```

Note: If you terminate your application in the debugger, it won't have an opportunity to remove the icon in the system tray. However, if you run the mouse cursor over this errant icon, the system will suddenly realize it no longer has support and remove it automatically.

Notes

The system tray is typically used by configuration applications to register an icon and then hide themselves. To create your own such application, use the AppWizard to create a Dialog Application. You can convert this into a tabbed Property Sheet by following Example 46, but not setting the Wizard mode. Then, once your application has registered an icon in the system tray, it can hide its window using ShowWindow(SW_HIDE). When someone then clicks on its icon in the system tray, you can put up a popup menu or unhide your application's dialog box with ShowWindow(SW_SHOW).

Note: Hiding the main window of an application also causes it to disappear from the task bar.

CD Notes

- When executing the project on the CD, a new icon will appear in the system tray. You can also put a break point on OnSystemTray() in Mainfrm.cpp and watch mouse clicking activity on that icon being reported to the application.

Example 6   Putting a Logo in the Main Window's Title Bar   **135**

Example 6   Putting a Logo in the Main Window's Title Bar

Objective

You would like to display a logo or other bitmap in your main window's caption bar, as seen in Figure 4.4.

II   4

Figure 4.4   A Logo in the Caption Bar

Instead of plain text,
highlight your
application's window
caption with a bitmap.

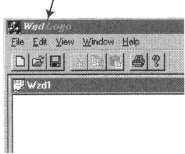

Strategy

The title bar is normally drawn by the system when a window receives the WM_NCPAINT message. However, the title bar is also used to reflect whether a window is active or not by changing its background color from bright to dull. Therefore, we need to intercept three windows messages to our main window, WM_NCPAINT, WM_ACTIVATE, and WM_NCACTIVATE, and draw our own bitmap then.

Steps

Create and Load the Bitmap

1. Create a long skinny bitmap that will become the new caption. Make sure the background is gray. Add this bitmap to your resources twice —

once as IDB_ACTIVE_CAPTION_BITMAP and then as IDB_INACTIVE_CAPTION_BITMAP. We can use the same bitmap for both captions because we will be substituting that gray background color later for each type of caption.

2. You will be loading this bitmap twice using a bitmap class, CWzdBitmap, which was featured in an example in the previous book. Essentially, what CWzdBitmap does is substitute the color gray in a bitmap for whatever color you specify when loading the bitmap. Embed two of these CWzdBitmap variables in your CMainFrame class.

```
CWzdBitmap m_bitmapActive;
CWzdBitmap m_bitmapInactive;
```

3. Now load the bitmap you created in step one twice, telling CWzdBitmap to substitute the gray color first for the current system caption bar color for an active window and then an inactve window.

```
m_bitmapActive.LoadBitmapEx( IDB_ACTIVE_CAPTION_BITMAP,
    ::GetSysColor( COLOR_ACTIVECAPTION ) );
m_bitmapInactive.LoadBitmapEx( IDB_INACTIVE_CAPTION_BITMAP,
    ::GetSysColor( COLOR_INACTIVECAPTION ) );
```

Intercept Non-Client Area Drawing Messages

1. Use the ClassWizard to add three message handlers to your CMainFrame class: WM_NCPAINT, WM_ACTIVATE, and WM_NCACTIVATE. This class owns your application's main window and, therefore, receives the messages that cause the main window to be drawn.

2. Fill in these message handlers as seen in the following code. Notice that you also need to add an m_bActive Boolean variable to your CMainFrame class to keep track of whether or not the main window is active. Also notice that you call a DrawTitle() helper function to do the actual drawing.

```
// WM_NCPAINT message handler
void CMainFrame::OnNcPaint()
{
    CMDIFrameWnd::OnNcPaint();

    // draw title
    DrawTitle();
```

Example 6   Putting a Logo in the Main Window's Title Bar   **137**

```
}

// WM_ACTIVE message handler
void CMainFrame::OnActivate( UINT nState, CWnd* pWndOther,
    BOOL bMinimized )
{
    CMDIFrameWnd::OnActivate( nState, pWndOther, bMinimized );

    // set state and draw title
    BOOL m_bActive;
    m_bActive = ( nState! = WA_INACTIVE );
    DrawTitle();
}

// WM_NCACTIVATE message handler
BOOL CMainFrame::OnNcActivate( BOOL bActive )
{
    BOOL b = CMDIFrameWnd::OnNcActivate( bActive );

    // set state and draw title
    m_bActive = bActive;
    DrawTitle();

    return b;
}
```

II   4

Draw the Title Bar

1. Add a `DrawTitle()` function to your `CMainFrame` class, which will start by determining if there's a title bar to draw.

```
void CMainFrame::DrawTitle()
{
    // if window isn't visible or is minimized, skip
    if ( !IsWindowVisible() || IsIconic())
        return;
```

2. Select the appropriate bitmap object into a memory device context, based on whether the window is active or not.

```
CDC memDC;
CDC* pDC = GetWindowDC();
memDC.CreateCompatibleDC( pDC );
memDC.SelectObject( m_bActive ? &m_bitmapActive:&m_bitmapInactive );
```

3. Calculate where to draw the bitmap. We want to avoid drawing over the icon at the far left and the window buttons (e.g., close, minimize, etc.) at the far right. We also want to avoid drawing over the window's border.

```
CRect rect, rectWnd;
GetWindowRect( &rect );
rect.top += GetSystemMetrics( SM_CYFRAME )+1;
    // for small caption use SM_CYDLGFRAME
rect.bottom = rect.top + GetSystemMetrics( SM_CYSIZE )-4;
    // for small caption use SM_CYSMSIZE
rect.left += GetSystemMetrics( SM_CXFRAME ) +
    // for small caption use SM_CXDLGFRAME
    GetSystemMetrics( SM_CXSIZE );
        // for small caption use SM_CXSMSIZE
rect.right -= GetSystemMetrics( SM_CXFRAME ) -
    // for small caption use SM_CXDLGFRAME
    ( 3 *
        // set to number of buttons already in caption + 1
    GetSystemMetrics( SM_CXSIZE ) )-1;
        // for small caption use SM_CXSMSIZE
GetWindowRect( rectWnd );
rect.OffsetRect( -rectWnd.left, -rectWnd.top );
```

4. Now just draw it and cleanup.

```
pDC->BitBlt( rect.left, rect.top, rect.Width(), rect.Height(),
    &memDC, 0, 0, SRCCOPY );


memDC.DeleteDC();
ReleaseDC( pDC );
}
```

Example 6   Putting a Logo in the Main Window's Title Bar   **139**

Notes

- Your old title will continue to be drawn just before this bitmap fills the title bar, which may cause a little flashing. You can minimize this flashing even more by preventing the document title from being included in the old title by adding the following to CMainFrame's PreCreateWindow() function.

```
BOOL CMainFrame::PreCreateWindow( CREATESTRUCT& cs )
{
    cs.style &= ~ FWS_ADDTOTITLE;
    return CMDIFrameWnd::PreCreateWindow( cs );
}
```

You could also eliminate the old title entirely. However, this title is the one that appears in the taskbar for your application.

CD Notes

- When executing the project on the CD, you will notice that the normal text in the application window's title bar has been replaced with a bitmap.

5

Chapter 5

Menus, Toolsbars, and Status Bars

Menus and control bars, such as toolbars, represent the primary way your user will interact with your application. The AppWizard will automatically add a generic menu, toolbar, and status bar to your application, but these menus and bars pale in comparison to the menu and bars that even your own Developer Studio uses. With a little manual effort, however, you can add this same look to your own application.

Example 7   Putting Icons in a Menu   We will emulate the icons you find in the Developer Studio's menus.

Example 8   Adjusting the Appearance of Your Command Bar   We will emulate the look of the toolbars found in the Developer Studio.

Example 9   Creating Programmable Toolbars   We will emulate the "feel" of the toolbars in the Developer Studio—namely, the ability for our user to configure their own set of toolbars from a pool of toolbar buttons.

Example 10 Putting a Toolbar, Menu, and Status Bar in Your Dialog Application We will manually add a toolbar and status bar to a dialog application.

Example 11 Adding a Bitmap Logo to a Popup Menu We will draw a bitmap along the side of a popup menu.

Example 12 Putting a Dropdown Button in a Toolbar We will create a double toolbar button that you use to create a popup menu, which appears to drop down from the toolbar.

Example 13 Putting an Icon in the Status Bar We will put a status icon in the status bar.

Example 14 Using a Rebar We will add a rebar to an application and fill it with toolbars and dialog bars.

Example 7 Putting Icons in a Menu

Objective

You would like to emulate the Developer Studio's menus, which display an icon next to each menu item that also has a toolbar button, as seen in Figure 5.1.

Example 7   Putting Icons in a Menu   **143**

Figure 5.1   Putting Icons in the Menus

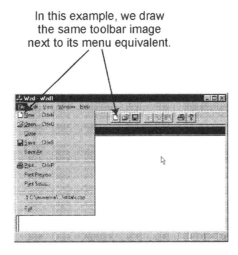

In this example, we draw
the same toolbar image
next to its menu equivalent.

II

5

Strategy

The menu seen in Figure 5.1 is an owner-drawn menu. To create an owner-drawn menu, we start by setting an option in each menu item that tells the system that it's owner-drawn. Because there are no provisions for this using the Menu Editor, we must do this dynamically. Owner-drawn menus then send two messages to their owner window (WM_MEASUREITEM and WM_DRAWITEM), which the owner must process by drawing the menu item. We will encapsulate all of this functionality in a Menu Class that we will derive from MFC's CMenu class. The InitMenu() function of our new class will not only mark every menu item as being owner-drawn, but it will also match up menu commands with their toolbar equivalents and use that matching toolbar button bitmap when drawing the menu item. Needless to say, it gets involved.

Steps

Create a New Class Derived from CMenu

1. Click on the Developer Studio's "Insert" and "New Class" menu commands to open the "New Class" dialog box. From the "Class type" combo box, pick "Generic Class". Enter a new class name and derive it

from `CMenu`. We aren't doing this through the ClassWizard because it doesn't currently support `CMenu`. For this reason, you will also be manually adding everything else to this class.

This example was designed to allow you to still use the Menu Editor to edit your application's menu. For this reason, the `InitMenu()` function of this new menu class does several things. First, it converts all of the menu items in your application to owner-drawn. Then, because owner-drawn menus don't save their text names, we will save each name in our own data array. We will also save in this array the approximate size that each menu item will require to be drawn in pixels to expedite drawing later. We will try to match each menu item command ID with an equivalent command ID in our application's toolbar. A pointer to this toolbar is passed as one of the arguments of `InitMenu()`. So, let's get started.

Add Your Own `InitMenu()` Function to This New Menu Class

1. Add an `InitMenu()` function to this class, passing it a pointer to your application's `CWnd` class and a `CMenu` pointer that wraps its menu.

```
void CWzdMenu::InitMenu( CWnd *pWnd, CMenu *pMenu, UINT idb,
    CToolBar *pToolBar )
{
```

2. Loop through all of the menu items found in this `CMenu` pointer.

```
CDC *pDC = pWnd -> GetDC();

// for all submenus
CMenu *pSubMenu = NULL;
for ( int i = 0; i < ( int )pMenu -> GetMenuItemCount(); i++ )
{
    pSubMenu = pMenu -> GetSubMenu( i );
    if ( pSubMenu )
    {
        for ( int j = 0;j < ( int )pSubMenu -> GetMenuItemCount(); j++ )
        {
            // if not a separator...
            UINT id = pSubMenu -> GetMenuItemID( j );
            if ( id )
            {
```

Example 7   Putting Icons in a Menu   **145**

```
                    // if already ownerdrawn, escape
                    if ( pSubMenu ->
                        GetMenuState( j,MF_BYPOSITION )&MF_OWNERDRAW )
                    {
                        pWnd -> ReleaseDC( pDC );
                        return;
                    }
```

Notice that we ignore separator menu items. If any item is already owner-drawn, we escape entirely! That's because this menu has already been converted at one point.

3. Unfortunately, once a menu item becomes owner-drawn, it no longer maintains a name string. Rather than maintaining a separate array of menu item names, we will simply save the names already in the menu that we created using the Menu Editor. For this purpose, we created a new structure called MENUITEM. In this step, we create a new instance of MENUITEM and copy the menu item's name string into it.

```
// fill in MENUITEM
MENUENTRY *pEntry = new MENUENTRY;
pEntry -> id = id;
pSubMenu -> GetMenuString( j,pEntry -> str,MF_BYPOSITION );
```

4. An owner-drawn menu also no longer processes quick key entries (the underlined letter you find in a menu that can be accessed using the keyboard). We will, therefore, be adding that functionality ourselves. We need to save the letter we selected using the Menu Editor when we originally created this menu by looking for the letter after the & character, if any. We save the resulting character in MENUITEM.

```
int k = pEntry -> str.Find( '&' );
pEntry -> chr = 0;
if ( k >= 0 )
    pEntry -> chr = pEntry -> str[k+1];
pEntry -> chr& = ~0x20;    // make upper case
```

5. We now calculate the size of our menu items in pixels and also save them in MENUITEM. Later, when we process the WM_MEASUREITEM message, we will need these values.

```
pEntry -> size = pDC -> GetTextExtent( pEntry -> str,
    pEntry -> str.GetLength() );
pEntry -> size.cx += BUTTON_WIDTH + 2;
pEntry-> size.cy = BUTTON_HEIGHT + 6;
```

6. We will now try to match up this menu command with its equivalent in the toolbar by scaning the toolbar for this command ID. If there is one, we'll save it for when we need to draw this item.

```
pEntry -> inx = -1;
for ( int m = 0;m < pToolBar -> GetToolBarCtrl().GetButtonCount();m++ )
{
    int inx;
    UINT idx,x;
    pToolBar -> GetButtonInfo( m,idx,x,inx );
    if ( id == idx )
    {
        pEntry -> inx = inx;
        break;
    }
}
```

7. We now go ahead and make this an owner drawn menu item.

```
// modify menu item to be owner drawn
pSubMenu -> ModifyMenu( id, MF_BYCOMMAND | MF_ENABLED | MF_OWNERDRAW,
    id, ( LPCTSTR )pEntry );
```

8. We also load up the toolbar's bitmap for use in drawing icons later. We will use the CWzdBitmap class created in the previous book, *Visual C++ MFC Programming by Example*, so that the background of our images will be translated to the current color of the menu, thus making them appear transparent.

```
m_bitmap.LoadBitmapEx(idb,TRUE);
```

The system sends your application two messages to allow you to draw an owner-drawn menu. The first is WM_MEASUREITEM, which you use to tell the system how big each item will be in the menu and from which the system can determine how big to make the entire menu. The second message,

Example 7   Putting Icons in a Menu   **147**

WM_DRAWITEM, is sent for each individual menu item to allow you to draw it using the standard drawing tools of Windows.

Add a MeasureItem() Function to This New Menu Class

1. Again, the ClassWizard doesn't work for CMenu, so manually add a WM_MEASUREITEM message handler to this class.

2. In this handler, return the size in pixels of the requested menu item. The system needs this value to determine how wide to make the entire menu. We already calculated these pixel values in InitMenu(), so we'll just whip those numbers out here.

```
// size of our menu item
void CWzdMenu::MeasureItem( LPMEASUREITEMSTRUCT lpMIS )
{
    MENUENTRY *pEntry = ( MENUENTRY * )lpMIS -> itemData;
    lpMIS -> itemWidth = pEntry -> size.cx;
    lpMIS -> itemHeight = pEntry -> size.cy;
}
```

Add a DrawItem() Function to This New Menu Class

1. Manually add a WM_DRAWITEM message handler to your new menu class.

```
void CWzdMenu::DrawItem( LPDRAWITEMSTRUCT lpDIS )
{
```

2. Start this handler by wrapping the information we get in the DRAWITEM-STRUCT stucture in MFC classes.

```
CDC dc;
dc.Attach( lpDIS -> hDC );          // device context to draw to
CRect rect( lpDIS -> rcItem );      // rectangular size of menu item
```

3. Next, draw the background of the menu item. The device context we get is ready to draw in the standard menu background color. However, if the

mouse is currently over our item, we need to invert the background and text colors.

```
// if our item is selected, then set colors accordingly
COLORREF bk = dc.GetBkColor();
COLORREF fg = dc.GetTextColor();
if ( lpDIS -> itemState & ODS_SELECTED )
{
    bk = ::GetSysColor( COLOR_HIGHLIGHT );
    fg = ::GetSysColor( COLOR_HIGHLIGHTTEXT );
}
dc.SetTextColor( fg );


// fill in background
CBrush brush( bk );
dc.FillRect( &rect, &brush );
```

4. If the menu item we are about to draw is disabled, we will need to gray out its text and any associated icon, using the `CDC::DrawState()` function. This function incorporates a callback function that does the actual drawing, so all we will be doing here in `DrawItem()` is setting this `DrawState()` function up to call our callback function.

```
// get enabled/disabled state and draw appropriately
UINT nState = DSS_NORMAL;
if ( lpDIS -> itemState & ODS_DISABLED )
{
    nState = DSS_DISABLED;
}
dc.DrawState( rect.TopLeft(),rect.Size(),DrawStateProc,
    ( LPARAM )lpDIS,nState,( HBRUSH )NULL );
```

5. To finish up `DrawItem()`, we cleanup the device context.

```
// cleanup
dc.SetTextColor( fg );
dc.SetBkMode( nBkMode );
dc.Detach();
```

Example 7   Putting Icons in a Menu   **149**

Add a CDC::DrawState() Callback Function to This New Menu Class

1. Now, we create the callback process required by CDC::DrawState() and again start by wrapping the device context handle and rectangle.

```
BOOL CALLBACK DrawStateProc( HDC hdc, LPARAM lData, WPARAM wData,
    int cx, int cy )
{
    CDC dc;
    LPDRAWITEMSTRUCT lpDIS = ( LPDRAWITEMSTRUCT )lData;
    dc.Attach( hdc );
    MENUENTRY *pEntry = ( MENUENTRY * )lpDIS -> itemData;
    CRect rect( 0,0,cx,cy );
```

2. If we found a toolbar match in InitMenu(), we will now draw its associated bitmap in our menu item CDC::BitBlt().

```
if ( pEntry -> inx != -1 )
    {
        CDC memDC;
        memDC.CreateCompatibleDC( &dc );
        memDC.SelectObject( pEntry -> pBitmap);
        dc.BitBlt( rect.left, rect.top, BUTTON_WIDTH, BUTTON_HEIGHT,
            &memDC, pEntry -> inx*BUTTON_WIDTH, 0, SRCCOPY );
        memDC.DeleteDC();
    }
```

3. Next, we check to see if this menu item is checked. If so, we draw a special icon check mark we created with the Icon Editor over the bitmap. If no bitmap was drawn in the last step, this check mark will be alone.

```
HICON hIcon;
if ( lpDIS -> itemState & ODS_CHECKED &&
    ( hIcon = AfxGetApp() -> LoadIcon( IDI_CHECK_ICON ) ) )
{
    dc.DrawIcon( rect.left, rect.top, hIcon );
}
```

II

5

4. Next, we draw the text string for this menu item using `DrawText()`. We use `DrawText()` to have the character behind the & sign underlined. `Draw-Text()` will also vertically align our text and expand any tabs.

```
rect.left += BUTTON_WIDTH + 2;
dc.DrawText( pEntry -> str, &rect,
    DT_LEFT|DT_EXPANDTABS|DT_VCENTER );


dc.Detach();
return TRUE;
}
```

Add a `MenuChar()` Function to This New Menu Class

1. As mentioned previously, owner-drawn menus no longer process quick key commands (the underlined character in the menu item). To manually process this command, you will need to add a `WM_MENUCHAR` message handler to this new menu class. We use the quick key information we stored previously in `InitMenu()` to determine if a menu item should be triggered in the following example.

```
LRESULT CWzdMenu::MenuChar( UINT nChar )
{
    nChar& = ~0x20;    // make uppercase
    // try to find char in current menu list
    for ( POSITION pos = m_CurrentMenuList.GetHeadPosition();pos; )
    {
        MENUENTRY *pEntry = ( MENUENTRY * )
            m_CurrentMenuList.GetNext( pos );
        if ( pEntry -> chr == nChar )
        {
            AfxGetMainWnd() -> SendMessage( WM_COMMAND,pEntry -> id );
            return( MAKELONG( 0,1 ) );
        }
    }
    return(0);
}
```

To see a complete and contiguous listing of this new Menu Class, please refer to the Menu Class for Example 7 (page 153).

Example 7   Putting Icons in a Menu   **151**

To add this new menu class to our application, we start by embedding it in our Main Frame Class. Then, we will add message handlers to the Main Frame Class that will call the member functions of this menu class.

Incorporate the New Menu Class into Your Application

1. Embed the new menu class in your CMainFrame class.

```
// add CWzdMenu to Mainfrm.h
private:
    CWzdMenu m_menu;
```

2. You will now need to connect the CWzdMenu class into the main menu by using the ClassWizard to add five message handlers. (Yes, you can finally use the ClassWizard again.) The window messages to add are: WM_INITMENU, WM_INITPOPUPMENU, WM_MEASUREITEM, WM_DRAWITEM, and WM_MENUCHAR.

3. Within each of the message handlers you added to CMainFrame, call a corresponding member function of the new menu class, as seen here.

```
void CMainFrame::OnInitMenu( CMenu* pMenu )
{

    CMDIFrameWnd::OnInitMenu( pMenu );

    m_menu.InitMenu( this,pMenu,IDR_MAINFRAME,&m_wndToolBar );

}

void CMainFrame::OnInitMenuPopup( CMenu* pPopupMenu, UINT nIndex,
    BOOL bSysMenu )
{
    CMDIFrameWnd::OnInitMenuPopup( pPopupMenu, nIndex, bSysMenu );

    m_menu.InitMenuPopup();

}

void CMainFrame::OnMeasureItem( int nIDCtl,
    LPMEASUREITEMSTRUCT lpMeasureItemStruct )
```

II

5

```
{
    if ( !nIDCtl ) m_menu.MeasureItem( lpMeasureItemStruct );

    CMDIFrameWnd::OnMeasureItem( nIDCtl, lpMeasureItemStruct );
}

void CMainFrame::OnDrawItem( int nIDCtl,
    LPDRAWITEMSTRUCT lpDrawItemStruct )
{
    if ( !nIDCtl ) m_menu.DrawItem( lpDrawItemStruct );

    CMDIFrameWnd::OnDrawItem( nIDCtl, lpDrawItemStruct );
}

LRESULT CMainFrame::OnMenuChar( UINT nChar, UINT nFlags, CMenu* pMenu )
{
    return m_menu.MenuChar( nChar );
}
```

Notes

- Due to the extent of this example, a few refinements were left out. The accelerator key descriptions in menu items should be drawn right-justified. When a menu item has the mouse cursor over it, the icon can be modified to indicate this by drawing it with a focus window (a thin border).

- Disabled menu items are drawn in a sort of embossed gray by using CDC::DrawState(). Bitmaps also appear embossed; however, in some cases their entire background is inverted, thus obscuring whatever image they had. If this is an undesirable effect, you can correct it by making the entire background of the bitmap white by loading the toolbar bitmap a second time with the background color translated into white. To do this, you will need to modify the CWzdBitmap class provided with this example.

CD Notes

- When executing the project on the accompanying CD, you will notice that any menu item that has a companion command in the toolbar will now have that toolbar's icon sitting next to it.

Example 7   Putting Icons in a Menu   **153**

Menu Class for Example 7

```
#if !defined( AFX_WZDMENU_H__18606914_D521_11D1_9B69_00AA003D8695__INCLUDED_ )
#define AFX_WZDMENU_H__18606914_D521_11D1_9B69_00AA003D8695__INCLUDED_

#if _MSC_VER >= 1000
#pragma once
#endif    // _MSC_VER >= 1000

// WzdMenu.h : header file
//
#include "afxtempl.h"

struct MENUENTRY
{
    CString str;
    UINT  id;
    UINT  chr;
    CSize size;
    int   inx;
};


/////////////////////////////////////////////////////////////////////////////
// CWzdMenu window

class CWzdMenu : public CMenu
{
// Construction
public:
    CWzdMenu();
    virtual ~CWzdMenu();

// Attributes
public:

// Operations
public:
    void InitMenu( CWnd *pWnd, CMenu *pMenu, UINT idb, CToolBar *pToolBar );
    void InitMenuPopup();
    void MeasureItem( LPMEASUREITEMSTRUCT lpMIS );
```

II

5

```cpp
        void DrawItem( LPDRAWITEMSTRUCT lpDIS );
        LRESULT MenuChar( UINT nChar );


// Implementation
public:

private:
    CBitmap m_bitmap;
    CList<MENUENTRY*,MENUENTRY*> m_FullMenuList;
    CList<MENUENTRY*,MENUENTRY*> m_CurrentMenuList;
};


/////////////////////////////////////////////////////////////////////////////


// {{AFX_INSERT_LOCATION}}
// Microsoft Developer Studio will insert additional declarations immediately
//      before the previous line.

#endif
    // !defined( AFX_WZDMENU_H__18606914_D521_11D1_9B69_00AA003D8695__INCLUDED_ )
// WzdMenu.cpp : implementation file
//
#include "stdafx.h"
#include "wzd.h"
#include "WzdMenu.h"
#include "WzdProject.h"


#ifdef _DEBUG
#define new DEBUG_NEW
#undef THIS_FILE
static char THIS_FILE[] = __FILE__;
#endif


BOOL CALLBACK DrawStateProc( HDC hdc, LPARAM lData, WPARAM wData, int cx,
    int cy );


/////////////////////////////////////////////////////////////////////////////
// CWzdMenu

CWzdMenu::CWzdMenu()
```

Example 7   Putting Icons in a Menu   **155**

```
{
}


CWzdMenu::~CWzdMenu()
{
    while ( !m_FullMenuList.IsEmpty() )
    {
        delete m_FullMenuList.RemoveHead();
    }
}

void CWzdMenu::InitMenu( CWnd *pWnd, CMenu *pMenu, UINT idb, CToolBar *pToolBar )
{
    CDC *pDC = pWnd -> GetDC();


    // for all submenus
    CMenu *pSubMenu = NULL;
    for (int i = 0; i < ( int )pMenu -> GetMenuItemCount(); i++ )
    {
        pSubMenu = pMenu -> GetSubMenu( i );
        if ( pSubMenu )
        {
            for ( int j = 0;j < ( int )pSubMenu -> GetMenuItemCount(); j++ )
            {
                // if not a separator...
                UINT id = pSubMenu -> GetMenuItemID( j );
                if ( id )
                {
                    // if already ownerdrawn, escape
                    if ( pSubMenu ->
                        GetMenuState( j,MF_BYPOSITION )&MF_OWNERDRAW )
                    {
                        pWnd -> ReleaseDC( pDC );
                        return;
                    }

                    // fill in MENUITEM
                    MENUENTRY *pEntry = new MENUENTRY;
                    pEntry -> id = id;
                    pSubMenu -> GetMenuString( j,pEntry -> str,MF_BYPOSITION );
```

```
                        int k = pEntry -> str.Find( '&' );
                        pEntry -> chr = 0;
                        if ( k >= 0 )
                            pEntry -> chr = pEntry -> str[k+1];
                        pEntry -> chr& = ~0x20;    // make upper case
                        pEntry -> size = pDC ->
                           GetTextExtent( pEntry -> str,pEntry -> str.GetLength() );
                        pEntry -> size.cx += BUTTON_WIDTH + 2;
                        pEntry -> size.cy = BUTTON_HEIGHT + 2;
                        pEntry -> inx = -1;
                        for ( int m = 0;m < pToolBar ->
                            GetToolBarCtrl().GetButtonCount();m++ )
                        {
                            int inx;
                            UINT idx,x;
                            pToolBar -> GetButtonInfo( m,idx,x,inx );
                            if ( id == idx )
                            {
                                pEntry -> inx = inx;
                                pEntry -> pBitmap = &m_bitmap;
                                break;
                            }
                        }

                        // add MENUITEM to full list
                        m_FullMenuList.AddTail( pEntry );

                        // modify menu item to be owner drawn
                        pSubMenu -> ModifyMenu( id, MF_BYCOMMAND | MF_ENABLED |
                            MF_OWNERDRAW, id, ( LPCTSTR )pEntry );
                    }
                }
            }
        }
    m_bitmap.LoadBitmapEx( idb,TRUE );
    pWnd -> ReleaseDC( pDC );
}


void CWzdMenu::InitMenuPopup()
{
```

Example 7   Putting Icons in a Menu   **157**

```
    // empty current menu list
    m_CurrentMenuList.RemoveAll();
}


// size of our menu item
void CWzdMenu::MeasureItem( LPMEASUREITEMSTRUCT lpMIS )
{
    MENUENTRY *pEntry = ( MENUENTRY * )lpMIS -> itemData;
    lpMIS -> itemWidth = pEntry -> size.cx;
    lpMIS -> itemHeight = pEntry -> size.cy;
}


// draw our menu item
void CWzdMenu::DrawItem( LPDRAWITEMSTRUCT lpDIS )
{
    // get our device context and rectangle to draw to
    CDC dc;
    dc.Attach( lpDIS -> hDC );
    CRect rect( lpDIS -> rcItem );

    // if our item is selected, then set colors accordingly
    COLORREF bk = dc.GetBkColor();
    COLORREF fg = dc.GetTextColor();
    if ( lpDIS -> itemState & ODS_SELECTED )
    {
        bk = ::GetSysColor( COLOR_HIGHLIGHT );
        fg = ::GetSysColor( COLOR_HIGHLIGHTTEXT );
    }
    dc.SetTextColor( fg );

    // fill in background
    CBrush brush( bk );
    dc.FillRect( &rect, &brush );

    // draw text withhout a background
    int nBkMode = dc.SetBkMode( TRANSPARENT );

    // get enabled/disabled state and draw appropriately
    UINT nState = DSS_NORMAL;
    if ( lpDIS -> itemState & ODS_DISABLED )
```

II

5

```
{
    nState = DSS_DISABLED;
}
dc.DrawState( rect.TopLeft(),rect.Size(),DrawStateProc,
    ( LPARAM )lpDIS,nState,( HBRUSH )NULL );

// add to current menu list
MENUENTRY *pEntry = ( MENUENTRY * )lpDIS -> itemData;
m_CurrentMenuList.AddTail( pEntry );

// cleanup
dc.SetTextColor( fg );
dc.SetBkMode( nBkMode );
dc.Detach();
}

BOOL CALLBACK DrawStateProc( HDC hdc, LPARAM lData, WPARAM wData, int cx, int cy )
{
    CDC dc;
    LPDRAWITEMSTRUCT lpDIS = ( LPDRAWITEMSTRUCT )lData;
    dc.Attach( hdc );
    MENUENTRY *pEntry = ( MENUENTRY * )lpDIS -> itemData;
    CRect rect( 0,0,cx,cy );

    // draw bitmap, if any
    if ( pEntry -> inx != -1 )
    {
        CDC memDC;
        memDC.CreateCompatibleDC( &dc );
        memDC.SelectObject( pEntry -> pBitmap );
        dc.BitBlt( rect.left, rect.top, BUTTON_WIDTH,BUTTON_HEIGHT, &memDC,
            pEntry -> inx*BUTTON_WIDTH, 0, SRCCOPY );
        memDC.DeleteDC();
    }

    // check it, if required
    HICON hIcon;
    if ( lpDIS -> itemState & ODS_CHECKED &&
        ( hIcon = AfxGetApp() -> LoadIcon( IDI_CHECK_ICON ) ) )
    {
```

Example 8   Adjusting the Appearance of Your Command Bar   **159**

```
            dc.DrawIcon( rect.left, rect.top, hIcon );
    }
    // draw text
    rect.left += BUTTON_WIDTH + 2;
    dc.DrawText( pEntry -> str, &rect, DT_LEFT|DT_EXPANDTABS|DT_VCENTER );

    dc.Detach();
    return TRUE;
}


// draw our menu item
LRESULT CWzdMenu::MenuChar( UINT nChar )
{
    nChar& = ~0x20;    // make uppercase
    // try to find char in current menu list
    for ( POSITION pos = m_CurrentMenuList.GetHeadPosition();pos; )
    {
        MENUENTRY *pEntry = ( MENUENTRY * )m_CurrentMenuList.GetNext( pos );
        if ( pEntry -> chr == nChar )
        {
            AfxGetMainWnd() -> SendMessage( WM_COMMAND,pEntry -> id );
            return( MAKELONG( 0,1 ) );
        }
    }
    return( 0 );
}
```

II

5

Example 8   Adjusting the Appearance of Your Command Bar

Objective

You would like to give your toolbar the same look as the Developer Studio's, as seen in Figure 5.2.

Figure 5.2   The Command Bar Look

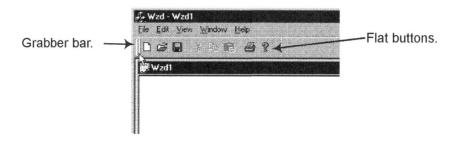

Grabber bar. ——→   ┌Flat buttons.

Strategy

We will use the TBSTYLE_FLAT toolbar style to give our toolbars a flat button look. We will then draw the grabber bars ourselves, when processing the WM_PAINT message sent to our toolbar. We will use the ClassWizard to encapsulate all of this functionality in our own toolbar class.

Note: The latest and greatest version of VC++ (v6.0) has added a toolbar style, CBRS_GRIPPER, that will automatically draw a gripper bar on your toolbar. However, CBRS_GRIPPER draws just a single gripper bar, not the double bar drawn in this example and the Developer Studio.

Steps

Create a New Toolbar Class

1. Use the ClassWizard to create a new toolbar class derived from CToolBarCtrl. Then, use the Text Editor to replace all instances of CToolBarCtrl with CToolBar, since the ClassWizard can't currently derive a new class from CToolBar.

2. Use the ClassWizard to add a WM_PAINT message handler to your new toolbar class. We will be drawing grabber bars at the front of the toolbar when the toolbar is horizontal, or at the top of the toolbar when the toolbar is vertically docked to the side of the your application's main window. We won't draw any grabber bars when the toolbar is floating.

Example 8   Adjusting the Appearance of Your Command Bar   **161**

We start by allowing the toolbar control to draw its buttons and border as usual.

```
void CWzdToolBar::OnPaint()
{
    CToolBar::OnPaint();
```

3. Next, if the toolbar is floating, we simply return

```
// if floating, no grabber bar
if ( IsFloating() )
    return;
```

4. We will create a window device context rather than a client device context since there's a nonclient area surrounding the toolbar.

```
// to draw to whole window!
CWindowDC dc( this );
```

5. The grabber bar will be drawn with colors defined by the system. Using hard coded numbers here won't work if our user changes the color scheme of their system in the Control Panel. There are two system colors we will be using for our 3D grabber lines: a highlight color and a shadow color. Create two pens with these colors.

```
CPen penHL( PS_SOLID,2,::GetSysColor( COLOR_3DHIGHLIGHT ) );
CPen penSH( PS_SOLID,3,::GetSysColor( COLOR_3DSHADOW ) );
```

6. Now, depending on your toolbar's orientation, draw two lines at the start or at the top of your toolbar.

```
CPen *pPen = dc.SelectObject( &penSH );
if ( GetBarStyle()&CBRS_ORIENT_HORZ )
{
    // draw shadow lines
    rect.OffsetRect( 4,4 );
    dc.MoveTo( rect.left,rect.top );
    dc.LineTo( rect.left,rect.bottom );
    dc.MoveTo( rect.left + 4,rect.top );
    dc.LineTo( rect.left + 4,rect.bottom );

    // draw highlight lines
    dc.SelectObject( &penHL );
```

II

5

```
    dc.MoveTo( rect.left,rect.top );
    dc.LineTo( rect.left,rect.bottom - 1 );
    dc.MoveTo( rect.left + 4,rect.top );
    dc.LineTo( rect.left + 4,rect.bottom - 1 );
}
else
{
    // draw shadow lines
    rect.OffsetRect( 4,2 );
    dc.MoveTo( rect.left,rect.top );
    dc.LineTo( rect.right,rect.top );
    dc.MoveTo( rect.left,rect.top + 4 );
    dc.LineTo( rect.right,rect.top + 4 );

    // draw highlight lines
    dc.SelectObject( &penHL );
    dc.MoveTo( rect.left,rect.top );
    dc.LineTo( rect.right - 1,rect.top );
    dc.MoveTo( rect.left,rect.top + 4 );
    dc.LineTo( rect.right - 1,rect.top + 4 );
}
dc.SelectObject(pPen);
```

7. To see a complete listing of this new toolbar class, please refer to the Toolbar Class for Example 8 (page 163).

Implement This New Toolbar Class

1. Substitute this new toolbar class for the current toolbar class in `CMain-Frame`. You can do this in the `MainFrm.h` file, as shown here. In this example, we named our new toolbar class `CWzdToolBar`.

```
CWzdToolBar    m_wndToolBar;
```

Example 8   Adjusting the Appearance of Your Command Bar   **163**

2. Apply the `TBSTYLE_FLAT` style to this toolbar in the `MainFrm.cpp` file, just after creating the toolbar, using the `CToolBar::ModifyStyle()` function.

```
if ( !m_wndToolBar.Create( this ) ||
    !m_wndToolBar.LoadToolBar( IDR_MAINFRAME ) )
{
    TRACE0( "Failed to create toolbar\n" );
    return -1;    // fail to create
}
m_wndToolBar.ModifyStyle( 0, TBSTYLE_FLAT );
```

Notes

- When using the flat button style with toolbars, it's harder to determine where the toolbar ends and the button begins. This makes it tougher when to grab a toolbar to drag—thus, the reason for the grabber bar.
- A rebar control resembles a Command Bar, but in fact isn't a toolbar at all. A rebar is a control window that can contain one or more other control windows, including a toolbar, and allows your user to move these control windows around within its control.

CD Notes

- When executing the project on the accompanying CD, you will notice that the toolbar looks similar to your Developer Studio's toolbar, including flat buttons and a grabber bar.

Toolbar Class for Example 8

```
#if !defined( AFX_WZDTOOLBAR_H__0939E911_E0EC_11D1_9B7A_00AA003D8695__INCLUDED_ )
#define AFX_WZDTOOLBAR_H__0939E911_E0EC_11D1_9B7A_00AA003D8695__INCLUDED_

#if _MSC_VER >= 1000
#pragma once
#endif    // _MSC_VER >= 1000
// WzdToolBar.h : header file
//

////////////////////////////////////////////////////////////////////////////
// CWzdToolBar window
```

```
class CWzdToolBar : public CToolBar
{
// Construction
public:
    CWzdToolBar();

// Attributes
public:

// Operations
public:

// Overrides
    // ClassWizard generated virtual function overrides
    // {{AFX_VIRTUAL( CWzdToolBar )
    // }}AFX_VIRTUAL

// Implementation
public:
    virtual ~CWzdToolBar();

// Generated message map functions
protected:
    // {{AFX_MSG( CWzdToolBar )
    afx_msg void OnPaint();
    // }}AFX_MSG

    DECLARE_MESSAGE_MAP()
};

//////////////////////////////////////////////////////////////////////////////

// {{AFX_INSERT_LOCATION}}
// Microsoft Developer Studio will insert additional declarations immediately
// before the previous line.

#endif
    // !defined( AFX_WZDTOOLBAR_H__0939E911_E0EC_11D1_9B7A_00AA003D8695__INCLUDED_ )
// WzdToolBar.cpp : implementation file
```

Example 8   Adjusting the Appearance of Your Command Bar   **165**

```
//

#include "stdafx.h"
#include "wzd.h"
#include "WzdToolBar.h"

#ifdef _DEBUG
#define new DEBUG_NEW
#undef THIS_FILE
static char THIS_FILE[] = __FILE__;
#endif

/////////////////////////////////////////////////////////////////////////////
// CWzdToolBar

CWzdToolBar::CWzdToolBar()
{
}


CWzdToolBar::~CWzdToolBar()
{
}



BEGIN_MESSAGE_MAP( CWzdToolBar, CToolBar )
    // {{AFX_MSG_MAP( CWzdToolBar )
    ON_WM_PAINT()
    // }}AFX_MSG_MAP
END_MESSAGE_MAP()

/////////////////////////////////////////////////////////////////////////////
// CWzdToolBar message handlers

void CWzdToolBar::OnPaint()
{
    CToolBar::OnPaint();

    // if floating, no grabber bar
    if ( IsFloating() )
        return;
```

```
// draw to whole window!
CWindowDC dc( this );

// draw horizontal or vertical grabber bar
CRect rect;
GetClientRect( &rect );

CPen penHL( PS_SOLID,2,::GetSysColor( COLOR_3DHIGHLIGHT ) );
CPen penSH( PS_SOLID,3,::GetSysColor( COLOR_3DSHADOW ) );

CPen *pPen = dc.SelectObject( &penSH );
if ( GetBarStyle()&CBRS_ORIENT_HORZ )
{
    // draw shadow lines
    rect.OffsetRect( 4,4 );
    dc.MoveTo( rect.left,rect.top );
    dc.LineTo( rect.left,rect.bottom );
    dc.MoveTo( rect.left + 4,rect.top );
    dc.LineTo( rect.left + 4,rect.bottom );

    // draw highlight lines
    dc.SelectObject( &penHL );
    dc.MoveTo( rect.left,rect.top );
    dc.LineTo( rect.left,rect.bottom - 1 );
    dc.MoveTo( rect.left + 4,rect.top );
    dc.LineTo( rect.left + 4,rect.bottom - 1 );
}
else
{
    // draw shadow lines
    rect.OffsetRect( 4,2 );
    dc.MoveTo( rect.left,rect.top );
    dc.LineTo( rect.right,rect.top );
    dc.MoveTo( rect.left,rect.top + 4 );
    dc.LineTo( rect.right,rect.top + 4 );

    // draw highlight lines
    dc.SelectObject( &penHL );
    dc.MoveTo( rect.left,rect.top );
    dc.LineTo( rect.right - 1,rect.top );
```

Example 9   Creating Programmable Toolbars   **167**

```
        dc.MoveTo( rect.left,rect.top + 4 );
        dc.LineTo( rect.right - 1,rect.top + 4 );
    }


    dc.SelectObject( pPen );
}
```

Example 9   Creating Programmable Toolbars

Objective

You would like to allow your user to create their own toolbars from a list of optional toolbar buttons, as seen in Figure 5.3.

Figure 5.3   Creating Toolbars from a Selection of Buttons

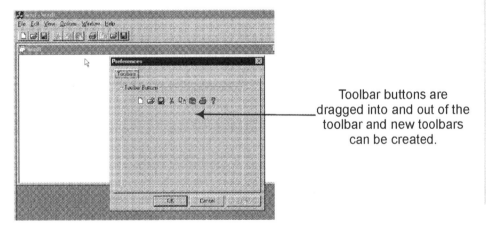

Toolbar buttons are dragged into and out of the toolbar and new toolbars can be created.

Strategy

First, we will create a new property sheet for our application's preferences that will configure our toolbars. This property page, when selected, will capture the mouse cursor so that clicking anywhere on the screen will send a mouse click to this page. This will allow us to determine if the user has clicked on a toolbar button outside of this page. Knowing this, we can use the regular functionality of CToolBar to modify that toolbar. The toolbar buttons in our property page will, in fact, simply be a list of bitmap images that represent toolbar buttons. We will also add functionality to our Main

Frame Class that will load and save any modified toolbars to the system registry so that they can be reloaded the next time our application is run. This is probably the most complicated of the examples in this book.

Steps

Create a Toolbar Property Page

1. Create a dialog box template using the Dialog Editor and give it a "Control" style. Put a simple group box in it for now.

2. Use the ClassWizard to create a toolbar property page class from this dialog box derived from CPropertyPage.

3. Embed a toolbar and bitmap class in this property page class.

```
CToolBar m_ToolBar;
CBitmap m_Bitmap;
```

4. Use the ClassWizard to add a WM_INITDIALOG message handler to this class. There, you will load whatever toolbars and toolbar bitmaps your application has. In this example, we are simply using the default toolbar. If there will be more in your application, simply load them all.

```
BOOL CToolBarPage::OnInitDialog()
{
    CPropertyPage::OnInitDialog();

    // create and load breeder toolbar
    m_ToolBar.Create( this );
    m_ToolBar.LoadToolBar( IDR_MAINFRAME );

    // load breeder toolbar bitmap

    m_bitmap.LoadBitmap( IDR_MAINFRAME );

    return TRUE;    // return TRUE unless you set the focus to a control
                    // EXCEPTION: OCX Property Pages should return FALSE
}
```

Example 9   Creating Programmable Toolbars   **169**

The toolbar(s) loaded here will not actually be visible to the user. Instead, we are using them as a reference for what buttons this property page can create.

5. Use the ClassWizard to add a WM_PAINT message handler to this property page class. Here, we will paint just the images of the buttons available for creating other toolbars within the group box. We will also use some extra processing to create some space between these button images.

```
void CToolBarPage::OnPaint()
{

    CPaintDC dc( this );    // device context for painting


    // display breeder toolbar bitmap
    CDC memDC;
    memDC.CreateCompatibleDC( &dc );
    memDC.SelectObject( &m_bitmap );


    int x = m_xBitmapStart;
    int y = m_yBitmapStart;
    for ( int i = 0; i < m_nButtonCount; i++ )
    {
        dc.BitBlt( x,y,BUTTON_WIDTH,BUTTON_HEIGHT, &memDC,
            i*BUTTON_WIDTH, 0, SRCCOPY );
        x += BUTTON_WIDTH + BUTTON_XSPACING;

    }
    memDC.DeleteDC();

}
```

Add a WM_SETACTIVE **Message Handler to This Property Page**

1. When this page is open, we need to grab control of the mouse cursor. If our user clicks on a toolbar button outside of this page, we need to be able to intercept that message so that we can either move or delete the button they click on. If we didn't grab control of the cursor, clicking a toolbar button would have no effect in our property page. Use the Class-Wizard to add a WM_SETACTIVE message handler to this property page

class. The WM_SETACTIVE message indicates this page is open. Then, use ::SetCapture to grab control of the cursor.

```
BOOL CToolBarPage::OnSetActive()
{
    SetCapture();
    return CPropertyPage::OnSetActive();
}
```

Add a WM_KILLACTIVE **Message Handler to This Property Page**

1. We also want to relinquish control of the mouse cursor when this page is closed. Use the ClassWizard to add a WM_KILLACTIVE message handler to release the cursor.

```
BOOL CToolBarPage::OnKillActive()
{
    ReleaseCapture();
    return CPropertyPage::OnKillActive();
}
```

So far, we have a property page that displays toolbar button images and that, when open, will know anytime the user has clicked on anything in the screen. We now need to process mouse messages that indicate the user has clicked on a button. It might indicate they are about to drag a button from an existing toolbar or an image from our property page. They might even have clicked on another tab in the property sheet that contains this property page. They might not have clicked on anything at all.

Add a Mouse Click Down Message Handler to This Property Page

1. Use the ClassWizard to add a WM_LBUTTONDOWN message handler to this class to process when the user has clicked something on the screen.

```
void CToolBarPage::OnLButtonDown( UINT nFlags, CPoint point )
{
```

Example 9   Creating Programmable Toolbars   **171**

2. Use the `CWnd::WindowFromPoint()` function to determine on which window the user has clicked.

```
m_bMoving = FALSE;
ClientToScreen( &point );
CWnd *pWnd = WindowFromPoint( point );
```

3. If the user has clicked on the property sheet or the tab of another property sheet, we must forward that message to that window.

```
// if point is parent or child of this property page
if ( pWnd != this && ( pWnd == GetParent() ||
    GetParent() -> IsChild( pWnd ) ) )


{
    CPoint pt( point );
    pWnd -> ScreenToClient( &pt );
    if ( pt.y >= 0 )
    {
        // if clicked on a cliet area, just send
        pWnd ->
            SendMessage( WM_LBUTTONDOWN,nFlags, MAKELONG( pt.x,pt.y ) );
    }
    *else
    {
        // if clicked on a non-client area, we must
        // perform a hit test before sending the click
        UINT ht = pWnd ->
            SendMessage( WM_NCHITTEST,0, MAKELONG( point.x,point.y ) );
        pWnd ->
            SendMessage( WM_NCLBUTTONDOWN,ht,
            MAKELONG( point.x,point.y ) );
    }
    return;
}
```

4. If the user has clicked on a button in an existing toolbar, we will go into a "move button" mode. We determine whether the user has clicked on such a button by simply checking to see if they clicked on a window that matches one of the toolbars maintained by our application. The "move button" mode simply means that we will determine and save the button

on which the user has clicked. We will also change the shape of the cursor to that of a button.

```
// see if this is a toolbar button
if ( m_pToolBar = GetToolBar( point ) )
{

    m_nButtonMoved = GetButtonIndex( m_pToolBar, point );
    m_pToolBar ->
        GetToolBarCtrl().GetButton( m_nButtonMoved,&m_tbbutton );
    m_bMoving = TRUE;
    ::SetCursor( m_hMoveCursor );

}
```

5. If the user clicks on one of the button images within this property page, we must determine on which button they clicked, save it, and again go into "move button" mode.

```
// else if this window
else if ( pWnd == this )
{
    ScreenToClient( &point );
    point.Offset( -m_xBitmapStart,-m_yBitmapStart );
    CRect rect( 0,0,(BUTTON_WIDTH+BUTTON_XSPACING )*
        m_nButtonCount,BUTTON_HEIGHT );
    if ( rect.PtInRect( point ) )
    {
        m_nButtonMoved = 0;
        int i = point.x/( BUTTON_WIDTH+BUTTON_XSPACING );
        for ( int j = 0;j < m_nButtonCount;j++ )
        {
            UINT k;
            int l;
            m_ToolBar.GetButtonInfo( j,k,k,l );
            if ( l == i )
            {
                m_nButtonMoved = j;
                break;
            }
        }
```

Example 9   Creating Programmable Toolbars   **173**

```
        }
        m_ToolBar.GetToolBarCtrl().GetButton(
            m_nButtonMoved,&m_tbbutton );
        m_bMoving = TRUE;
        ::SetCursor( m_hMoveCursor );
        SetCapture();
    }
```

6. Any other clicks are ignored. For a complete listing, please refer to the
   Property Page Class for Example 9 (page 191).

The location at which the user releases a dragged button will determine
the action your application should take next. If the user releases it on
another toolbar, your application should add that button to that toolbar. If
the user releases the button in the property page itself, it should cancel the
operation—unless the button was dragged from an existing toolbar, in
which case, your application should delete the button. If the user releases a
toolbar in empty space, your application should create a new floating or
docked toolbar.

Add a Mouse Click Release Message Handler to This Property Page

1. Use the ClassWizard to add a WM_LBUTTONUP message handler to this class.

```
void CToolBarPage::OnLButtonUp( UINT nFlags, CPoint point )
{
```

2. If we are not currently in "move button" mode, we will simply ignore
   this message.

```
if ( m_bMoving )
{
```

3. If we started the "move button" mode with a button from an existing
   toolbar, delete that button now.

```
// delete button from source toolbar
if ( m_pToolBar )
{
    m_pToolBar-> GetToolBarCtrl().DeleteButton( m_nButtonMoved );
}
```

4. Check to see if the user released the button on this property page. If so, we're done. If we were dragging an existing button, it has already been deleted. If we were dragging an image out of the property page, that move should now be canceled.

```
// if dropped anywhere but toolbar property page
CRect rect;
ClientToScreen (&point );
GetWindowRect( &rect );
if ( !rect.PtInRect( point ) )
{
```

5. Check to see if the user released the button over an existing toolbar and, if so, add the button to that toolbar at the location the user made the release.

```
// if dropped on existing toolbar, add button to it
CToolBar *pToolBar;
if ( pToolBar = GetToolBar( point ) )
{
    int i = GetButtonIndex( pToolBar,point );
    pToolBar -> GetToolBarCtrl().InsertButton( i,&m_tbbutton );
    UpdateToolBar( pToolBar );
}
```

GetToolBar() is actually a little helper function also created for this class. You will find it in the Property Page Class for Example 9 (page 191).

6. If the user didn't release the button over an existing toolbar or over the property page itself, we need to create a brand new floating or docked toolbar and insert this button into that.

```
// else create a new toolbar and add our button to it
else
{
    pToolBar = new CToolBar;
    CList<CToolBar*,CToolBar*> *pList =
        ( ( CMainFrame* )AfxGetMainWnd() ) -> GetToolBarList();
    pList -> AddTail( pToolBar );
    pToolBar -> Create( GetParentFrame(),
        WS_CHILD|WS_VISIBLE|CBRS_TOP|CBRS_TOOLTIPS );
    SIZE sizeButton, sizeImage;
```

Example 9   Creating Programmable Toolbars   **175**

```
sizeImage.cx = BUTTON_WIDTH;
sizeImage.cy = BUTTON_HEIGHT;
sizeButton.cx = sizeImage.cx + BUTTON_XSPACING;
sizeButton.cy = sizeImage.cy + BUTTON_YSPACING;
pToolBar -> SetSizes( sizeButton, sizeImage );
pToolBar -> EnableDocking( CBRS_ALIGN_ANY | CBRS_FLOAT_MULTI );

// add all possible tool button bitmaps
pToolBar ->
    GetToolBarCtrl().AddBitmap( m_nButtonCount,IDR_MAINFRAME );
// add new button to this new toolbar
pToolBar -> GetToolBarCtrl().InsertButton( 0,&m_tbbutton );
```

7. Now we must determine if the toolbar we have created can be docked to the frame or if it is a floating toolbar. We use a helper function called Get-ToolBarNear(), which essentially looks at all docked control bars to see if this toolbar is in line with it and, if so, whether it is vertically aligned or horizontally aligned. If the new toolbar can't be docked, then we simply float it.

```
UINT nMode;
BOOL bHorz;
if ( GetToolBarNear( point,nMode,bHorz ) )
{
    // dock toolbar centered on cursor
    CSize size = pToolBar -> CalcFixedLayout( FALSE,bHorz );
    if ( bHorz )
        point.x -= size.cx/2;
    else
        point.y -= size.cy/2;
        CRect rectx( point,size );
        GetParentFrame() -> DockControlBar( pToolBar,nMode,&rectx );
        pToolBar -> Invalidate();
        GetParentFrame() -> RecalcLayout();
}
```

```
else
{
    GetParentFrame() ->
        FloatControlBar( pToolBar, point, CBRS_ALIGN_TOP );
}
```

8. If this button has come from an existing toolbar, we now check to see if that toolbar is now devoid of buttons. If it has no buttons left, we delete it.

```
// else update source toolbar if any
else if ( m_pToolBar )
{
    UpdateToolBar( m_pToolBar );
    // if no buttons left on toolbar, delete it
    if ( m_pToolBar && m_pToolBar ->
        GetToolBarCtrl().GetButtonCount() == 0 )
    {
        POSITION pos;
        CList<CToolBar*,CToolBar*> *pList =
            ( ( CMainFrame* )AfxGetMainWnd() ) -> GetToolBarList();
        if ( pos = pList -> Find( m_pToolBar ) )
        {
            pList -> RemoveAt( pos );
            m_pToolBar -> m_bAutoDelete = TRUE;
            if ( !m_pToolBar -> IsFloating() )
            {
                m_pToolBar -> SendMessage( WM_CLOSE );
            }
            else
            {
                CDockBar* pDockBar = m_pToolBar -> m_pDockBar;
                CMiniDockFrameWnd* pDockFrame =
                    ( CMiniDockFrameWnd* )pDockBar -> GetParent();
                // won't close till page is gone so hide till then
                pDockFrame -> ShowWindow( SW_HIDE );
                pDockFrame -> SendMessage( WM_CLOSE );
            }
        }
    }
```

Example 9   Creating Programmable Toolbars   **177**

Add a Toolbar Property Page to Your Application's Property Sheet Options

1. Add this toolbar property page to your application's preferences.

```
void CMainFrame::OnOptionsPreferences()
{
    CPropertySheet sheet( _T( "Preferences" ),this );

    m_pToolBarPage = new CToolBarPage;

    sheet.AddPage( m_pToolBarPage );

    sheet.DoModal();

    delete m_pToolBarPage;
}
```

2. For a complete list of this property sheet class, please refer to the Property Page Class for Example 9 (page 191).

Although our user can now create their very own toolbars, this functionality means nothing if they can't then save their toolbars for the next time they run your application. To this end, we need to add two new member functions to the Main Frame Class. The first function will save our toolbar configuration when our application is about to terminate. The other will load the configuration when our application is first created.

Create a SaveToolbar() Function

1. Create a new function in CMainFrame called SaveToolbars(). There, we will start by saving the number of custom toolbars in your application. We will use CWinApp's WriteProfileInt() function to save this value.

```
void CMainFrame::SaveToolbars()
{
    int nNumToolBars = m_ToolBarList.GetCount();
    AfxGetApp() ->
        WriteProfileInt( TOOLBAR_SUMMARY_KEY,TOOLBAR_NUM_KEY,
        nNumToolBars );
```

II

5

2. If there aren't any toolbars to save, we return immediately.

```
if ( nNumToolBars )
{
```

3. Next, we loop through each toolbar in our application and create a new system registry key for it.

```
int i = 0;
int idc = IDC_CUSTOM_TOOLBARS;
for ( POSITION pos = m_ToolBarList.GetHeadPosition();pos;i++ )
{
    CToolBar *pToolBar = m_ToolBarList.GetNext( pos );
    // create key for this toolbar
    CString key;
    key.Format( TOOLBAR_KEY,i );

    // give toolbar a unique, sequential ID for
    //      SaveBarState/LoadBarState
    pToolBar -> SetDlgCtrlID( idc++ );
```

4. Then, we loop through each button in each toolbar, saving its information to the system registry.

```
// write number of buttons in this toolbar
int nButtonCount = pToolBar -> GetToolBarCtrl().GetButtonCount();
AfxGetApp() -> WriteProfileInt( key,NUM_OF_BUTTONS_KEY, nButtonCount );
// write info on each button
TBBUTTON *ptbbutton = new TBBUTTON[nButtonCount];
for ( int j = 0;j < nButtonCount;j++ )
{
    pToolBar -> GetToolBarCtrl().GetButton( j,ptbbutton+j );
}
AfxGetApp() -> WriteProfileBinary( key,BUTTON_INFO_KEY,
    ( BYTE* )ptbbutton,nButtonCount*sizeof( TBBUTTON ) );
delete []ptbbutton;
```

5. Use the ClassWizard to add a WM_CLOSE message handler to your Main Frame Class and call this SaveToolBars() function from that handler.

Example 9   Creating Programmable Toolbars   **179**

Also, call `CMainFrame::SaveBarState()` to save the current position and state of every bar in your application.

```
void CMainFrame::OnClose()
{
    // save all custom toolbars
    SaveToolbars();

    // save state of all control bars
    SaveBarState( "Control Bar States" );

    CMDIFrameWnd::OnClose();
}
```

6. For a full listing of `SaveToolbars()`, please see the Main Frame Class for Example 9 (page 184).

Add a `LoadToolbars()` Function to the Main Frame Class

1. Create another new function in `CMainFrame` called `LoadToolbars()`. We will start there by retrieving the number of toolbars we will be creating, and. if none, we return empty handed.

```
int CMainFrame::LoadToolbars()
{
    int nNumToolBars = AfxGetApp() ->
        GetProfileInt( TOOLBAR_SUMMARY_KEY, TOOLBAR_NUM_KEY,0 );
    if ( nNumToolBars )
    {
```

2. If there are toolbars to create, loop through each one creating it. Don't worry about which one is docked and which one is floating where because we will be using `CMainFrame::LoadBarState()` later to restore the state of all of our control bars.

```
int idc = IDC_CUSTOM_TOOLBARS;
for ( int i = 0;i < nNumToolBars;i++ )
{
    // create empty toolbar
    CToolBar *pToolBar = new CToolBar;
    m_ToolBarList.AddTail( pToolBar );
```

II

5

```
pToolBar ->
    Create( this, WS_CHILD|WS_VISIBLE|CBRS_TOP|CBRS_TOOLTIPS, idc++ );
// set button sizes
SIZE sizeButton, sizeImage;
sizeImage.cx = BUTTON_WIDTH;
sizeImage.cy = BUTTON_HEIGHT;
sizeButton.cx = sizeImage.cx + BUTTON_XSPACING;
sizeButton.cy = sizeImage.cy + BUTTON_YSPACING;
pToolBar -> SetSizes( sizeButton, sizeImage );
pToolBar -> EnableDocking( CBRS_ALIGN_ANY | CBRS_FLOAT_MULTI );
// add all possible tool button bitmaps to this empty toolbar
//      after first finding out how many buttons are in this bitmap
BITMAP bm;
CBitmap bitmap;
bitmap.LoadBitmap( IDR_MAINFRAME );
bitmap.GetObject( sizeof(BITMAP), &bm );
pToolBar -> GetToolBarCtrl().AddBitmap(
    bm.bmWidth/BUTTON_WIDTH,IDR_MAINFRAME );


// create key for this toolbar
CString key;
key.Format( TOOLBAR_KEY,i );
```

3. Then, add buttons to these toolbars based on their stored configuration in the system registry.

```
// get number of buttons in this toolbar
int nNumButtons = AfxGetApp() ->
    GetProfileInt(key,NUM_OF_BUTTONS_KEY,0);


// get button info and insert buttons into created toolbar
UINT k;
TBBUTTON *ptbbutton =;
AfxGetApp()->
    GetProfileBinary( key,BUTTON_INFO_KEY, ( BYTE ** )&ptbbutton,&k );
for ( int j = 0;j < nNumButtons;j++ )
{
```

Example 9   Creating Programmable Toolbars   **181**

```
        pToolBar -> GetToolBarCtrl().InsertButton(j,ptbbutton+j);
    }
    delete []ptbbutton;
}
```

Modify the Main Frame Class's `OnCreate()` Function

1. When your application first starts up, there will be no toolbars in the system registry, so `LoadToolbars()` will return with zero toolbars loaded and your application will initially have no toolbars. Therefore, you will want to initially load a default toolbar from your application's resources if there are no toolbars saved in the registry. You can use the following code to do this.

```
// create and load custom toolbars
EnableDocking( CBRS_ALIGN_ANY );
if ( !LoadToolbars() )
{
    // if none, load standard toolbar(s)
    CToolBar *pToolBar = new CToolBar;
    if ( !pToolBar -> Create( this ) ||
        !pToolBar -> LoadToolBar( IDR_MAINFRAME ) )
    {
        TRACE0( "Failed to create toolbar\n" );
        return -1;    // fail to create
    }
    :    :    :
```

2. At the end of the `OnCreate()` function, call `LoadBarState()` to restore the position, size, and state of each of your toolbars.

```
// reload all bar states
LoadBarState( "Control Bar States" );

return 0;
```

3. For a full listing of the Main Frame Class, please see the Main Frame Class for Example 9 (page 184).

One last item remains. What happens when the user clicks on the close button of a floating toolbar? The default action is to simply hide the toolbar.

However, in this example, we will actually delete the toolbar. Normally, we could create our own toolbar class and process the WM_CLOSE message to destroy the window. However, when a toolbar is floating, it's actually sitting in what's called a Mini Dock Frame Window. So, we need to create our own class derived from CMiniDockFrameWnd and add a WM_CLOSE message handler to it that will destroy the window.

Destroy Floating Toolbars

1. Use the ClassWizard to create a new class derived from CMiniFrameWnd. Then, replace all occurrences of CMiniFrameWnd with CMiniDockFrameWnd. Unfortunately, CMiniDockFrameWnd is undocumented and unsupported by ClassWizard.

2. Use the ClassWizard to add a WM_CLOSE message handler to this class where we will destroy this window, rather than hide it.

```
void CWzdMiniDockFrameWnd::OnClose()
{
    DestroyWindow();
    CMiniDockFrameWnd::OnClose();
}
```

3. To force your application to use this new class, you will need to add the following line after the EnableDock() line in your CMainFrame::OnCreate() function.

```
m_pFloatingFrameClass = RUNTIME_CLASS( CWzdMiniDockFrameWnd );
```

We can't simply substitute our new class name for the original because MFC creates this floating mini frame using CObject::CreateObject(). So, this method is also back-door and undocumented.

4. When a toolbar is destroyed, we also want it to be removed from the list of toolbars this application maintains. To do this, we will use the Class-Wizard to create a new class from the CToolBarCtrl class. Use the Text Editor to substitute all occurrences of CToolBarCtrl with CToolBar.

Example 9   Creating Programmable Toolbars   **183**

5. Then, use the ClassWizard to add a WM_DESTROY message handler to this new toolbar class. There you will access the list of toolbars in the Main Frame Class to delete this toolbar from it.

```
void CWzdToolBar::OnDestroy()
{
    CToolBar::OnDestroy();
    CList<CToolBar*,CToolBar*> *pList =
        ( ( CMainFrame* )AfxGetMainWnd() ) -> GetToolBarList();
    POSITION pos = pList -> Find( this );
    if ( pos )
    {
        pList -> RemoveAt( pos );
    }

}
```

Notes

- Due to the size of this example, there are many refinements left out. For example, when the user selects a toolbar bitmap "button" in the property page, you can put a descriptive message up in the page to describe the button. When the mouse is in "move button" mode, you can draw a line on the toolbar to indicate where the button will be inserted. (Currently, the button is inserted at the point at which the user releases the mouse button.) As with the programmable toolbar buttons in the Developer Studio, you could add toolbar categories to this example, displaying only certain toolbar bitmaps at a time.

- When the user clicks the close button on a floating toolbar, the current action in this example is to delete the toolbar from the application's list of toolbars kept in CMainFrame. You might, instead, maintain a dynamic list of toolbars under your application's "View" menu command which is what the Developer Studio does.

CD Notes

- When executing the project on the accompanying CD, click on the "Options" and then "Preferences" menu commands to invoke the preferences property sheet. To delete a button from the existing toolbar, drag it to the preferences sheet. To add it back, drag the appropriate image from the preferences sheet back to the toolbar. To create a new toolbar, drag a button from anywhere into open space. To delete a toolbar, cause it to float and then click on its close button.

Main Frame Class for Example 9

```
// MainFrm.h : interface of the CMainFrame class
//
/////////////////////////////////////////////////////////////////////////////

#if !defined( AFX_MAINFRM_H__CA9038EA_B0DF_11D1_A18C_DCB3C85EBD34__INCLUDED_ )
#define AFX_MAINFRM_H__CA9038EA_B0DF_11D1_A18C_DCB3C85EBD34__INCLUDED_

#if _MSC_VER >= 1000
#pragma once
#endif    // _MSC_VER >= 1000

#include "ToolBarPage.h"
#include <afxtempl.h>

class CMainFrame : public CMDIFrameWnd
{
    DECLARE_DYNAMIC( CMainFrame )
public:
    CMainFrame();

// Attributes
public:

// Operations
public:
    void LoadToolBars();
    void SaveToolBars();
    CList<CToolBar*,CToolBar*> *GetToolBarList(){return &m_ToolBarList;};

// Overrides
    // ClassWizard generated virtual function overrides
    // {{AFX_VIRTUAL( CMainFrame )
    virtual BOOL PreCreateWindow( CREATESTRUCT& cs );
    // }}AFX_VIRTUAL
```

Example 9   Creating Programmable Toolbars   **185**

```
// Implementation
public:
    virtual ~CMainFrame();
#ifdef _DEBUG
    virtual void AssertValid() const;
    virtual void Dump( CDumpContext& dc ) const;
#endif

protected:    // control bar embedded members
    CStatusBar  m_wndStatusBar;

// Generated message map functions
protected:
    // {{AFX_MSG(CMainFrame)
    afx_msg int OnCreate( LPCREATESTRUCT lpCreateStruct );
    afx_msg void OnClose();
    afx_msg void OnOptionsPreferences();
    // }}AFX_MSG
    DECLARE_MESSAGE_MAP()
private:
    CToolBarPage *m_pToolBarPage;

    int LoadToolbars();
    void SaveToolbars();
    CList<CToolBar*,CToolBar*> m_ToolBarList;

};

/////////////////////////////////////////////////////////////////////////////

// {{AFX_INSERT_LOCATION}}
// Microsoft Developer Studio will insert additional declarations immediately
//    before the previous line.

#endif
    // !defined( AFX_MAINFRM_H__CA9038EA_B0DF_11D1_A18C_DCB3C85EBD34__INCLUDED_ )
// MainFrm.cpp : implementation of the CMainFrame class
//

#include "stdafx.h"
```

```cpp
#include "Wzd.h"
#include "WzdProject.h"
#include "WzdToolBar.h"
#include "WzdMiniDockFrameWnd.h"

#include "MainFrm.h"

#include <afxpriv.h>

#ifdef _DEBUG
#define new DEBUG_NEW
#undef THIS_FILE
static char THIS_FILE[] = __FILE__;
#endif

/////////////////////////////////////////////////////////////////////////////
// CMainFrame

IMPLEMENT_DYNAMIC( CMainFrame, CMDIFrameWnd )

BEGIN_MESSAGE_MAP( CMainFrame, CMDIFrameWnd )
    // {{AFX_MSG_MAP( CMainFrame )
    ON_WM_CREATE()
    ON_WM_CLOSE()
    ON_COMMAND( ID_OPTIONS_PREFERENCES, OnOptionsPreferences )
    ON_WM_PALETTECHANGED()
    ON_WM_QUERYNEWPALETTE()
    // }}AFX_MSG_MAP
END_MESSAGE_MAP()

static UINT indicators[] =
{
    ID_SEPARATOR,           // status line indicator
    ID_INDICATOR_CAPS,
    ID_INDICATOR_NUM,
    ID_INDICATOR_SCRL,
};

/////////////////////////////////////////////////////////////////////////////
// CMainFrame construction/destruction
```

Example 9   Creating Programmable Toolbars   **187**

```
CMainFrame::CMainFrame()
{
}


CMainFrame::~CMainFrame()
{
    while ( !m_ToolBarList.IsEmpty() )
    {
        delete m_ToolBarList.RemoveHead();
    }
}


int CMainFrame::OnCreate( LPCREATESTRUCT lpCreateStruct )
{
    if ( CMDIFrameWnd::OnCreate( lpCreateStruct ) == -1 )
        return -1;

    // create and load custom toolbars
    EnableDocking( CBRS_ALIGN_ANY );
    m_pFloatingFrameClass = RUNTIME_CLASS( CWzdMiniDockFrameWnd );
    if ( !LoadToolbars() )
    {
        // if none, load standard toolbar(s)
        CToolBar *pToolBar = ( CWzdToolBar* )new CWzdToolBar;
        if ( !pToolBar -> Create( this ) ||
            !pToolBar -> LoadToolBar( IDR_MAINFRAME ) )
        {
            TRACE0( "Failed to create toolbar\n" );
            return -1;    // fail to create
        }
        m_ToolBarList.AddTail( pToolBar );

        // TODO: Remove this if you don't want tool tips or a resizeable toolbar
        pToolBar -> SetBarStyle( pToolBar -> GetBarStyle() |
            CBRS_TOOLTIPS | CBRS_FLYBY | CBRS_SIZE_DYNAMIC );

        // TODO: Delete these three lines if you don't want the toolbar to
        //       be dockable
        pToolBar -> EnableDocking( CBRS_ALIGN_ANY );
```

II

5

```
        DockControlBar( pToolBar );
    }


    if ( !m_wndStatusBar.Create( this ) ||
        !m_wndStatusBar.SetIndicators( indicators,
        sizeof( indicators )/sizeof( UINT ) ) )
    {
        TRACE0( "Failed to create status bar\n" );
        return -1;    // fail to create
    }


    // reload all bar states
    LoadBarState( "Control Bar States" );
    return 0;
}


BOOL CMainFrame::PreCreateWindow( CREATESTRUCT& cs )
{
    // TODO: Modify the Window class or styles here by modifying
    //     the CREATESTRUCT cs

    return CMDIFrameWnd::PreCreateWindow( cs );
}


/////////////////////////////////////////////////////////////////////////////
// CMainFrame diagnostics

#ifdef _DEBUG
void CMainFrame::AssertValid() const
{
    CMDIFrameWnd::AssertValid();
}


void CMainFrame::Dump( CDumpContext& dc ) const
{
    MDIFrameWnd::Dump( dc );
}


#endif    //_DEBUG
```

Example 9   Creating Programmable Toolbars   **189**

```
/////////////////////////////////////////////////////////////////////////////
// CMainFrame message handlers

void CMainFrame::OnClose()
{
    // save all custom toolbars
    SaveToolbars();

    // save state of all control bars
    SaveBarState( "Control Bar States" );

    CMDIFrameWnd::OnClose();
}

void CMainFrame::OnOptionsPreferences()
{
    CPropertySheet sheet( _T( "Preferences" ),this );
    m_pToolBarPage = new CToolBarPage;
    sheet.AddPage( m_pToolBarPage );
    sheet.DoModal();
    delete m_pToolBarPage;
}

int CMainFrame::LoadToolbars()
{
    int nNumToolBars = AfxGetApp() ->
        GetProfileInt( TOOLBAR_SUMMARY_KEY,TOOLBAR_NUM_KEY,0 );
    if ( nNumToolBars )
    {
        int idc = IDC_CUSTOM_TOOLBARS;
        for ( int i = 0;i < nNumToolBars;i++ )
        {
            // create empty toolbar
            CToolBar *pToolBar = ( CToolBar* )new CWzdToolBar;
            m_ToolBarList.AddTail( pToolBar );
            pToolBar ->
                Create( this, WS_CHILD|WS_VISIBLE|CBRS_TOP|CBRS_TOOLTIPS, idc++ );
            SIZE sizeButton, sizeImage;
            sizeImage.cx = BUTTON_WIDTH;
            sizeImage.cy = BUTTON_HEIGHT;
```

II

5

```
            sizeButton.cx = sizeImage.cx + BUTTON_XSPACING;
            sizeButton.cy = sizeImage.cy + BUTTON_YSPACING;
            pToolBar -> SetSizes( sizeButton, sizeImage );
            pToolBar -> EnableDocking( CBRS_ALIGN_ANY | CBRS_FLOAT_MULTI );


            // add all possible tool button bitmaps to this empty toolbar
            //     after first finding out how many buttons are in this bitmap
            BITMAP bm;
            CBitmap bitmap;
            bitmap.LoadBitmap( IDR_MAINFRAME );
            bitmap.GetObject( sizeof(BITMAP), &bm );
            pToolBar ->
               GetToolBarCtrl().AddBitmap( bm.bmWidth/BUTTON_WIDTH,IDR_MAINFRAME );


            // create key for this toolbar
            CString key;
            key.Format( TOOLBAR_KEY,i );


            // get number of buttons in this toolbar
            int nNumButtons = AfxGetApp() ->
               GetProfileInt( key,NUM_OF_BUTTONS_KEY,0 );


            // get button info and insert buttons into created toolbar
            UINT k;
            TBBUTTON *ptbbutton;
            AfxGetApp() ->
               GetProfileBinary( key,BUTTON_INFO_KEY,( BYTE ** )&ptbbutton,&k );
            for ( int j = 0;j < nNumButtons;j++ )
            {
                pToolBar -> GetToolBarCtrl().InsertButton( j,ptbbutton+j );
            }
            delete []ptbbutton;
        }
    }
return(nNumToolBars);
}


void CMainFrame::SaveToolbars()
{
    int nNumToolBars = m_ToolBarList.GetCount();
```

Example 9   Creating Programmable Toolbars   **191**

```
AfxGetApp() ->
    WriteProfileInt( TOOLBAR_SUMMARY_KEY,TOOLBAR_NUM_KEY,nNumToolBars );
if ( nNumToolBars )
{
    int i = 0;
    int idc = IDC_CUSTOM_TOOLBARS;
    for ( POSITION pos = m_ToolBarList.GetHeadPosition();pos;i++ )
    {
        CToolBar *pToolBar = m_ToolBarList.GetNext( pos );

        // create key for this toolbar
        CString key;
        key.Format( TOOLBAR_KEY,i );

        // give toolbar a unique, sequential ID for SaveBarState/LoadBarState
        pToolBar -> SetDlgCtrlID( idc++ );

        // write number of buttons in this toolbar
        int nButtonCount = pToolBar -> GetToolBarCtrl().GetButtonCount();
        AfxGetApp() -> WriteProfileInt( key,NUM_OF_BUTTONS_KEY,nButtonCount );

        // write info on each button
        TBBUTTON *ptbbutton = new TBBUTTON[nButtonCount];
        for ( int j = 0;j < nButtonCount;j++ )
        {
            pToolBar -> GetToolBarCtrl().GetButton( j,ptbbutton+j );
        }
        AfxGetApp() ->
            WriteProfileBinary( key,BUTTON_INFO_KEY,
            ( BYTE* )ptbbutton,nButtonCount*sizeof( TBBUTTON ) );
        delete []ptbbutton;
    }
}
}
```

Property Page Class for Example 9

```
#if !defined TOOLBARPAGE_H
#define TOOLBARPAGE_H
```

```
// ToolBarPage.h : header file
//

#include "WzdProject.h"

/////////////////////////////////////////////////////////////////////////////
// CToolBarPage dialog

class CToolBarPage : public CPropertyPage
{
    DECLARE_DYNCREATE( CToolBarPage )

// Construction
public:
    CToolBarPage();
    ~CToolBarPage();

// Dialog Data
    // {{AFX_DATA( CToolBarPage )
    enum { IDD = IDD_TOOLBAR_PAGE };
    // }}AFX_DATA

// Overrides
    // ClassWizard generate virtual function overrides
    // {{AFX_VIRTUAL( CToolBarPage )
public:
    virtual BOOL OnSetActive();
    virtual BOOL OnKillActive();
protected:
    virtual void DoDataExchange( CDataExchange* pDX );    // DDX/DDV support
    // }}AFX_VIRTUAL

// Implementation
protected:
    // Generated message map functions
    // {{AFX_MSG( CToolBarPage )
    virtual BOOL OnInitDialog();
    afx_msg void OnLButtonDown( UINT nFlags, CPoint point );
    afx_msg void OnLButtonUp( UINT nFlags, CPoint point );
    afx_msg void OnPaint();
```

Example 9   Creating Programmable Toolbars   **193**

```
    // }}AFX_MSG
    DECLARE_MESSAGE_MAP()
private:
    CToolBar    m_ToolBar;
    CBitmap     m_bitmap;
    int         m_xBitmapStart;
    int         m_yBitmapStart;
    int         m_nButtonCount;

    BOOL        m_bMoving;
    int         m_nButtonMoved;
    TBBUTTON    m_tbbutton;
    HICON       m_hMoveCursor;
    CToolBar    *m_pToolBar;
    CToolBar    *GetToolBar( CPoint point );
    int         GetButtonIndex( CToolBar *pToolBar, CPoint point );
    void        UpdateToolBar( CToolBar *pToolBar );
    BOOL        GetToolBarNear( CPoint &point,UINT &nMode,BOOL &bHorz );

};

#endif
// ToolBarPage.cpp : implementation file
//

#include "stdafx.h"
#include "wzd.h"
#include "ToolBarPage.h"
#include "WzdToolBar.h"
#include "WzdProject.h"
#include "MainFrm.h"

#include <afxcmn.h>
#include <afxpriv.h>

#ifdef _DEBUG
#define new DEBUG_NEW
#undef THIS_FILE
static char THIS_FILE[] = __FILE__;
#endif
```

II

5

```
///////////////////////////////////////////////////////////////////////////////
// CToolBarPage property page

IMPLEMENT_DYNCREATE( CToolBarPage, CPropertyPage )

CToolBarPage::CToolBarPage() : CPropertyPage( CToolBarPage::IDD )
{
    // {{AFX_DATA_INIT( CToolBarPage )
    // }}AFX_DATA_INIT
    m_bMoving = FALSE;
    m_nButtonCount = 0;
    m_pToolBar = NULL;
    m_hMoveCursor = AfxGetApp() -> LoadCursor( IDI_MOVING_BUTTON );
}


CToolBarPage::~CToolBarPage()
{
}


void CToolBarPage::DoDataExchange( CDataExchange* pDX )
{
    CPropertyPage::DoDataExchange( pDX );
    // {{AFX_DATA_MAP( CToolBarPage )
    // }}AFX_DATA_MAP
}


BEGIN_MESSAGE_MAP( CToolBarPage, CPropertyPage )
    // {{AFX_MSG_MAP(CToolBarPage)
    ON_WM_LBUTTONDOWN()
    ON_WM_LBUTTONUP()
    ON_WM_PAINT()
    // }}AFX_MSG_MAP
END_MESSAGE_MAP()


///////////////////////////////////////////////////////////////////////////////
// CToolBarPage message handlers

BOOL CToolBarPage::OnInitDialog()
```

Example 9 Creating Programmable Toolbars **195**

```
{
    CPropertyPage::OnInitDialog();

    // create and load breeder toolbar
    m_ToolBar.Create( this );
    m_ToolBar.LoadToolBar( IDR_MAINFRAME );
    // load breeder toolbar bitmap
    m_bitmap.LoadBitmap( IDR_MAINFRAME );

    // get window and button count
    m_nButtonCount = m_ToolBar.GetToolBarCtrl().GetButtonCount();


    CRect rect;
    GetClientRect( &rect );
    m_xBitmapStart = rect.Width()/6;
    m_yBitmapStart = rect.Height()/6;


    return TRUE;      // return TRUE unless you set the focus to a control
                      // EXCEPTION: OCX Property Pages should return FALSE
}

void CToolBarPage::OnPaint()
{
    CPaintDC dc( this );      // device context for painting

    // display breeder toolbar bitmap
    CDC memDC;
    memDC.CreateCompatibleDC( &dc );
    memDC.SelectObject( &m_bitmap );

    int x = m_xBitmapStart;
    int y = m_yBitmapStart;
    for ( int i = 0; i < m_nButtonCount; i++ )
    {
        dc.BitBlt( x,y,BUTTON_WIDTH,BUTTON_HEIGHT, &memDC, i*BUTTON_WIDTH,
            0, SRCCOPY );
        x += BUTTON_WIDTH + BUTTON_XSPACING;
    }
    memDC.DeleteDC();
```

II

5

```
}

BOOL CToolBarPage::OnSetActive()
{

    SetCapture();
    return CPropertyPage::OnSetActive();
}


BOOL CToolBarPage::OnKillActive()
{

    ReleaseCapture();
    return CPropertyPage::OnKillActive();
}

void CToolBarPage::OnLButtonDown( UINT nFlags, CPoint point )
{
    m_bMoving = FALSE;
    ClientToScreen( &point );
    CWnd *pWnd = WindowFromPoint( point );

    // if point is parent or child of
    if ( pWnd != this && ( pWnd == GetParent() ||
        GetParent() -> IsChild( pWnd ) ) )
    {
        CPoint pt( point );
        pWnd -> ScreenToClient( &pt );
        if ( pt.y >= 0 )
        {
            pWnd -> SendMessage( WM_LBUTTONDOWN,nFlags,MAKELONG( pt.x,pt.y ) );
        }
        else
        {
            UINT ht = pWnd ->
                SendMessage( WM_NCHITTEST,0,MAKELONG( point.x,point.y ) );
            pWnd ->
                SendMessage( WM_NCLBUTTONDOWN,ht,MAKELONG( point.x,point.y ) );
        }
        return;
    }
```

Example 9   Creating Programmable Toolbars   **197**

```
// see if this is a toolbar button
if ( m_pToolBar = GetToolBar( point ) )
{

    m_nButtonMoved = GetButtonIndex( m_pToolBar, point );
    m_pToolBar -> GetToolBarCtrl().GetButton( m_nButtonMoved,&m_tbbutton );
    m_bMoving = TRUE;
    ::SetCursor( m_hMoveCursor );
}


// else if this window
else if ( pWnd == this )
{
    ScreenToClient( &point );
    point.Offset( -m_xBitmapStart,-m_yBitmapStart );
    CRect rect( 0,0,
        ( BUTTON_WIDTH+BUTTON_XSPACING )*m_nButtonCount,BUTTON_HEIGHT );
    if ( rect.PtInRect( point ) )
    {
        m_nButtonMoved = 0;
        int i = point.x/( BUTTON_WIDTH + BUTTON_XSPACING );
        for ( int j = 0;j < m_nButtonCount;j++ )
        {
            UINT k;
            int l;
            m_ToolBar.GetButtonInfo( j,k,k,l );
            if ( l == i )
            {
                m_nButtonMoved = j;
                break;
            }
        }
        m_ToolBar.GetToolBarCtrl().GetButton( m_nButtonMoved,&m_tbbutton );
        m_bMoving = TRUE;
        ::SetCursor( m_hMoveCursor );
        SetCapture();
    }
}

CPropertyPage::OnLButtonDown( nFlags, point );
```

```
}

void CToolBarPage::OnLButtonUp( UINT nFlags, CPoint point )
{
    if ( m_bMoving )
    {

        // delete button from source toolbar
        if ( m_pToolBar )
        {
            m_pToolBar -> GetToolBarCtrl().DeleteButton( m_nButtonMoved );
        }

        // if dropped anywhere but toolbar property page
        CRect rect;
        ClientToScreen( &point );
        GetWindowRect( &rect );
        if ( !rect.PtInRect( point ) )
        {
            // if dropped on existing toolbar, add button to it
            CToolBar *pToolBar;
            if ( pToolBar = GetToolBar( point ) )
            {
                int i = GetButtonIndex( pToolBar,point );
                pToolBar -> GetToolBarCtrl().InsertButton( i,&m_tbbutton );
                UpdateToolBar( pToolBar );
            }
            // else create a new toolbar and add our button to it
            else
            {
                pToolBar = ( CWzdToolBar* )new CWzdToolBar;
                CList<CToolBar*,CToolBar*> *pList =
                    ( ( CMainFrame* )AfxGetMainWnd() ) -> GetToolBarList();
                pList -> AddTail( pToolBar );
                pToolBar ->
                    Create( GetParentFrame(),
                    WS_CHILD|WS_VISIBLE|CBRS_TOP|CBRS_TOOLTIPS );
                SIZE sizeButton, sizeImage;
                sizeImage.cx = BUTTON_WIDTH;
                sizeImage.cy = BUTTON_HEIGHT;
```

Example 9   Creating Programmable Toolbars   **199**

```
            sizeButton.cx = sizeImage.cx + BUTTON_XSPACING;
            sizeButton.cy = sizeImage.cy + BUTTON_YSPACING;
            pToolBar -> SetSizes( sizeButton, sizeImage );
            pToolBar -> EnableDocking( CBRS_ALIGN_ANY | CBRS_FLOAT_MULTI );

            // add all possible tool button bitmaps
            pToolBar ->
                GetToolBarCtrl().AddBitmap( m_nButtonCount,IDR_MAINFRAME );

            // add new button to this new toolbar
            pToolBar -> GetToolBarCtrl().InsertButton( 0,&m_tbbutton );

            UINT nMode;
            BOOL bHorz;
            if ( GetToolBarNear( point,nMode,bHorz ) )
            {
                // dock toolbar centered on cursor
                CSize size = pToolBar -> CalcFixedLayout( FALSE,bHorz );
                if ( bHorz )
                    point.x -= size.cx/2;
                else
                    point.y -= size.cy/2;
                CRect rectx( point,size );
                GetParentFrame() -> DockControlBar( pToolBar,nMode,&rectx );
                pToolBar -> Invalidate();
                GetParentFrame() -> RecalcLayout();
            }
            else
            {
                GetParentFrame() ->
                    FloatControlBar( pToolBar, point, CBRS_ALIGN_TOP );
            }
        }
    }
}
// else update source toolbar if any
else if ( m_pToolBar )
{
    UpdateToolBar( m_pToolBar );

    // if no buttons left on toolbar, delete it
```

```
            if ( m_pToolBar && m_pToolBar ->
                GetToolBarCtrl().GetButtonCount() == 0 )
            {
                POSITION pos;
                CList<CToolBar*,CToolBar*> *pList =
                    ( ( CMainFrame* )AfxGetMainWnd() ) -> GetToolBarList();
                if ( pos = pList -> Find( m_pToolBar ) )
                {
                    pList -> RemoveAt( pos );
                    m_pToolBar -> m_bAutoDelete = TRUE;
                    if ( !m_pToolBar -> IsFloating() )
                    {
                        m_pToolBar -> SendMessage( WM_CLOSE );
                    }
                    else
                    {
                        CDockBar* pDockBar = m_pToolBar -> m_pDockBar;
                        CMiniDockFrameWnd* pDockFrame =
                            ( CMiniDockFrameWnd* )pDockBar -> GetParent();
                        pDockFrame -> ShowWindow( SW_HIDE );
                        pDockFrame -> SendMessage( WM_CLOSE );
                    }
                }
            }
        m_bMoving = FALSE;
        ::SetCursor( AfxGetApp() -> LoadStandardCursor( IDC_ARROW ) );
    }


    CPropertyPage::OnLButtonUp( nFlags, point );
}

///////////////////////////////////////////////////////////////////////////////
///////////////////////////////////////////////////////////////////////////////
// Helper Toolbar Functions

CToolBar *CToolBarPage::GetToolBar( CPoint point )
{
    CRect rect;
    CList<CToolBar*,CToolBar*> *pList =
        ( ( CMainFrame* )AfxGetMainWnd() ) -> GetToolBarList();
```

Example 9   Creating Programmable Toolbars   **201**

```
        if ( !pList -> IsEmpty() )
        {
            for ( POSITION pos = pList -> GetHeadPosition(); pos; )
            {
                CToolBar *pToolBar = pList -> GetNext( pos );
                pToolBar -> GetWindowRect( &rect );
                if ( rect.PtInRect( point ) )
                {
                    return pToolBar;
                }
            }
        }
        return( NULL );
}


int CToolBarPage::GetButtonIndex( CToolBar *pToolBar, CPoint point )
{
    CRect rect;
    pToolBar -> ScreenToClient( &point );
    int nButtons = pToolBar ->GetToolBarCtrl().GetButtonCount();
    for ( int i = 0; i < nButtons; i++ )
    {
        pToolBar -> GetItemRect( i,&rect );
        if ( rect.PtInRect( point ) )
        {
            return( i );
        }
    }
    return( -1 );
}


void CToolBarPage::UpdateToolBar( CToolBar *pToolBar )
{
    if ( pToolBar )
    {
        if ( pToolBar -> IsFloating() )
        {
            CDockBar* pDockBar = pToolBar -> m_pDockBar;
            CMiniDockFrameWnd* pDockFrame =
                ( CMiniDockFrameWnd* )pDockBar -> GetParent();
```

II

5

```
                pDockFrame -> RecalcLayout( TRUE );
                pDockFrame -> UpdateWindow();
            }
            else
            {
                pToolBar -> Invalidate();
                GetParentFrame() -> RecalcLayout();
            }
        }
    }
}


BOOL CToolBarPage::GetToolBarNear( CPoint &point,UINT &nMode,BOOL &bHorz )
{
    CRect rect;
    BOOL bDock = FALSE;
    CDockState dockstate;
    ( ( CMainFrame * )AfxGetMainWnd() ) -> GetDockState( dockstate );
    for ( int i = 0; i < dockstate.m_arrBarInfo.GetSize(); i++ )
    {
        CControlBarInfo *pBarInfo =
            ( CControlBarInfo * )dockstate.m_arrBarInfo[i];
        CControlBar *pBar = ( CControlBar * )pBarInfo -> m_pBar;
        // if this bar is visible, can be docked and isn't floating, check it out
        if ( pBarInfo - >m_bVisible && pBarInfo ->
            m_bDocking && !pBar -> IsFloating() )
        {
            pBarInfo -> m_pBar -> GetWindowRect( &rect );
            DWORD dwStyle = pBar -> GetBarStyle();
            if ( bHorz = ( dwStyle & CBRS_ORIENT_HORZ ) )
            {
                // if user clicked in this region, dock new toolbar here
                if ( point.y >= rect.top && point.y < rect.bottom)
                {
                    bDock = TRUE;
                    point.y = rect.top;
                    if ( dwStyle & CBRS_ALIGN_TOP )
                        nMode = AFX_IDW_DOCKBAR_TOP;
                    else
                        nMode = AFX_IDW_DOCKBAR_BOTTOM;
                    break;
```

```
            }
        }
        else
        {
            // else if user clicked in this region, dock here
            if ( point.x >= rect.left && point.x < rect.right )
            {
                bDock = TRUE;
                point.x = rect.left;
                if ( dwStyle & CBRS_ALIGN_LEFT )
                    nMode = AFX_IDW_DOCKBAR_LEFT;
                else
                    nMode = AFX_IDW_DOCKBAR_RIGHT;
                break;
            }
        }
    }
    return( bDock );
}
```

II

5

Example 10   Putting a Toolbar, Menu, and Status Bar in Your Dialog App

Objective

You would like to add a menu, toolbar, and status bar to your Dialog Application, as seen in Figure 5.4.

Figure 5.4   A Dialog Box with a Toolbar and Status Bar

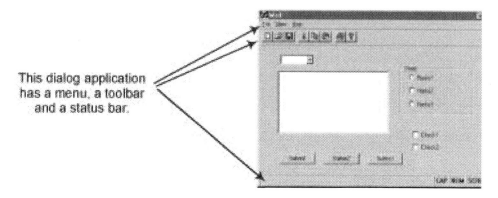

This dialog application has a menu, a toolbar and a status bar.

Strategy

Unlike an SDI or MDI application, the AppWizard does not create a Dialog Application with a menu, toolbar, or status bar. That doesn't mean a Dialog Application can't have these user input methods—after all, any popup window can have a menu, and toolbars and status bars are just control windows that can be added to any other window. The difference, then, is that we need to manually add these items, as needed.

Steps

Add a Menu

1. Use the Menu Editor to create a new menu, or copy one from another project. If you're the daring type, you can create a new SDI application just to steal its menu from its .rc file and paste it into your Dialog App's .rc file.

2. Use the Dialog Editor to add this new menu to your application's dialog template.

3. Use the ClassWizard to add command handlers to your dialog class to handle these menu items.

Add a Toolbar

1. Use the Toolbar Editor to create a new toolbar, or copy one from another project, as described in the last section.

2. Since the Dialog Editor doesn't allow you to place a toolbar control in a template, you will need to create one dynamically with your dialog class. However, a good way to specify the location at which to create this toolbar is by using the Dialog Editor to create a static control "placeholder" where you want to put the toolbar, then transfer the position and size of this control to your toolbar when creating it. Therefore, use the Dialog Editor to add a static control to your dialog's template where you want the toolbar to go. Make it long and skinny.

3. Embed a `CToolbar` variable in your dialog class.

```
CToolbar m_wndToolBar;
```

4. In the dialog class's `OnInitDialog()` message handler, create the toolbar.

```
if ( !m_wndToolBar.Create( this ) ||
    !m_wndToolBar.LoadToolBar( IDR_MAINFRAME ) )
{
    TRACE0( "Failed to create toolbar\n" );
    return -1;    // fail to create
}
```

5. Position the toolbar window inside the static control we created with the dialog editor.

```
CRect rect;
GetDlgItem( IDC_TOOLBAR_STATIC ) -> GetWindowRect( &rect );
ScreenToClient( &rect );
m_wndToolBar.MoveWindow( &rect );
```

Add a Status Bar

1. Use the String Table Editor to define your status bar panes.

```
ID_INDICATOR_ONE    "xxx"
ID_INDICATOR_TWO    "yyy"
```

2. Again, we will use the Dialog Editor to add a static control to our dialog template in the location at which you would like to place the status bar.

3. Embed a status bar control in your dialog class.

```
CStatusBar   m_wndStatusBar;
```

4. Add a list of the status pane definitions you created with the String Editor to the top of your dialog class.

```
static UINT indicators[] =
{
    ID_SEPARATOR,           // status line indicator
    ID_INDICATOR_ONE,
    ID_INDICATOR_TWO,
};
```

5. Create the status bar and position it over the static control.

```
if ( !m_wndStatusBar.Create(this,WS_CHILD|WS_VISIBLE|WS_BORDER) ||
    !m_wndStatusBar.SetIndicators( indicators,
    sizeof( indicators )/sizeof( UINT ) ) )
{
    TRACE0( "Failed to create status bar\n" );
    return -1;    // fail to create
}

GetDlgItem( IDC_STATUSBAR_STATIC ) -> GetWindowRect( &rect );
ScreenToClient( &rect );
m_wndStatusBar.MoveWindow( &rect );
```

Notes

- The first pane of the status bar created with CStatusBar is created without a border. Although this is okay for an SDI or MDI application since the status bar typically sits by itself below the view, this pane in fact disappears into a dialog window. To give this pane a defining border, you can use the following right after the status bar is created.

```
UINT nID,nStyle;
int nWidth;
m_wndStatusBar.GetPaneInfo( 0, nID, nStyle, nWidth );
m_wndStatusBar.SetPaneInfo( 0, nID, SBPS_STRETCH, nWidth );
```

- Just as the toolbar and status bar had to be manually added, you are also responsible for updating their status. For a toolbar, this means manually

Example 11   Adding a Bitmap Logo to a Popup Menu   **207**

graying out or checking indiviudal buttons. For a status bar, this means manually setting the text in each pane.

- If you impliment the menu, you can remove the help menu item from your Dialog App's system menu. You can then also remove the supporting logic in your dialog class.

CD Notes

- When executing the project on the accompanying CD, you will notice that this Dialog Application has a menu, toolbar, and status bar.

Example 11   Adding a Bitmap Logo to a Popup Menu

Objective

You would like to add a bitmap logo to a popup menu, as seen in Figure 5.5.

Figure 5.5   Popup Menu with a Logo

This popup menu has a bitmap along its side.

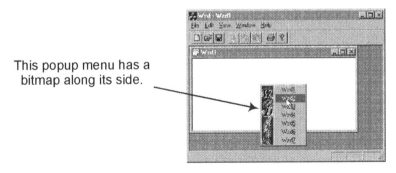

Strategy

Actually, we will be using a bit of a trick to add this bitmap. In fact, we won't be adding a bitmap to the menu at all. Instead, we will be opening a dialog box with a vertical, but thin, bitmap. We will open a popup menu next to the bitmap. Another more complicated approach would involve drawing the menu yourself, for which you can refer to Example 7.

Steps

Create a Popup Menu Dialog Class

1. Create a popup menu resource with the menu items you require.

2. Create a dialog box resource with a picture control on one side and a static control on the other. Configure the picture control to hold a bitmap to appear along the side of the static control.

3. Next, size the static control and dialog template to conform to the size of the popup menu. You will need to perform the next steps to be able to see the final result. For now, you can only make a rough estimate.

4. Use the ClassWizard to create a dialog class that uses this template.

5. Use the ClassWizard to add a WM_INITDIALOG message handler to this dialog class. There, you will find out where the cursor is so you can reposition the dialog box there.

```
BOOL CWzdDlg::OnInitDialog()
{
CDialog::OnInitDialog();

// position where mouse button was clicked
CPoint pt;
GetCursorPos( &pt );
SetWindowPos( NULL,pt.x,pt.y,0,0,SWP_NOSIZE );

return TRUE;      // return TRUE unless you set the focus to a control
                  // EXCEPTION: OCX Property Pages should return FALSE
}
```

6. Use the ClassWizard to add a WM_PAINT message handler to this class to take care of opening the popup menu over the static control.

```
void CWzdDlg::OnPaint()
{
    CDialog::OnPaint();

    // load up menu
    CMenu menu;
    menu.LoadMenu( IDR_WZD_MENU );
```

Example 11   Adding a Bitmap Logo to a Popup Menu   **209**

```
CMenu* pPopup = menu.GetSubMenu( 0 );

// get location of static control and display popup menu there
CRect rect;
GetDlgItem( IDC_MENU_STATIC ) -> GetWindowRect( &rect );
int nLeft = rect.right + 2;
GetWindowRect( &rect );
pPopup ->
    TrackPopupMenu( TPM_RIGHTBUTTON, nLeft, rect.top, GetParent() );

// cancel this dialog
PostMessage( WM_CLOSE );
}
```

Notice that we also send a message to ourselves to immediately close this dialog box once the user has selected a menu item.

Implement the Popup Menu Dialog Class

1. Use the ClassWizard to add a WM_RBUTTONDOWN message handler to the class in which you want to implement this menu. In this example, this menu is implemented in the View Class. There you will create the dialog with DoModal().

```
void CWzdView::OnRButtonDown( UINT nFlags, CPoint point )
{
    CWzdDlg dlg;
    dlg.DoModal();

    CView::OnRButtonDown( nFlags, point );
}
```

Notes

- As mentioned previously, you can also try drawing the menu yourself, positioning your menu item text to the left or right and drawing some portion of the bitmap when drawing that item.

II

5

CD Notes

- When executing the project on the accompanying CD, right-click the mouse on the view to open a popup menu, which will have a bitmap displayed on the left side.

Example 12   Putting a Dropdown Button in a Toolbar

Objective

You would like to create a double button in a toolbar, as seen in Figure 5.6.

Figure 5.6   A Drop-down Button in a Toolbar

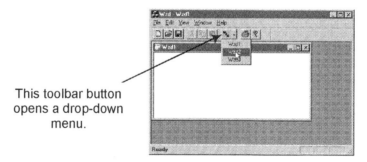

This toolbar button opens a drop-down menu.

Strategy

We will be using two toolbar styles that are new to MFC v6.0. Your application must also run on Windows 98 or with a system that has Internet Explorer 4.01 or better installed. The first, TBSTYLE_DROPDOWN, causes a toolbar button to issue a TBN_DROPDOWN notification message when it's pressed. (Normally, a toolbar button issues a WM_COMMAND message along with the ID of the button.) This notification can then be used to display a popup menu underneath the button to give it a menu-like feel. The second style, TBSTYLE_EX_DRAWDDARROWS, causes a button with the TBSTYLE_ drop-down style to be drawn with a drop-down arrow, as seen in Figure 5.6.

To keep our CMainFrame class from getting cluttered with code applying these new styles, we will encapsulate it in our very own new toolbar class derived from CToolbar.

Example 12   Putting a Dropdown Button in a Toolbar   **211**

Steps

Create a New Toolbar Class

1. Use the ClassWizard to create a new class derived from `CToolbarCtrl`. Then, use the Text Editor to change all references to `CToolbarCtrl` to `CToolbar` (You can't create a class derived from `CToolbar` with Class Wizard).

2. Add a new function to this class that will load our toolbar, as before; however, it will also apply two new toolbar styles to a selected toolbar button that will cause it to display a drop-down button and to issue a `TBN_DROPDOWN` notification message whenever it's pressed.

```
BOOL CWzdToolBar::LoadToolBarEx( UINT id )
{
    // load toolbar info
    BOOL bRet;
    bRet = CToolBar::LoadToolBar( id );

    // find where our dropdown button will go
    int pos = CommandToIndex( IDC_WZD_DROPBUTTON );

    // set this button to be a dropdown button
    int iImage;
    UINT nID,nStyle;
    GetButtonInfo( pos,nID,nStyle,iImage );
    SetButtonInfo( pos,nID,nStyle|TBSTYLE_DROPDOWN,iImage );

    // ask toolbar to draw a down arrow next to this and
    //     any button with TBSTYLE_DROPDOWN
    SendMessage( TB_SETEXTENDEDSTYLE,0,TBSTYLE_EX_DRAWDDARROWS );
    return bRet;
}
```

Notice that we need to set the `TBSTYLE_EX_DRAWDDARROWS` style using the new toolbar message, `TB_SETEXTENDEDSTYLE`.

3. If this drop-down button is sitting in a floating toolbar, the buttons on the end of the toolbar may be clipped because `CToolbar`'s `CalcDynamicLayout()` function doesn't yet seem to account for the fact that drop-down

buttons are a bit longer than normal buttons. To compensate for this, you will also need to override the `CalcDynamicLayout()` function in your toolbar class to add extra pixels to the size of your toolbar. Therefore, declare this override and add this implementation.

```
// in your toolbar class's declaration file:
CSize CalcDynamicLayout( int nLength, DWORD dwMode );
:    :    :
// in your toolbar class's implementation file:
CSize CWzdToolBar::CalcDynamicLayout( int nLength, DWORD dwMode )
{
    CSize size = CToolBar::CalcDynamicLayout( nLength,dwMode );
    // make room on non-fixed toolbars for additional down arrows
    if ( dwMode&LM_HORZ ) size.cx += 16*1;    // 16 * number of buttons
                                              //        with arrows

    return size;
}
```

4. Now, manually add a `TBN_DROPDOWN` message handler to this class. In the declaration file, add the following.

```
// {{AFX_MSG( CWzdToolBar )
// }}AFX_MSG
// add the following
void OnDropdownButton( LPNMTOOLBAR pNotifyStruct, LRESULT* result );
DECLARE_MESSAGE_MAP()
```

5. To the message map, add

```
BEGIN_MESSAGE_MAP( CWzdToolBar, CToolBar )
    // {{AFX_MSG_MAP( CWzdToolBar )
    // }}AFX_MSG_MAP
    ON_NOTIFY_REFLECT( TBN_DROPDOWN,OnDropdownButton ) // <<< add
END_MESSAGE_MAP()
```

Example 12   Putting a Dropdown Button in a Toolbar   **213**

6. Process this notification by opening a popup menu right below the tool-bar button.

```
void CWzdToolBar::OnDropdownButton( LPNMTOOLBAR lpnmtb,
    LRESULT *result )
{
    // get location of clicked button
    CRect rect;
    GetItemRect( CommandToIndex( lpnmtb -> iItem ),&rect );
    ClientToScreen( &rect );

    // putup a popup menu there
    CMenu menu;
    menu.LoadMenu( IDR_WZD_MENU );
    CMenu* pPopup = menu.GetSubMenu( 0 );

    pPopup ->
        TrackPopupMenu( TPM_RIGHTBUTTON, rect.left, rect.bottom, this );

    *result = TBDDRET_DEFAULT;     // drop-down was handled, also
    // TBDDRET_NODEFAULT drop-down was not handled.
    // TBDDRET_TREATPRESSED generate a WM_COMMAND message
}
```

7. For a complete listing of this class, please refer to the Toolbar Class for Example 12 (page 214).

Implement This New Toolbar Class

1. Use this new toolbar class in CMainFrame and load your toolbar resource using LoadBitmapEx().

Notes

- If your toolbar will be fixed, you don't need to worry about overriding the CalcDynamicLayout() function to resize the toolbar—the toolbar will always be as big as it can be. Please see Chapter 2 for more on toolbars and control bars. If your toolbar can be docked vertically, you will notice that the drop-down arrow will be clipped off by the toolbar. This can also be repaired in CalcDynamicLayout(); however, this seems to be a feature of Microsoft and not a bug. (In other words, drop-down toolbar buttons in other Microsoft applications also clip the drop-down arrow.)

CD Notes

- When executing the project on the accompanying CD, you will notice a drop-down button in the toolbar that, when clicked, causes a menu to appear beneath the toolbar button.
- Since this project requires functionality only available in v6.0 of the Developer Studio, you will only find a v6.0 version of this project on the CD.

Toolbar Class for Example 12

```
#if !defined( AFX_WZDTOOLBAR_H__27649E31_C807_11D1_9B5D_00AA003D8695__INCLUDED_ )
#define AFX_WZDTOOLBAR_H__27649E31_C807_11D1_9B5D_00AA003D8695__INCLUDED_

#if _MSC_VER >= 1000
#pragma once
#endif    // _MSC_VER >= 1000
// WzdToolBar.h : header file
//


/////////////////////////////////////////////////////////////////////////////
// CWzdToolBar window

class CWzdToolBar : public CToolBar
{
// Construction
public:
    CWzdToolBar();

    BOOL LoadToolBarEx( UINT id );

// Attributes
public:

// Operations
public:
```

Example 12   Putting a Dropdown Button in a Toolbar   **215**

```
// Overrides
    // ClassWizard generated virtual function overrides
    // {{AFX_VIRTUAL( CWzdToolBar )
    // }}AFX_VIRTUAL
    CSize CalcDynamicLayout( int nLength, DWORD dwMode );

// Implementation
public:
    virtual ~CWzdToolBar();

    // Generated message map functions
protected:
    // {{AFX_MSG( CWzdToolBar )
    // }}AFX_MSG
    void OnDropdownButton( LPNMTOOLBAR pNotifyStruct, LRESULT* result );
    DECLARE_MESSAGE_MAP()
};


/////////////////////////////////////////////////////////////////////////////

// {{AFX_INSERT_LOCATION}}
// Microsoft Developer Studio will insert additional declarations immediately
//      before the previous line.

#endif
    // !defined( AFX_WZDTOOLBAR_H__27649E31_C807_11D1_9B5D_00AA003D8695__INCLUDED_ )
// WzdToolBar.cpp : implementation file
//

#include "stdafx.h"
#include "wzd.h"
#include "WzdTlBar.h"

#ifdef _DEBUG
#define new DEBUG_NEW
#undef THIS_FILE
static char THIS_FILE[] = __FILE__;
#endif
```

```
//////////////////////////////////////////////////////////////////////////
// CWzdToolBar

CWzdToolBar::CWzdToolBar()
{
}


CWzdToolBar::~CWzdToolBar()
{
}



BEGIN_MESSAGE_MAP( CWzdToolBar, CToolBar )
    // {{AFX_MSG_MAP( CWzdToolBar )
    // }}AFX_MSG_MAP
    ON_NOTIFY_REFLECT( TBN_DROPDOWN,OnDropdownButton )
END_MESSAGE_MAP()

//////////////////////////////////////////////////////////////////////////
// CWzdToolBar message handlers

BOOL CWzdToolBar::LoadToolBarEx( UINT id )
{
    // load toolbar info
    BOOL bRet;
    bRet = CToolBar::LoadToolBar( id );
    // find where our dropdown button will go
    int pos = CommandToIndex( IDC_WZD_DROPBUTTON );

    // set this button to be a dropdown button
    int iImage;
    UINT nID,nStyle;
    GetButtonInfo( pos,nID,nStyle,iImage );
    SetButtonInfo( pos,nID,nStyle|TBSTYLE_DROPDOWN,iImage );

    // ask toolbar to draw a down arrow next to this and any button with
    //      TBSTYLE_DROPDOWN
    SendMessage( TB_SETEXTENDEDSTYLE,0,TBSTYLE_EX_DRAWDDARROWS );

    return bRet;
```

Example 13   Putting an Icon in the Status Bar   **217**

```
}

CSize CWzdToolBar::CalcDynamicLayout( int nLength, DWORD dwMode )
{
    CSize size = CToolBar::CalcDynamicLayout( nLength,dwMode );
    // make room on non-fixed toolbars for additional down arrows
    if (dwMode&LM_HORZ) size.cx += 16*1;    // 16 * number of buttons
                                      //      with arrows
    return size;
}


void CWzdToolBar::OnDropdownButton( LPNMTOOLBAR lpnmtb, LRESULT *result )
{
    // get location of clicked button
    CRect rect;
    GetItemRect( CommandToIndex( lpnmtb -> iItem ),&rect );
    ClientToScreen( &rect );

    // putup a popup menu there
    CMenu menu;
    menu.LoadMenu( IDR_WZD_MENU );
    CMenu* pPopup = menu.GetSubMenu( 0 );
    pPopup -> TrackPopupMenu( TPM_RIGHTBUTTON, rect.left, rect.bottom, this );

    *result = TBDDRET_DEFAULT;    //drop-down was handled, also
    // TBDDRET_NODEFAULT drop-down was not handled.
    // TBDDRET_TREATPRESSED treat click as a button press
}
```

Example 13   Putting an Icon in the Status Bar

Objective

You would like to put an icon to indicate status in your status bar, as seen in Figure 5.7.

Figure 5.7   An Icon in the Status Bar

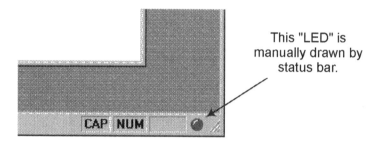

This "LED" is manually drawn by status bar.

Strategy

Just as a menu can be owner-drawn, so can the panes in a status bar. We will, therefore, add a new pane to our status bar and change its style to owner-drawn. In this example, we will draw this pane with one of two LED icons, depending on a status flag.

Steps

Create New Resources

1. Use the Icon Editor to create two or more icons indicating the differing status of some application feature. In this example, a red and a green LED icon were created.

2. Use the String Table Editor to create a new string filled with blanks. The number of blanks in this string will determine the width of your status pane. In this example, we use ID_INDICATOR_RED to identify this string.

Create a New Status Bar Class

1. Use the ClassWizard to create a new class derived from CStatusBarCtrl. Use the Text Editor to change all references from CStatusBarCtrl to CStatusBar. (The ClassWizard doesn't support CStatusBar.)

Example 13 Putting an Icon in the Status Bar **219**

2. Embed HICON variables in this new class and load the new icons in the constructor.

```
CWzdStatusBar::CWzdStatusBar()
{
    // load any graphics
    m_hRedLedIcon = AfxGetApp() -> LoadIcon( IDI_RED_LED );
    m_hGreenLedIcon = AfxGetApp() -> LoadIcon( IDI_GREEN_LED );
}
```

3. We need to apply an owner-drawn style to the pane we wish to draw. We will encapsulate this functionality in a class function called InitDrawing().

```
void CWzdStatusBar::InitDrawing()
{
    UINT nID,nStyle;
    int nPane,nWidth;

    // for each pane that we will be drawing
    nPane = CommandToIndex( ID_INDICATOR_LED );
    GetPaneInfo( nPane, nID, nStyle, nWidth );
    SetPaneInfo( nPane, nID, SBPS_OWNERDRAW|SBPS_NOBORDERS, nWidth );
}
```

4. Now, override the DrawItem() function of this class to draw this pane. Start this function by getting the device context, pane ID, and pane area in which to draw from the DRAWITEMSTRUCT calling argument.

```
void CWzdStatusBar::DrawItem( LPDRAWITEMSTRUCT lpDrawItemStruct )
{
    // get graphic context to draw to
    CDC* pDC = CDC::FromHandle( lpDrawItemStruct -> hDC) ;

    // get the pane's rectangle
    CRect rect( lpDrawItemStruct -> rcItem );

    // get the pane's id
    UINT nID,nStyle;
    int nWidth;
    GetPaneInfo( lpDrawItemStruct -> itemID, nID, nStyle, nWidth );
```

5. Since more than one pane can be drawn in this class, create a switch statement on the pane ID and draw one or the other icon based on some condition. Notice we start by filling the background area with the current color for a button face so that what we draw will blend in with the rest of the status bar.

```
// draw to that pane based on a status
switch ( nID )
{
    case ID_INDICATOR_LED:
    // draw the background
    CBrush bkColor( GetSysColor( COLOR_3DFACE ) );
    CBrush* pOldBrush = ( CBrush* )pDC -> SelectObject( &bkColor );
    pDC -> FillRect( rect, &bkColor );
    pDC -> SelectObject( pOldBrush );

    // draw the appropriate LED
    HICON hicon = ( ( CMainFrame* )AfxGetMainWnd() ) -> m_bTest?
    m_hRedLedIcon:m_hGreenLedIcon;
    pDC -> DrawIcon( rect.left,rect.top,hicon );
    break;

}
}    // end of DrawItem()
```

6. To see a complete listing of this status bar class, please refer to the Status Bar Class for Example 13 (page 222).

Implement the New Status Bar Class

1. Substitute this new class name for CStatusBar in Mainfrm.h.
2. Add the new status pane ID to the table in Mainfrm.cpp in the location at which you want this pane to appear.

```
static UINT indicators[] =
{
    ID_SEPARATOR,            // status line indicator
    ID_INDICATOR_CAPS,
```

Example 13   Putting an Icon in the Status Bar   **221**

```
    ID_INDICATOR_NUM,
    ID_INDICATOR_SCRL,
    ID_INDICATOR_LED,      <<<<<< ADD HERE
};
```

3. Initialize the new status bar class in CMainFrame's OnCreate() function by calling that InitDrawing() function.

```
// change selected pane style(s) to self-draw
m_wndStatusBar.InitDrawing();
```

4. This status bar is now open for business. When you want to change the status of the LED, you should also invalidate the status bar to cause it to redraw.

```
// change status
m_bTest = FALSE;

// force icon to be re-drawn
m_wndStatusBar.Invalidate();
```

Notes

- You may have to play with the size of the icon to make it fit in the status bar. Make sure to use the 32 by 32 icon size, but use only the top-left corner for your icon. Obviously, you can also use a bitmap here, but an icon lends itself more to this solution since it is typically small and automatically has a transparent color that will allow the status bar to bleed through and give the effect that the icon is part of the status bar.

CD Notes

- When executing the project on the accompanying CD, you will notice a green or red icon in the status bar. To change its color, click on the "Test" and then the "Wzd" menu commands.

II

5

Status Bar Class for Example 13

```
#if !defined( AFX_WZDSTATB_H__5DF01360_876F_11D2_A18D_D6622706D73F__INCLUDED_ )
#define AFX_WZDSTATB_H__5DF01360_876F_11D2_A18D_D6622706D73F__INCLUDED_

#if _MSC_VER > 1000
#pragma once
#endif    // _MSC_VER > 1000
// WzdStatB.h : header file
//


/////////////////////////////////////////////////////////////////////////////
// CWzdStatusBar window

class CWzdStatusBar : public CStatusBar
{
// Construction
public:
    CWzdStatusBar();

// Attributes
public:

// Operations
public:

    void InitDrawing();

// Overrides
    // ClassWizard generated virtual function overrides
    // {{AFX_VIRTUAL( CWzdStatusBar )
    // }}AFX_VIRTUAL
    virtual void DrawItem( LPDRAWITEMSTRUCT lpDrawItemStruct );

// Implementation
public:
    virtual ~CWzdStatusBar();


    // Generated message map functions
protected:
```

Example 13   Putting an Icon in the Status Bar   **223**

```
    // {{AFX_MSG( CWzdStatusBar )
    // NOTE - the ClassWizard will add and remove member functions here.
    // }}AFX_MSG

    DECLARE_MESSAGE_MAP()
private:
    HICON    m_hRedLedIcon;
    HICON    m_hGreenLedIcon;
};


///////////////////////////////////////////////////////////////////////////

// {{AFX_INSERT_LOCATION}}
// Microsoft Visual C++ will insert additional declarations immediately before
//    the previous line.

#endif
    // !defined( AFX_WZDSTATB_H__5DF01360_876F_11D2_A18D_D6622706D73F__INCLUDED_ )
// WzdStatB.cpp : implementation file
//

#include "stdafx.h"
#include "wzd.h"
#include "WzdStatB.h"
#include "MainFrm.h"

#ifdef _DEBUG
#define new DEBUG_NEW
#undef THIS_FILE
static char THIS_FILE[] = __FILE__;
#endif


///////////////////////////////////////////////////////////////////////////
// CWzdStatusBar

CWzdStatusBar::CWzdStatusBar()
{
    // load any graphics
    m_hRedLedIcon = AfxGetApp() -> LoadIcon( IDI_RED_LED );
    m_hGreenLedIcon = AfxGetApp() -> LoadIcon( IDI_GREEN_LED );
```

```
}

CWzdStatusBar::~CWzdStatusBar()
{
}


BEGIN_MESSAGE_MAP( CWzdStatusBar, CStatusBar )
    // {{AFX_MSG_MAP( CWzdStatusBar )
    // NOTE - the ClassWizard will add and remove mapping macros here.
    // }}AFX_MSG_MAP
END_MESSAGE_MAP()

///////////////////////////////////////////////////////////////////////////
// CWzdStatusBar message handlers

void CWzdStatusBar::InitDrawing()
{
    UINT nID,nStyle;
    int nPane,nWidth;

    // for each pane that we will be drawing
    nPane = CommandToIndex( ID_INDICATOR_LED );
    GetPaneInfo( nPane, nID, nStyle, nWidth );
    SetPaneInfo( nPane, nID, SBPS_OWNERDRAW|SBPS_NOBORDERS, nWidth );

}

void CWzdStatusBar::DrawItem( LPDRAWITEMSTRUCT lpDrawItemStruct )
{
    // get graphic context to draw to
    CDC* pDC = CDC::FromHandle( lpDrawItemStruct -> hDC );

    // get the pane's rectangle
    CRect rect( lpDrawItemStruct -> rcItem );

    // get the pane's id
    UINT nID,nStyle;
    int nWidth;
```

Example 14   Using a Rebar   **225**

```
GetPaneInfo( lpDrawItemStruct -> itemID, nID, nStyle, nWidth );

// draw to that pane based on a status
switch ( nID )
{
    case ID_INDICATOR_LED:
        // draw the background
        CBrush bkColor( GetSysColor( COLOR_3DFACE ) );
        CBrush* pOldBrush = ( CBrush* )pDC -> SelectObject( &bkColor );
        pDC -> FillRect( rect, &bkColor );
        pDC -> SelectObject( pOldBrush );

        // draw the appropriate LED
        HICON hicon = ( ( CMainFrame* )AfxGetMainWnd() ) -> m_bTest?
            m_hRedLedIcon:m_hGreenLedIcon;
        pDC -> DrawIcon( rect.left,rect.top,hicon );
        break;
}
}
```

Example 14   Using a Rebar

Objective

You would like to add a rebar to your application, as seen in Figure 5.8.

Figure 5.8   A ReBar

This application
has two rebars.

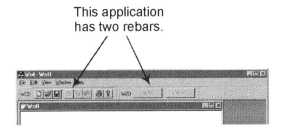

Strategy

MFC's new `CReBar` and `CReBarCtrl` classes take most of the work out of adding rebars to your application. Unfortunately, these classes are only available with v6.0 and rebars are only available with Windows 98 systems or systems with Internet Explorer v4.01 or better installed.

Steps

Create a Rebar

1. Embed a rebar class variable in your `CMainFrame` class.

```
CReBar    m_wndReBar;
```

2. Create this rebar's window in `CMainFrame`'s `OnCreate()` function.

```
if ( !m_wndReBar.Create( this ) )
{
    TRACE0( "Failed to create rebar\n" );
    return -1;    // fail to create
}
```

Add Controls to a Rebar

1. To add any toolbars to this rebar, use the `AddBar()` function.

```
m_wndReBar.AddBar( &m_wndToolBar,
    "WZD",                                  // rebar title
    NULL,                                   // a CBitmap background
    RBBS_GRIPPERALWAYS | RBBS_FIXEDBMP      // style
);
```

Don't enable docking for this toolbar or suffer the fate of an ASSERT. This toolbar doesn't need docking anyway—a rebar has the automatic ability to allow your user to move bars around its insides.

2. To add other types of controls to your rebar such as buttons or edit boxes, put them in a dialog template, alone or with one or two others. Then, create a dialog bar class for this template and embed that dialog

Example 14   Using a Rebar   **227**

bar class variable in CMainFrame. You can then create and add this dialog bar to your rebar.

```
if ( !m_WzdDialogBar.Create( this, IDD_WZD_DIALOG,
    WS_CHILD|WS_VISIBLE,-1 ) || !m_WzdDialogBar.InitDialog() )
{
    TRACE0( "Failed to create dialog bar\n" );
    return -1;    // fail to create
}

m_wndReBar.AddBar( &m_WzdDialogBar,
    "WZD",                                  // rebar title
    NULL,                                   // a CBitmap background
    RBBS_GRIPPERALWAYS | RBBS_FIXEDBMP      // style
);
```

Notes

- The biggest advantage of the rebar is to allow a user to customize the positions of their toolbars and other controls. MFC applications already had this ability using docking bars. Rebars are, therefore, probably Microsoft's way of providing this functionality to non-MFC applications. This is also why you can't make a toolbar dockable and still include it in a rebar. As an MFC application writer, you might, in fact, eschew using rebars for docking bars unless you're attempting to emulate a particular look.

CD Notes

- When executing the project on the accompanying CD, you will see a toolbar and dialog bar in a rebar at the top of the application. You will also notice that these bars can be moved by dragging their grabber bars at the front of each.

6

Views

The View in an SDI or MDI application is the primary method your user has to interact with your application and, in particular, the document that your application is editing. All of the examples in this chapter relate to the view, from creating a view out of a Property Sheet to printing a view to dragging files into your view.

Example 15   Creating Tabbed Form Views   We will create a view out of a property sheet.

Example 16   Creating a View From Any Common Control   We will look at how to create a view out of any control (buttons, list boxes, and edit boxes are controls). In this example, we will create a view out of a simple combo box.

Example 17   Printing a Report   We will use the capabilities built into the MFC `CView` class to print reports.

Example 18   Printing the View   We will dynamically capture your application's screen image and print it.

Example 19   Drawing to the MDI Client View   We will look at a way to add color and pattern to the otherwise plain backdrop of an MDI application. The backdrop is the MDI Client View over which the individual views of an MDI application hover.

Example 20   Dragging and Dropping Files into Your View   We will open files that have been dropped into the view.

Example 15   Creating Tabbed Form Views

Objective

You would like to create a tabbed view that can contain several dialog templates, as seen in Figure 6.1.

Figure 6.1   A Tabbed Form View

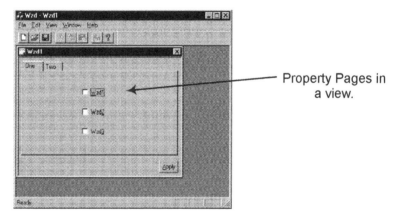

Property Pages in a view.

Strategy

We will create our own View Class, deriving it from `CScrollView`. In this view, we will simply create a property sheet control window.

Example 15   Creating Tabbed Form Views   **231**

Steps

Create the Pages for the Tabbed Form View

1. Use the Dialog Editor to create one or more dialog templates that will become the pages in your tabbed view. Set their style(s) as child windows with a thin frame. The window title you give each of these pages will become the names that will appear in their tabs.

2. Use the ClassWizard to create classes for these pages, deriving them from `CPropertyPage`.

Create a Tabbed Form View Class

1. Use the ClassWizard to create a new class derived from `CScrollView`.

2. Embed a property sheet variable into this new View Class.

```
// WzdTabbedView.h
private:
    CPropertySheet m_sheet;
```

3. Also, embed a button control in this class. This will become our "Apply" button.

```
CButton m_button;
```

Property sheets normally create their own "Apply" button, but only when created as a modal dialog box. Therefore, we have to dynamically add our own Apply button to this view.

4. Use the ClassWizard to add a `WM_CREATE` message handler to this class. Create the property sheet and Apply button here and add the property page classes we created previously to the property sheet.

```
int CWzdTabbedView::OnCreate( LPCREATESTRUCT lpCreateStruct )
{
    if ( CScrollView::OnCreate( lpCreateStruct ) == -1 )
        return -1;

    // create property sheet
    m_sheet.AddPage( &m_pageOne );
    m_sheet.AddPage( &m_pageTwo );
```

II

6

```
    m_sheet.Create( this,WS_CHILD|WS_VISIBLE );

    // create apply button
    CRect rect ( 0,0,10,10 );
    CFont *pFont =
        CFont::FromHandle( ( HFONT )::GetStockObject( ANSI_VAR_FONT ) );
    m_button.Create( "&Apply", WS_VISIBLE | WS_CHILD, rect, this,
        IDC_WZD_APPLY );
    m_button.SetFont( pFont );


    return 0;
}
```

Notice that we initially don't care where or how big these controls are since we will be moving them shortly.

5. Use the ClassWizard to override the OnInitUpdate() of the new View Class. We will start here by moving and sizing the property sheet and the "Apply" button. We then shrink the view around these two items using CScrollView::ResizeParentToFit(), having first told CScrollView the size of our view using SetScrollSizes().

```
void CWzdTabbedView::OnInitialUpdate()
{
    CScrollView::OnInitialUpdate();

    // home the property sheet
    CRect rect;
    m_sheet.GetClientRect( &rect );
    m_sheet.MoveWindow( rect );
    rect.bottom += BUTTON_HEIGTH + 10;

    // move apply button into place
    CRect brect( rect.right-BUTTON_WIDTH-5,
        rect.bottom-BUTTON_HEIGTH-5,
        rect.right-5,
        rect.bottom-5 );
    m_button.MoveWindow( brect );
```

Example 15   Creating Tabbed Form Views   **233**

```
   // size child frame around property sheet
   SIZE size = { rect.Width(), rect.Height() };
   SetScrollSizes( MM_TEXT, size );
   ResizeParentToFit( FALSE );

   // make sure the scroll bars are gone
   SetScrollSizes( MM_TEXT, CSize( 20,20 ) );

   // copy document into property pages
   m_pageOne.m_bWzd1 = ( ( CWzdDoc* )GetDocument() ) -> m_bWzd1;
   m_pageTwo.m_sEdit = ( ( CWzdDoc* )GetDocument() ) -> m_sEdit;

}
```

The last thing we do here is initialize the values in our property pages with values from the document. This can obviously be a lot more involved than what's shown here.

6. Because the property sheet and button controls don't cover all of the view, we need to paint the view's background. To do this, use the Class-Wizard to add a WM_ERASEBKGND message handler to this class and paint the background using the system button face color, as seen here.

```
BOOL CWzdTabbedView::OnEraseBkgnd(CDC* pDC)
{
   CPen pen( PS_SOLID,0,GetSysColor( COLOR_BTNFACE ) );
   CPen *pPen = pDC -> SelectObject( &pen );
   CBrush brush( GetSysColor( COLOR_BTNFACE ) );
   CBrush *pBrush = pDC -> SelectObject( &brush );

   CRect rect;
   GetClientRect( &rect );
   pDC -> Rectangle( rect );
   pDC -> SelectObject( pPen );
   pDC -> SelectObject( pBrush );
   return TRUE;
}
```

7. Add a message handler for the "Apply" button. You can do this manually, or you can create an Apply menu item using the same command ID

as the button and then use the ClassWizard to automatically add a message handler to your View Class. In this handler, you tell the property sheet to update its pages. Even though the property sheet has no Apply button when used modelessly, you can access the functionality of that apply button by using CPropertySheet's PressButton() function.

```
void CWzdTabbedView::OnWzdApply()
{
    m_sheet.PressButton( PSBTN_APPLYNOW );

    ( ( CWzdDoc* )GetDocument() ) -> m_bWzd1 = m_pageOne.m_bWzd1;
    ( ( CWzdDoc* )GetDocument() ) -> m_sEdit = m_pageTwo.m_sEdit;
}
```

The last thing we do here is copy any variable values from our property pages back into the document. Again, this can be a lot more involved than seen here.

8. To see a complete listing of the Tabbed Form View Class, please refer to the Tabbed Form View Class for Example 15 (page 235).

Implement the New Tabbed Form View Class

1. Substitute the use of this new class in your Application Class's InitInstance() function.

```
CMultiDocTemplate* pDocTemplate;
pDocTemplate = new CMultiDocTemplate(
    IDR_WZDTYPE,
    RUNTIME_CLASS( CWzdDoc ),
    RUNTIME_CLASS( CChildFrame ),
    RUNTIME_CLASS( CWzdTabbedView ) );     <<<<<<<< ADD HERE
AddDocTemplate( pDocTemplate );
```

You can delete any View Class created for your application by the App-Wizard.

Example 15   Creating Tabbed Form Views   **235**

2. To prevent your user from resizing this view, modify the `PreCreateWin-`
`dow()` function in your Main Frame or Child Frame Class, as seen here.

```
BOOL CChildFrame::PreCreateWindow( CREATESTRUCT& cs )
{
    // removes min/max boxes
    cs.style &= ~( WS_MAXIMIZEBOX|WS_MINIMIZEBOX );

    // makes dialog box unsizable
    cs.style &= ~WS_THICKFRAME;

    return CMDIChildWnd::PreCreateWindow( cs );
}
```

Notes

- Why did we derive our new view class from `CScrollView` instead of `CView`? So that we could use the `CScrollView::ResizeParentToFit()` function to shrink-wrap the view around our property sheet and button.

CD Notes

- When executing the project on the accompanying CD, you will notice that the view is actually a property sheet containing property pages.

Tabbed Form View Class for Example 15

```
#if !defined( AFX_WZDTABBEDVIEW_H__9A0B9504_E043_11D1_9B77_00AA003D8695__INCLUDED_ )
#define AFX_WZDTABBEDVIEW_H__9A0B9504_E043_11D1_9B77_00AA003D8695__INCLUDED_

#if _MSC_VER >= 1000
#pragma once
#endif    // _MSC_VER >= 1000
// WzdTabbedView.h : header file
//

#include "PageOne.h"
#include "PageTwo.h"
```

```
/////////////////////////////////////////////////////////////////////////
// CWzdTabbedView view

class CWzdTabbedView : public CScrollView
{
protected:
    CWzdTabbedView();     // protected constructor used by dynamic creation
    DECLARE_DYNCREATE( CWzdTabbedView )

// Attributes
public:

// Operations
public:

// Overrides
    // ClassWizard generated virtual function overrides
    // {{AFX_VIRTUAL( CWzdTabbedView )
protected:
    virtual void OnDraw( CDC* pDC );        // overridden to draw this view
    virtual void OnInitialUpdate();         // first time after construct
    // }}AFX_VIRTUAL

// Implementation
protected:
    virtual ~CWzdTabbedView();
#ifdef _DEBUG
    virtual void AssertValid() const;
    virtual void Dump( CDumpContext& dc ) const;
#endif

    // Generated message map functions
    // {{AFX_MSG(CWzdTabbedView)
    afx_msg int OnCreate(LPCREATESTRUCT lpCreateStruct);
    afx_msg void OnWzdApply();
    afx_msg BOOL OnEraseBkgnd(CDC* pDC);
    // }}AFX_MSG
    DECLARE_MESSAGE_MAP()
private:
    CButton m_button;
```

Example 15   Creating Tabbed Form Views   **237**

```
    CPageOne m_pageOne;
    CPageTwo m_pageTwo;
    CPropertySheet m_sheet;
};


/////////////////////////////////////////////////////////////////////////////

// {{AFX_INSERT_LOCATION}}
// Microsoft Developer Studio will insert additional declarations immediately
//     before the previous line.

#endif
    // !defined( AFX_WZDTABBEDVIEW_H__9A0B9504_E043_11D1_9B77_00AA003D8695__INCLUDED_ )
// WzdTabbedView.cpp : implementation file
//

#include "stdafx.h"
#include "wzd.h"
#include "WzdDoc.h"
#include "WzdTabbedView.h"

#ifdef _DEBUG
#define new DEBUG_NEW
#undef THIS_FILE
static char THIS_FILE[] = __FILE__;
#endif

#define BUTTON_WIDTH 40
#define BUTTON_HEIGTH 25


/////////////////////////////////////////////////////////////////////////////
// CWzdTabbedView

IMPLEMENT_DYNCREATE( CWzdTabbedView, CScrollView )

CWzdTabbedView::CWzdTabbedView()
{
}
```

II

6

```
CWzdTabbedView::~CWzdTabbedView()
{
}


BEGIN_MESSAGE_MAP( CWzdTabbedView, CScrollView )
    // {{AFX_MSG_MAP( CWzdTabbedView )
    ON_WM_CREATE()
    ON_COMMAND( IDC_WZD_APPLY, OnWzdApply )
    ON_WM_ERASEBKGND()
    // }}AFX_MSG_MAP
END_MESSAGE_MAP()

////////////////////////////////////////////////////////////////////////////
// CWzdTabbedView drawing

void CWzdTabbedView::OnInitialUpdate()
{
    CScrollView::OnInitialUpdate();

    // home the property sheet
    CRect rect;
    m_sheet.GetClientRect( &rect );
    m_sheet.MoveWindow( rect );
    rect.bottom += BUTTON_HEIGTH + 10;

    // move apply button into place
    CRect brect( rect.right-BUTTON_WIDTH-5,
        rect.bottom-BUTTON_HEIGTH-5,
        rect.right-5,
        rect.bottom-5);
    m_button.MoveWindow( brect );

    // size child frame around property sheet
    SIZE size = {rect.Width(), rect.Height()};
    SetScrollSizes( MM_TEXT, size );
    ResizeParentToFit( FALSE );

    // make sure the scroll bars are gone
    SetScrollSizes( MM_TEXT, CSize( 20,20 ) );
```

Example 15   Creating Tabbed Form Views   **239**

```
    // copy document into property pages
    m_pageOne.m_bWzd1 = ( ( CWzdDoc* )GetDocument() ) -> m_bWzd1;
    m_pageTwo.m_sEdit = ( ( CWzdDoc* )GetDocument() ) -> m_sEdit;

}


void CWzdTabbedView::OnDraw( CDC* pDC )
{
    CDocument* pDoc = GetDocument();
    // TODO: add draw code here
}


/////////////////////////////////////////////////////////////////////////////
// CWzdTabbedView diagnostics

#ifdef _DEBUG
void CWzdTabbedView::AssertValid() const
{
    CScrollView::AssertValid();
}


void CWzdTabbedView::Dump(CDumpContext& dc) const
{
    CScrollView::Dump( dc );
}
#endif    //_DEBUG


/////////////////////////////////////////////////////////////////////////////
// CWzdTabbedView message handlers

int CWzdTabbedView::OnCreate( LPCREATESTRUCT lpCreateStruct )
{
    if ( CScrollView::OnCreate( lpCreateStruct ) == -1 )
    return -1;

    // create property sheet
    m_sheet.AddPage( &m_pageOne );
    m_sheet.AddPage( &m_pageTwo );
    m_sheet.Create( this,WS_CHILD|WS_VISIBLE );
```

II

6

```
    // create apply button
    CRect rect ( 0,0,10,10 );
    CFont *pFont = CFont::FromHandle( ( HFONT )::GetStockObject( ANSI_VAR_FONT ) );
    m_button.Create( "&Apply", WS_VISIBLE | WS_CHILD, rect, this, IDC_WZD_APPLY );
    m_button.SetFont( pFont );

    return 0;
}


void CWzdTabbedView::OnWzdApply()
{
    m_sheet.PressButton( PSBTN_APPLYNOW );

    ( ( CWzdDoc* )GetDocument() ) -> m_bWzd1 = m_pageOne.m_bWzd1;
    ( ( CWzdDoc* )GetDocument() ) -> m_sEdit = m_pageTwo.m_sEdit;
}


BOOL CWzdTabbedView::OnEraseBkgnd( CDC* pDC )
{
    CPen pen( PS_SOLID,0,GetSysColor( COLOR_BTNFACE ) );
    CPen *pPen = pDC -> SelectObject( &pen );
    CBrush brush( GetSysColor( COLOR_BTNFACE ) );
    CBrush *pBrush = pDC -> SelectObject( &brush );

    CRect rect;
    GetClientRect( &rect );
    pDC -> Rectangle( rect );
    pDC -> SelectObject( pPen );
    pDC -> SelectObject( pBrush );
    return TRUE;
}
```

Example 16   Creating a View From Any Common Control   **241**

Example 16   Creating a View From Any Common Control

Objective

You would like to create a view out of any common control such as a button or a listbox. In this example, we will create a view out of a simple combo box, as seen in Figure 6.2.

Figure 6.2   A Combo Box as a View

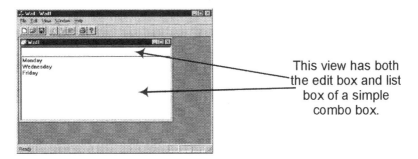

This view has both the edit box and list box of a simple combo box.

Strategy

We will embed the MFC class of a common control in the standard CView class. (In this example, we will be using CComboBox.) Then, we will add two message handlers to force this control window to take over our view. The WM_CREATE message handler will create the control window. The WM_SIZE handler will expand the control window to fill the view. We will override CView's OnInitialUpdate() and OnUpdate() functions to put document values into our control.

Steps

Create a Combo Box View

1. Create your application as usual with the AppWizard taking the default CView class to create your view.

2. Embed a combo box control in your View Class.

```
CComboBox m_combobox;
```

3. Use the ClassWizard to add a WM_CREATE message handler to your View Class where you will create a simple combo box with the following styles.

```
int CWzdView::OnCreate( LPCREATESTRUCT lpCreateStruct )
{
    if ( CView::OnCreate( lpCreateStruct ) == -1 )
        return -1;

    CRect rect( 0,0,0,0 );
    m_combobox.Create( WS_CHILD| WS_VISIBLE| CBS_SIMPLE|
        CBS_NOINTEGRALHEIGHT| WS_VSCROLL,
        rect, this, IDC_WZD_COMBOBOX );

    return 0;
}
```

Any control window you would like to turn into a view should have the WS_CHILD and WS_VISIBLE styles set. The other styles will vary from control to control.

We are creating a simple combo box using the CBS_SIMPLE style because the list box of a simple combo box is always visible. Otherwise, our view would always be an edit box with a drop-down button.

4. Use the ClassWizard to add a WM_SIZE message handler to your View Class. Use CWnd::MoveWindow to expand the size of the control to fill the view.

```
void CWzdView::OnSize( UINT nType, int cx, int cy )
{
    CView::OnSize( nType, cx, cy );

    m_combobox.MoveWindow( 0,0,cx,cy );
}
```

Example 16   Creating a View From Any Common Control   **243**

Update a Combo Box View

1. Use the ClassWizard to override the `OnInitialUpdate()` and `OnUpdate()` member functions of `CView`. There, you will fill the control with data from your document.

```
void CWzdView::OnInitialUpdate()
{
    CView::OnInitialUpdate();

    // get initial data from document
    m_combobox.AddString( "Monday" );
    m_combobox.AddString( "Wednesday" );
    m_combobox.AddString( "Friday" );


}

void CWzdView::OnUpdate( CView* pSender, LPARAM lHint, CObject* pHint )
{
    // in response to UpdateAllViews from document
}
```

2. Manually add handlers for control notifications from your control (e.g., notifications to your view when the selection in the list box changes or when the user types something in the edit box). An easy way to determine what to add would be to start by creating a dialog box and then add a simple combo box to it. Then, use the ClassWizard to create a dialog class and add any handlers you need in your view to this dialog class. Then, simply cut and paste the message macros it creates into your View Class.

3. To see a complete listing of this Combo View Class, please refer to the Combo Box View Class for Example 16 (page 244).

Notes

- Another variation of this theme is to create a view using two or more control windows. For example, you could create a static control window at the top to display some sort of status with an edit window at the bottom to view data. You can do this by embedding both control classes in the view class, creating both in the `WM_CREATE` message handler, and then determining how much each control window gets of the view in the WM_SIZE message handler.

CD Notes

- When executing the project on the accompanying CD, you will notice that the view is filled with a simple combo box.

Combo Box View Class for Example 16

```
// WzdView.h : interface of the CWzdView class
//
/////////////////////////////////////////////////////////////////////////////

#if !defined( AFX_WZDVIEW_H__CA9038F0_B0DF_11D1_A18C_DCB3C85EBD34__INCLUDED_ )
#define AFX_WZDVIEW_H__CA9038F0_B0DF_11D1_A18C_DCB3C85EBD34__INCLUDED_

#if _MSC_VER >= 1000
#pragma once
#endif    // _MSC_VER >= 1000

class CWzdView : public CView
{
protected: // create from serialization only
    CWzdView();
    DECLARE_DYNCREATE( CWzdView )

// Attributes
public:
    CWzdDoc* GetDocument();

// Operations
public:

// Overrides
    // ClassWizard generated virtual function overrides
    // {{AFX_VIRTUAL( CWzdView )
public:
    virtual void OnDraw( CDC* pDC );    // overridden to draw this view
    virtual BOOL PreCreateWindow( CREATESTRUCT& cs );
```

Example 16   Creating a View From Any Common Control   **245**

```
        virtual void OnInitialUpdate();
protected:
    virtual BOOL OnPreparePrinting( CPrintInfo* pInfo );
    virtual void OnBeginPrinting( CDC* pDC, CPrintInfo* pInfo );
    virtual void OnEndPrinting( CDC* pDC, CPrintInfo* pInfo );
    virtual void OnUpdate( CView* pSender, LPARAM lHint, CObject* pHint );
    // }}AFX_VIRTUAL

// Implementation
public:
    virtual ~CWzdView();
#ifdef _DEBUG
    virtual void AssertValid() const;
    virtual void Dump( CDumpContext& dc ) const;
#endif

protected:

// Generated message map functions
protected:
    // {{AFX_MSG( CWzdView )
    afx_msg int OnCreate( LPCREATESTRUCT lpCreateStruct );
    afx_msg void OnSize( UINT nType, int cx, int cy );
    // }}AFX_MSG
    afx_msg void OnEditchangeCombo();
    afx_msg void OnEditupdateCombo();
    afx_msg void OnSelchangeCombo();
    DECLARE_MESSAGE_MAP()
private:
    CComboBox m_combobox;
};

#ifndef _DEBUG    // debug version in WzdView.cpp
inline CWzdDoc* CWzdView::GetDocument()
    { return ( CWzdDoc* )m_pDocument; }
#endif

/////////////////////////////////////////////////////////////////////////////

// {{AFX_INSERT_LOCATION}}
```

II

6

```
// Microsoft Developer Studio will insert additional declarations immediately
//      before the previous line.

#endif
    // !defined( AFX_WZDVIEW_H__CA9038F0_B0DF_11D1_A18C_DCB3C85EBD34__INCLUDED_ )
// WzdView.cpp : implementation of the CWzdView class
//

#include "stdafx.h"
#include "Wzd.h"

#include "WzdDoc.h"
#include "WzdView.h"

#ifdef _DEBUG
#define new DEBUG_NEW
#undef THIS_FILE
static char THIS_FILE[] = __FILE__;
#endif

/////////////////////////////////////////////////////////////////////////////
// CWzdView

IMPLEMENT_DYNCREATE( CWzdView, CView )

BEGIN_MESSAGE_MAP( CWzdView, CView )
    // {{AFX_MSG_MAP( CWzdView )
    ON_WM_CREATE()
    ON_WM_SIZE()
    // }}AFX_MSG_MAP
    ON_CBN_EDITCHANGE( IDC_WZD_COMBOBOX, OnEditchangeCombo )
    ON_CBN_EDITUPDATE( IDC_WZD_COMBOBOX, OnEditupdateCombo )
    ON_CBN_SELCHANGE( IDC_WZD_COMBOBOX, OnSelchangeCombo )
    // Standard printing commands
    ON_COMMAND( ID_FILE_PRINT, CView::OnFilePrint )
    ON_COMMAND( ID_FILE_PRINT_DIRECT, CView::OnFilePrint )
    ON_COMMAND( ID_FILE_PRINT_PREVIEW, CView::OnFilePrintPreview )
END_MESSAGE_MAP()

/////////////////////////////////////////////////////////////////////////////
// CWzdView construction/destruction
```

Example 16   Creating a View From Any Common Control   **247**

```
CWzdView::CWzdView()
{
    // TODO: add construction code here

}

CWzdView::~CWzdView()
{
}

BOOL CWzdView::PreCreateWindow( CREATESTRUCT& cs )
{
    // TODO: Modify the Window class or styles here by modifying
    //      the CREATESTRUCT cs

    return CView::PreCreateWindow( cs );
}

///////////////////////////////////////////////////////////////////////////
// CWzdView drawing

void CWzdView::OnDraw( CDC* pDC )
{
    CWzdDoc* pDoc = GetDocument();
    ASSERT_VALID( pDoc );

    // TODO: add draw code for native data here
}

///////////////////////////////////////////////////////////////////////////
// CWzdView printing

BOOL CWzdView::OnPreparePrinting( CPrintInfo* pInfo )
{
    // default preparation
    return DoPreparePrinting( pInfo );
}
```

II

6

```
void CWzdView::OnBeginPrinting( CDC* /*pDC*/, CPrintInfo* /*pInfo*/ )
{
    // TODO: add extra initialization before printing
}

void CWzdView::OnEndPrinting( CDC* /*pDC*/, CPrintInfo* /*pInfo*/ )
{
    // TODO: add cleanup after printing
}

/////////////////////////////////////////////////////////////////////////
// CWzdView diagnostics

#ifdef _DEBUG
void CWzdView::AssertValid() const
{
    CView::AssertValid();
}

void CWzdView::Dump( CDumpContext& dc ) const
{
    CView::Dump( dc );
}

CWzdDoc* CWzdView::GetDocument()     // non-debug version is inline
{
    ASSERT( m_pDocument -> IsKindOf( RUNTIME_CLASS( CWzdDoc ) ) );
    return ( CWzdDoc* )m_pDocument;
}
#endif    //_DEBUG

/////////////////////////////////////////////////////////////////////////
// CWzdView message handlers

int CWzdView::OnCreate( LPCREATESTRUCT lpCreateStruct )
{
    if ( CView::OnCreate( lpCreateStruct ) == -1 )
        return -1;

    CRect rect( 0,0,0,0 );
```

Example 16   Creating a View From Any Common Control   **249**

```
    m_combobox.Create( WS_CHILD|WS_VISIBLE|CBS_SIMPLE|CBS_NOINTEGRALHEIGHT|WS_VSCROLL,
        rect, this, IDC_WZD_COMBOBOX );

    return 0;
}

void CWzdView::OnSize( UINT nType, int cx, int cy )
{
    CView::OnSize( nType, cx, cy );

    m_combobox.MoveWindow( 0,0,cx,cy );
}

void CWzdView::OnInitialUpdate()
{
    CView::OnInitialUpdate();

    // get initial data from document
    m_combobox.AddString( "Monday" );
    m_combobox.AddString( "Wednesday" );
    m_combobox.AddString( "Friday" );

}

void CWzdView::OnUpdate( CView* pSender, LPARAM lHint, CObject* pHint )
{
    // in response to UpdateAllViews from document
}

void CWzdView::OnEditchangeCombo()
{
    // TODO: Add your control notification handler code here
}

void CWzdView::OnEditupdateCombo()
{
    // TODO: Add your control notification handler code here
}
```

II

6

```
void CWzdView::OnSelchangeCombo()
{
    // TODO: Add your control notification handler code here
}
```

Example 17   Printing a Report

Objective

You would like to print the information in your document in a report, as seen in Figure 6.3.

Figure 6.3   Preview of Printed Report

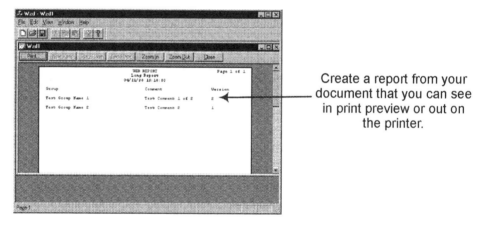

Create a report from your document that you can see in print preview or out on the printer.

Strategy

Printing reports in an MFC application is, by convention, performed in the View Class. I suppose it can be considered just another way to view the document. The "File/Print" and "File/Print Preview" menu commands are automatically processed by the View Class. To take advantage of this automation, however, we must override two View Class functions. These include OnBeginPrint(), which is called just before printing to initialize our own printing variables, and OnPrint(), which is called every time a new page needs to be printed. Here we will loop through the data collection in our document and draw it to the printer's device context. In this example, we will also be prompting the user for the type of report they want to print.

Example 17   Printing a Report   **251**

Steps

Prompt the User for the Type of Report to Print

1. Start by writing a small helper function that prompts your user for a report type. Add this function to your View Class. The following example simply creates a dialog box that asks the user if they want a short or long report. Yours can obviously ask your user for much more detail.

```
BOOL CWzdView::GetReportOptions()
{
    CWzdDialog dlg;
    dlg.m_nReportType = m_nReportType;
    if( dlg.DoModal() == IDOK )
    {
        m_nReportType = dlg.m_nReportType;
        return TRUE;
    }
    return FALSE;
}
```

The best place to prompt our user with this dialog box would be right after they click on the "Print" or "Print Preview" menu commands. However, if we were to directly intercept these commands, we run into a problem when our user is previewing a printed page. One of the buttons on a preview page allows your user to directly print that page. This sends out yet another ID_FILE_PRINT command, which would again open our dialog box prompt. To avoid this problem, we will change the main menu to send out two new command messages that will prompt our user for the report type they want before sending out the normal print commands.

2. Use the Menu Editor to change the command ID's for the "File/Print" and "File/Print Preview" commands to IDC_FILE_PRINT and IDC_FILE_PRINT_PREVIEW, respectively. (Add a "C" to the "ID_".)

3. Use the ClassWizard to add a command message handler to the View Class for these two command messages. There, you will call GetReportOptions(), which itself will open our dialog prompt for a report type. We

then generate the command ID's that will cause the View Class to go into print mode.

```
void CWzdView::OnFilePrint()
{
    if ( GetReportOptions() )
    {
        AfxGetMainWnd() -> SendMessage( WM_COMMAND, ID_FILE_PRINT );
    }
}


void CWzdView::OnFilePrintPreview()
{
    if ( GetReportOptions() )
    {
        AfxGetMainWnd() ->
            SendMessage( WM_COMMAND, ID_FILE_PRINT_PREVIEW );
    }
}
```

The listing for `GetReportOptions()` can be found in the View Class for Example 17 (page 255).

Initialize the View Class for Printing

1. Fill in the `OnBeginPrint()` member function of the View Class. You will not need the ClassWizard because the AppWizard has already added this function. `OnBeginPrint()` is called once before printing starts. In this function, you will calculate any values you will need to print, including the average printed character size, number of characters allowed on a printed line, number of lines possible per page, and the total number of pages required for this report.

```
void CWzdView::OnBeginPrinting( CDC* pDC, CPrintInfo* pInfo )
{
    // get printed character height and width
    TEXTMETRIC tm;
    pDC -> GetTextMetrics( &tm );
    m_nPrintCharHeight = tm.tmHeight;
    m_nPrintCharWidth = tm.tmAveCharWidth;
```

Example 17   Printing a Report   **253**

```
    // get number of characters per line
    int nPageWidth = pDC -> GetDeviceCaps( HORZRES );
    m_nPageWidth = nPageWidth/m_nPrintCharWidth;

    // get number of lines per page (with and w/o title on each page)
    int nPageHeight = pDC -> GetDeviceCaps( VERTRES );
    int nPrintLinesPerPage = nPageHeight / m_nPrintCharHeight;
    m_nPrintableLinesPerPage =
        nPrintLinesPerPage-LINES_IN_REPORT_TITLE;

    // determine number of total pages in this document
    int nLines = GetDocument() -> GetWzdInfoList() -> GetCount();
    int nPages = ( nLines + m_nPrintableLinesPerPage - 1 )/
        m_nPrintableLinesPerPage;
    if ( nPages <= 0 ) nPages = 1;
    pInfo -> SetMaxPage( nPages );
    pInfo -> m_nCurPage = 1;
}
```

II

6

Print Each Page

Use the ClassWizard to override the OnPrint() member function of your View Class. The OnPrint() function is called by the View Class every time a new page is ready to be printed until there are no more pages to print. Therefore, one of the things OnPrint() must do is determine where in the document its page starts. To print the report in this example, OnPrint() calls three helper functions: PrintTitle() to print the report title, PrintColumn-Headers() to print the column headers for the report, and PrintLine() to print each report line.

```
void CWzdView::OnPrint( CDC* pDC, CPrintInfo* pInfo )
{
    // print title
    int y = 0;
    PrintTitle( pDC,&y,pInfo );

    // print column headers
    PrintColumnHeaders( pDC,&y );
```

```
    // determine part of document to print on this page
    int nWzdInfoListInx =
        ( pInfo -> m_nCurPage-1 )*m_nPrintableLinesPerPage;
    int nWzdInfoListEnd =
        GetDocument() -> GetWzdInfoList() -> GetCount();
    if ( nWzdInfoListEnd > nWzdInfoListInx + m_nPrintableLinesPerPage )
    {
        nWzdInfoListEnd = nWzdInfoListInx + m_nPrintableLinesPerPage;
    }


    // print report lines
    for ( ; nWzdInfoListInx < nWzdInfoListEnd; nWzdInfoListInx++ )
    {
        POSITION pos = GetDocument() ->
            GetWzdInfoList() -> FindIndex(nWzdInfoListInx);
        CWzdInfo *pInfo =
            GetDocument() -> GetWzdInfoList() -> GetAt( pos );
        PrintLine( pDC,&y,pInfo );
        y += m_nPrintCharHeight;
    }


    CView::OnPrint( pDC, pInfo );
}
```

2. Add these three helper functions to your View Class: `PrintTitle()`, `Print-ColumnHeaders()` and `PrintLine()`. `PrintTitle()` should print the title using the `TextOut()` member function of the `CDC` class. The `PrintColumnHeader()` and `PrintLine()` functions can use the `TabbedTextOut()` function to automatically create columns. For an example of these helper functions, please see the View Class for Example 17 (page 255).

Notes

*   If your view contains graphics instead of a listing or other text, you don't need to add anything to the View Class to print the view. The `CView` class will call its own `OnDraw()` member function with a printer device context instead of a screen device context, which will cause your view to be automatically drawn to the printer.

Example 17　　Printing a Report　**255**

- To create a report of any other type of view, please refer to the next example.

CD Notes

- When executing the project on the accompanying CD, click on the "File" then "Print" or "Print Preview" menu command. A dialog appears asking if you want to print the short or long report. Select one and press "OK" and watch a report being printed to paper or the screen.

View Class for Example 17

```
// WzdView.h : interface of the CWzdView class
//
///////////////////////////////////////////////////////////////////////////

#if !defined( AFX_WZDVIEW_H__CA9038F0_B0DF_11D1_A18C_DCB3C85EBD34__INCLUDED_ )
#define AFX_WZDVIEW_H__CA9038F0_B0DF_11D1_A18C_DCB3C85EBD34__INCLUDED_

#if _MSC_VER >= 1000
#pragma once
#endif    // _MSC_VER >= 1000

class CWzdView : public CView
{
protected:    // create from serialization only
    CWzdView();
    DECLARE_DYNCREATE( CWzdView )

// Attributes
public:
    CWzdDoc* GetDocument();

    enum {
        SHORTREPORT,
        LONGREPORT
    };

// Operations
public:
```

```
// Overrides
    // ClassWizard generated virtual function overrides
    // {{AFX_VIRTUAL( CWzdView )
public:
    virtual void OnDraw( CDC* pDC );    // overridden to draw this view
    virtual BOOL PreCreateWindow( CREATESTRUCT& cs );
protected:
    virtual BOOL OnPreparePrinting( CPrintInfo* pInfo );
    virtual void OnBeginPrinting( CDC* pDC, CPrintInfo* pInfo );
    virtual void OnEndPrinting( CDC* pDC, CPrintInfo* pInfo );
    virtual void OnPrint( CDC* pDC, CPrintInfo* pInfo );
    // }}AFX_VIRTUAL

// Implementation
public:
    virtual ~CWzdView();
#ifdef _DEBUG
    virtual void AssertValid() const;
    virtual void Dump( CDumpContext& dc ) const;
#endif

protected:

// Generated message map functions
protected:
    // {{AFX_MSG(CWzdView)
    afx_msg void OnFilePrint();
    afx_msg void OnFilePrintPreview();
    // }}AFX_MSG
    DECLARE_MESSAGE_MAP()
private:
    int m_nPrintCharHeight;
    int m_nPrintCharWidth;
    int m_nPageWidth;
    int m_nPrintableLinesPerPage;
    int m_nReportType;
    static int m_reportTabs[];
```

Example 17   Printing a Report   **257**

```
    BOOL GetReportOptions();
    void PrintTitle( CDC *pDC, int *y, CPrintInfo* pInfo );
    void PrintColumnHeaders( CDC *pDC, int *y );
    void PrintLine( CDC *pDC, int *y, CWzdInfo *pInfo );


};


#ifndef _DEBUG     // debug version in WzdView.cpp
inline CWzdDoc* CWzdView::GetDocument()
   { return ( CWzdDoc* )m_pDocument; }
#endif


/////////////////////////////////////////////////////////////////////////////


// {{AFX_INSERT_LOCATION}}
// Microsoft Developer Studio will insert additional declarations immediately
//     before the previous line.


#endif
    // !defined( AFX_WZDVIEW_H__CA9038F0_B0DF_11D1_A18C_DCB3C85EBD34__INCLUDED_ )
// WzdView.cpp : implementation of the CWzdView class
//


#include "stdafx.h"
#include "Wzd.h"


#include "WzdDoc.h"
#include "WzdView.h"
#include "WzdDialog.h"


#ifdef _DEBUG
#define new DEBUG_NEW
#undef THIS_FILE
static char THIS_FILE[] = __FILE__;
#endif


int CWzdView::m_reportTabs[] = {1200,2000};
#define NUM_TABS sizeof(m_reportTabs)/sizeof(int)
```

II

6

```
///////////////////////////////////////////////////////////////////////////
// CWzdView

IMPLEMENT_DYNCREATE( CWzdView, CView )

BEGIN_MESSAGE_MAP( CWzdView, CView )
    // {{AFX_MSG_MAP( CWzdView )
    ON_COMMAND( IDC_FILE_PRINT, OnFilePrint )
    ON_COMMAND( IDC_FILE_PRINT_PREVIEW, OnFilePrintPreview )
    // }}AFX_MSG_MAP
    // Standard printing commands
    ON_COMMAND( ID_FILE_PRINT, CView::OnFilePrint )
    ON_COMMAND( ID_FILE_PRINT_DIRECT, CView::OnFilePrint )
    ON_COMMAND( ID_FILE_PRINT_PREVIEW, CView::OnFilePrintPreview )
END_MESSAGE_MAP()


///////////////////////////////////////////////////////////////////////////
// CWzdView construction/destruction

CWzdView::CWzdView()
{
    m_nReportType = 0;


}


CWzdView::~CWzdView()
{
}


BOOL CWzdView::PreCreateWindow( CREATESTRUCT& cs )
{
    // TODO: Modify the Window class or styles here by modifying
    //      the CREATESTRUCT cs

    return CView::PreCreateWindow( cs );
}


///////////////////////////////////////////////////////////////////////////
// CWzdView drawing
```

Example 17   Printing a Report   **259**

```
void CWzdView::OnDraw( CDC* pDC )
{
    CWzdDoc* pDoc = GetDocument();
    ASSERT_VALID( pDoc );

    // TODO: add draw code for native data here
}

//////////////////////////////////////////////////////////////////////////////
// CWzdView printing

#define LINES_IN_REPORT_TITLE 7

BOOL CWzdView::OnPreparePrinting( CPrintInfo* pInfo )
{
    // default preparation
    return DoPreparePrinting( pInfo );
}

void CWzdView::OnBeginPrinting( CDC* pDC, CPrintInfo* pInfo )
{
    // get printed character height and width
    TEXTMETRIC tm;
    pDC -> GetTextMetrics( &tm );
    m_nPrintCharHeight = tm.tmHeight;
    m_nPrintCharWidth = tm.tmAveCharWidth;

    // get number of characters per line
    int nPageWidth = pDC -> GetDeviceCaps( HORZRES );
    m_nPageWidth = nPageWidth/m_nPrintCharWidth;

    // get number of lines per page (with and w/o title on each page)
    int nPageHeight = pDC -> GetDeviceCaps( VERTRES );
    int nPrintLinesPerPage = nPageHeight / m_nPrintCharHeight;
    m_nPrintableLinesPerPage = nPrintLinesPerPage-LINES_IN_REPORT_TITLE;

    // determine number of total pages in this document
    int nLines = GetDocument() -> GetWzdInfoList() -> GetCount();
    int nPages = ( nLines + m_nPrintableLinesPerPage - 1 )/
        m_nPrintableLinesPerPage;
```

```
    if ( nPages <= 0 ) nPages = 1;
    pInfo -> SetMaxPage( nPages );
    pInfo -> m_nCurPage = 1;
}

void CWzdView::OnEndPrinting( CDC* /*pDC*/, CPrintInfo* /*pInfo*/ )
{
    // TODO: add cleanup after printing
}


/////////////////////////////////////////////////////////////////////////
// CWzdView diagnostics

#ifdef _DEBUG
void CWzdView::AssertValid() const
{
    CView::AssertValid();
}

void CWzdView::Dump( CDumpContext& dc ) const
{
    CView::Dump( dc );
}

CWzdDoc* CWzdView::GetDocument()    // non-debug version is inline
{
    ASSERT( m_pDocument -> IsKindOf( RUNTIME_CLASS( CWzdDoc ) ) );
    return ( CWzdDoc* )m_pDocument;
}
#endif //_DEBUG


/////////////////////////////////////////////////////////////////////////
// CWzdView message handlers

void CWzdView::OnFilePrint()
{
    if ( GetReportOptions() )
    {
        AfxGetMainWnd() -> SendMessage( WM_COMMAND, ID_FILE_PRINT );
    }
```

Example 17   Printing a Report   **261**

```
}

void CWzdView::OnFilePrintPreview()
{

    if ( GetReportOptions() )
    {
        AfxGetMainWnd() -> SendMessage( WM_COMMAND, ID_FILE_PRINT_PREVIEW );
    }
}

BOOL CWzdView::GetReportOptions()
{
    CWzdDialog dlg;
    dlg.m_nReportType = m_nReportType;
    if( dlg.DoModal() == IDOK )
    {
        m_nReportType = dlg.m_nReportType;
        return TRUE;
    }
    return FALSE;
}

void CWzdView::OnPrint( CDC* pDC, CPrintInfo* pInfo )
{
    // print title
    int y = 0;
    PrintTitle( pDC,&y,pInfo );

    // print column headers
    PrintColumnHeaders( pDC,&y );

    // determine part of document to print on this page
    int nWzdInfoListInx = ( pInfo -> m_nCurPage-1 )*m_nPrintableLinesPerPage;
    int nWzdInfoListEnd = GetDocument() -> GetWzdInfoList() -> GetCount();
    if ( nWzdInfoListEnd > nWzdInfoListInx + m_nPrintableLinesPerPage )
    {
        nWzdInfoListEnd = nWzdInfoListInx + m_nPrintableLinesPerPage;
    }
```

II

6

```
    // print report lines
    for ( ; nWzdInfoListInx < nWzdInfoListEnd; nWzdInfoListInx++ )
    {
        POSITION pos =
            GetDocument() -> GetWzdInfoList() -> FindIndex( nWzdInfoListInx );
        CWzdInfo *pInfo = GetDocument() -> GetWzdInfoList() -> GetAt( pos );
        PrintLine( pDC,&y,pInfo );
        y += m_nPrintCharHeight;
    }


    CView::OnPrint( pDC, pInfo );
}


void CWzdView::PrintTitle( CDC *pDC, int *y, CPrintInfo* pInfo )
{
    // title
    int x = pDC -> GetDeviceCaps( HORZRES );
    pDC -> SetTextAlign( TA_CENTER );
    pDC -> TextOut( x/2, *y, "WZD REPORT" );

    // page #
    CString str;
    str.Format( "Page %d of %d",pInfo -> m_nCurPage,pInfo -> GetMaxPage() );
    pDC -> SetTextAlign( TA_RIGHT );
    pDC -> TextOut( x, *y, str );

    // report type
    *y += m_nPrintCharHeight;
    pDC -> SetTextAlign( TA_CENTER );
    switch ( m_nReportType )
    {
        case CWzdView::SHORTREPORT:
            str = "Short Report";
            break;

        case CWzdView::LONGREPORT:
            str = "Long Report";
            break;
```

Example 17   Printing a Report   **263**

```
    }
    pDC -> TextOut( x/2, *y, str );

    // date
    *y += m_nPrintCharHeight;
    COleDateTime dt( COleDateTime::GetCurrentTime() );
    pDC -> TextOut( x/2, *y, dt.Format( "%c" ) );
    *y += m_nPrintCharHeight;
    *y += m_nPrintCharHeight;    // leave space between title and column headers
}


void CWzdView::PrintColumnHeaders( CDC *pDC, int *y )
{
    CString str;
    switch ( m_nReportType )
    {
        case CWzdView::SHORTREPORT:
            str = "Group\tVersion";
            break;

        case CWzdView::LONGREPORT:
            str = "Group\tComment\tVersion";
            break;
    }
    pDC -> SetTextAlign( TA_LEFT );
    pDC -> TabbedTextOut( 0,*y,str,NUM_TABS,m_reportTabs,0 );
    *y += m_nPrintCharHeight;
    *y += m_nPrintCharHeight;    // leave space between column headers and report
}


void CWzdView::PrintLine( CDC *pDC, int *y, CWzdInfo *pInfo )
{
    CString str;
    switch ( m_nReportType )
    {
        case CWzdView::SHORTREPORT:
            str.Format( "%s\t%d",pInfo -> m_sGroupName, pInfo -> m_nVersion );
            break;
```

```
        case CWzdView::LONGREPORT:
            str.Format( "%s\t%s\t%d",pInfo -> m_sGroupName,
                pInfo -> m_sComment, pInfo -> m_nVersion );
            break;
    }

    pDC -> SetTextAlign( TA_LEFT );
    pDC -> TabbedTextOut( 0,*y,str,NUM_TABS,m_reportTabs,0 );
    *y += m_nPrintCharHeight;
}
```

Example 18   Printing the View

Objective

You would like to dynamically capture your application's screen image and print it.

Strategy

In the previous example, we printed out a report of our document rather than what was appearing in the view at the time. In this example, we will literally do a screen capture and print that. MFC does provide limited support for printing the view, but only if you draw your view yourself using the device context that MFC provides in CView::OnDraw(). When you want to print the view, MFC simply calls OnDraw() with the device context of a printer. Clever, huh? Unfortunately, if you aren't drawing the view yourself, as is the case when you fill your view with one or more control windows, nothing will be printed. Each of your controls will use its very own device context to print itself out to—you guessed it—the screen. Therefore, the only way to print the view is to actually capture the screen (i.e., copy its contents to a bitmap object) and print that to the printer. Since this functionality is all related to a bitmap, we will encapsulate it in a new bitmap class.

Example 18   Printing the View   **265**

Steps

Create a New Bitmap Class

1. Use the ClassWizard to create a new class derived from CBitmap.

2. To this class, we will add two functions that are covered in other examples in this book. The Capture() function is used to copy the contents of an area of the screen into a bitmap object and is shown in Example 33. CreateDIB() is used to convert a device dependant bitmap (DDB) into a device independent bitmap (DIB) and is covered in Example 34.

3. The last function we will add to this class is a Print() function, which, when provided with a printer device context, will print the contained bitmap to the printer. Before this function is called, it is assumed that the view has already been captured into this bitmap using Capture(). We, therefore, start this function by calling CreateDIB() to get a DIB version of this captured screen bitmap.

```
void CWzdBitmap::Print( CDC *pDC )
{
    // get DIB version of bitmap
    int bmData;
    HANDLE hDIB = CreateDIB( &bmData );
```

4. We could now just copy this DIB bitmap to the printer context and be done. However, the printer typically has a lot more resolution than the screen and, therefore, our bitmap needs to be stretched to fill that space. We continue by determining just how big a printer area we need to fill.

```
    // stretch bitmap to fill printed page with 1/4 inch borders
    int cxBorder = pDC -> GetDeviceCaps(LOGPIXELSX)/4;
    int cyBorder = pDC -> GetDeviceCaps(LOGPIXELSY)/4;
    int cxPage = pDC -> GetDeviceCaps(HORZRES) - (cxBorder*2);
    int cyPage = ( int )( ( ( double )cxPage/
        ( double )m_nWidth ) * ( double )m_nHeight );
```

5. We will be using StretchDIBits() to actually do the stretching of our bitmap to the printer device context. StretchDIBits() requires us to provide

two pointers to our DIB bitmap: one to the bitmap header and one to the data.

```
LPBITMAPINFOHEADER lpDIBHdr = ( LPBITMAPINFOHEADER )::GlobalLock( hDIB );
LPSTR lpDIBBits = ( LPSTR )lpDIBHdr+bmData;
```

6. Finally, we do the stretch and cleanup.

```
    // stretch the bitmap for the best fit on the printed page
    pDC -> SetStretchBltMode( COLORONCOLOR );
    ::StretchDIBits( pDC -> m_hDC,
        cxBorder,cyBorder,cxPage,cyPage,     // destination dimensions
        0,0,m_nWidth,m_nHeight,
        // source bitmap dimensions (use all of bitmap)
        lpDIBBits,                    // bitmap picture data
        ( LPBITMAPINFO )lpDIBHdr,     // bitmap header info
        DIB_RGB_COLORS,               // specify color table has
                                      //      RGB values
        SRCCOPY                   // simple source to destination copy
    );


    // cleanup
    ::GlobalUnlock( hDIB );
    ::GlobalFree( hDIB );
    return;
}
```

To see a complete listing of this class, please refer to the Bitmap Class for Example 18 (page 268).

We could now use this bitmap class "as-is" to capture the view and print it to the printer. However, a more elegant way to use this class would be to work it into our View Class so that when the user clicks on File/Print or File/Print Preview, our view appears on the printer. We will implement such an approach next.

Implement the New Bitmap Class

1. Embed this new class in your view class.

```
CWzdBitmap m_bitmap;
```

Example 18   Printing the View   **267**

2. Comment out the default `ID_FILE_PRINT` and `ID_FILE_PRINT_PREVIEW` command handlers and use the ClassWizard to add your own.

```
BEGIN_MESSAGE_MAP( CWzdView, CView )
    // {{AFX_MSG_MAP( CWzdView )
    ON_COMMAND( ID_FILE_PRINT, OnFilePrint )
    ON_COMMAND( ID_FILE_PRINT_PREVIEW, OnFilePrintPreview )
    // }}AFX_MSG_MAP
    // Standard printing commands
    // ON_COMMAND( ID_FILE_PRINT, CView::OnFilePrint )
    ON_COMMAND( ID_FILE_PRINT_DIRECT, CView::OnFilePrint )
    // ON_COMMAND( ID_FILE_PRINT_PREVIEW, CView::OnFilePrintPreview )
END_MESSAGE_MAP()
```

3. Fill in the `OnFilePrint()` and `OnFilePrintPreview()` functions to use the `Capture()` function of our new bitmap class to capture the view.

```
void CWzdView::OnFilePrint()
{
    // capture our view
    CRect rect;
    GetWindowRect( &rect );
    m_bitmap.Capture( rect );

    CView::OnFilePrint();
}

void CWzdView::OnFilePrintPreview()
{
    // capture our view
    CRect rect;
    GetWindowRect( &rect );
    m_bitmap.Capture( rect );

    CView::OnFilePrintPreview();
}
```

4. Then, use the ClassWizard to override the OnPrint() function and fill it in to call our bitmap class's Print() function.

```
void CWzdView::OnPrint( CDC* pDC, CPrintInfo* pInfo )
{
    // print captured bitmap to pDC
    m_bitmap.Print( pDC );

    // CView::OnPrint( pDC, pInfo );
}
```

Notes

- Why are we using ::StretchDIBits() and not CDC::StretchBlt() to stretch our bitmap to the printer? Why do we need to convert our bitmap to a DIB before we can print it out? Why can't we just select our screen palette into the printer device context and let the device context sort it out for true interoperability? Well, it just doesn't work. The functionality to convert from any device palette to any other device palette is apparently more costly for an operating system programmer to implement than to start with a given format (DIB) and convert from there.

CD Notes

- When executing the project on the accompanying CD, click on the "File" then "Print" or "Print Preview" menu command. The view (which happens to be empty) will print to the printer or the print preview dialog.

Bitmap Class for Example 18

```
#ifndef WZDBITMAP_H
#define WZDBITMAP_H


class CWzdBitmap : public CBitmap
{
public:
    DECLARE_DYNAMIC( CWzdBitmap )

// Constructors
    CWzdBitmap();
```

Example 18   Printing the View   **269**

```
    void Capture( CRect &rect );
    CPalette *GetPalette(){return m_pPalette;};
    HANDLE CreateDIB( int *pbmData = NULL );
    void Print( CDC *pDC );

// Implementation
public:
    virtual ~CWzdBitmap();

// Attributes
    int m_nWidth;
    int m_nHeight;
// Operations

private:
    CPalette *m_pPalette;
};
#endif
// WzdBitmap.cpp : implementation of the CWzdBitmap class
//

#include "stdafx.h"
#include "WzdBtmap.h"


/////////////////////////////////////////////////////////////////////////////
// CWzdBitmap

IMPLEMENT_DYNAMIC( CWzdBitmap, CBitmap )


CWzdBitmap::CWzdBitmap()
{
    m_pPalette = NULL;
}

CWzdBitmap::~CWzdBitmap()
{
    if ( m_pPalette )
    {
```

```
            delete m_pPalette;
    }
}


void CWzdBitmap::Capture(CRect &rect)
{
    // cleanup from last capture
    if (m_pPalette)
    {
        delete m_pPalette;
        DeleteObject();
    }

    // save width and height
    m_nWidth = rect.Width();
    m_nHeight = rect.Height();

    /////////////////////////////////////////
    // copy screen image into a bitmap object
    /////////////////////////////////////////

    // create a device context that accesses the whole screen
    CDC dcScreen;
    dcScreen.CreateDC( "DISPLAY", NULL, NULL, NULL );

    // create an empty bitmap in memory
    CDC dcMem;
    dcMem.CreateCompatibleDC( &dcScreen );
    CreateCompatibleBitmap( &dcScreen, m_nWidth, m_nHeight );
    dcMem.SelectObject( this );

    // copy screen into empty bitmap
    dcMem.BitBlt( 0,0,m_nWidth,m_nHeight,&dcScreen,rect.left,rect.top,SRCCOPY );

    // this bitmap is worthless without the current system palette, so...

    /////////////////////////////////////////
    // save system palette in this bitmap's palette
    /////////////////////////////////////////
```

Example 18   Printing the View   **271**

```
    // create an empty logical palette that's big enough to hold all the colors
    int nColors = ( 1 << ( dcScreen.GetDeviceCaps( BITSPIXEL ) *
        dcScreen.GetDeviceCaps( PLANES ) ) );
    LOGPALETTE *pLogPal = ( LOGPALETTE * )new BYTE[
        sizeof( LOGPALETTE ) + ( nColors * sizeof( PALETTEENTRY ) )];

    // initialize this empty palette's header
    pLogPal -> palVersion    = 0x300;
    pLogPal -> palNumEntries = nColors;

    // load this empty palette with the system palette's colors
    ::GetSystemPaletteEntries( dcScreen.m_hDC, 0, nColors,
        ( LPPALETTEENTRY )( pLogPal -> palPalEntry ) );

    // create the palette with this logical palette
    m_pPalette = new CPalette;
    m_pPalette -> CreatePalette( pLogPal );

    // clean up
    delete []pLogPal;
    dcMem.DeleteDC();
    dcScreen.DeleteDC();
}


HANDLE CWzdBitmap::CreateDIB( int *pbmData )
{
    /////////////////////////////////////////////
    // create DIB header from our BITMAP header
    /////////////////////////////////////////////

    BITMAPINFOHEADER bi;
    memset( &bi, 0, sizeof( bi ) );
    bi.biSize = sizeof( BITMAPINFOHEADER );
    bi.biPlanes = 1;
    bi.biCompression = BI_RGB;

    // get and store dimensions of bitmap
    BITMAP bm;
    GetObject( sizeof( bm ),( LPSTR )&bm );
```

II

6

```
bi.biWidth = bm.bmWidth;
bi.biHeight = bm.bmHeight;

// get number of bits required per pixel
int bits = bm.bmPlanes * bm.bmBitsPixel;
if (bits <= 1)
    bi.biBitCount = 1;
else if ( bits <= 4 )
    bi.biBitCount = 4;
else if ( bits <= 8 )
    bi.biBitCount = 8;
else
    bi.biBitCount = 24;


// calculate color table size
int biColorSize = 0;
if ( bi.biBitCount! = 24 ) biColorSize = ( 1 << bi.biBitCount );
biColorSize* = sizeof( RGBQUAD );

// calculate picture data size
bi.biSizeImage = ( DWORD )bm.bmWidth * bi.biBitCount;       // bits per row
bi.biSizeImage = ( ( ( bi.biSizeImage ) + 31 ) / 32 ) * 4;  // DWORD aligned
bi.biSizeImage* = bm.bmHeight;                     // bytes required for whole bitmap

// return size to caler in case they want to save to file
if ( pbmData )
    *pbmData = bi.biSize + biColorSize;

/////////////////////////////////////////////
// get DIB color table and picture data
/////////////////////////////////////////////

// allocate a hunk of memory to hold header, color table and picture data
HANDLE hDIB = ::GlobalAlloc( GHND, bi.biSize + biColorSize + bi.biSizeImage );

// get a memory pointer to this hunk by locking it
LPBITMAPINFOHEADER lpbi = ( LPBITMAPINFOHEADER )::GlobalLock( hDIB );
```

Example 18   Printing the View   **273**

```
    // copy our header structure into hunk
    *lpbi = bi;

    // get a device context and select our bitmap's palette into it
    CDC dc;
    dc.Attach( ::GetDC( NULL ) );
    CPalette *pPal = dc.SelectPalette( m_pPalette,FALSE );
    dc.RealizePalette();

    // load our memory hunk with the color table and picture data
    ::GetDIBits( dc.m_hDC, ( HBITMAP )m_hObject, 0, ( UINT )bi.biHeight,
        ( LPSTR )lpbi + (WORD)lpbi->biSize + biColorSize, (LPBITMAPINFO)lpbi,
        DIB_RGB_COLORS);

    // clean up
    ::GlobalUnlock( hDIB );
    dc.SelectPalette( pPal,FALSE );
    dc.RealizePalette();

    // return handle to the DIB
    return hDIB;
}

void CWzdBitmap::Print( CDC *pDC )
{
    // get DIB version of bitmap
    int bmData;
    HANDLE hDIB = CreateDIB( &bmData );

    // get memory pointers to the DIB's header and data bits
    LPBITMAPINFOHEADER lpDIBHdr = ( LPBITMAPINFOHEADER )::GlobalLock( hDIB );
    LPSTR lpDIBBits = ( LPSTR )lpDIBHdr+bmData;

    // stretch bitmap to fill printed page with 1/4 inch borders
    int cxBorder = pDC -> GetDeviceCaps(LOGPIXELSX)/4;
    int cyBorder = pDC -> GetDeviceCaps(LOGPIXELSY)/4;
    int cxPage = pDC -> GetDeviceCaps(HORZRES) - (cxBorder*2);
    int cyPage = ( int )( ( ( double )cxPage/
        ( double )m_nWidth ) * ( double )m_nHeight );
```

```
// stretch the bitmap for the best fit on the printed page
pDC -> SetStretchBltMode( COLORONCOLOR );
int i = ::StretchDIBits( pDC -> m_hDC,
    cxBorder,cyBorder,cxPage,cyPage,      // destination dimensions
    0,0,m_nWidth,m_nHeight,               // source bitmap dimensions
                                          //     (use all of bitmap)

    lpDIBBits,                            // bitmap picture data
    (LPBITMAPINFO)lpDIBHdr,               // bitmap header info
    DIB_RGB_COLORS,                       // specify color table
                                          //     has RGB values

    SRCCOPY                               // simple source to destination copy
);


// cleanup
::GlobalUnlock( hDIB );
::GlobalFree( hDIB );
return;
}
```

Example 19   Drawing to the MDI Client View

Objective

In an MDI application, you would like to draw something decorative in the MDI Client view, as seen in Figure 6.4.

Figure 6.4   Filling the MDI Client

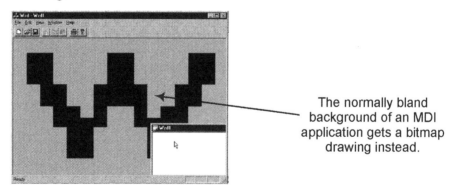

The normally bland background of an MDI application gets a bitmap drawing instead.

Example 19   Drawing to the MDI Client View   **275**

Strategy

An MDI application is potentially filled with several Child Frame Windows, each filled with a View Class's window. The area these Child Frame Windows cover, however, is not simply the client area of the Main Frame Window. Instead, this area is covered by a child window created using the MDI Client Window Class. Therefore, to draw to this background, we can't simply add a `WM_PAINT` message handler to our `CMainframe` class and paint the background. Instead, we have to subclass this MDI Client window and intercept its `WM_PAINT` window message.

Steps

Create a New MDI Client Window Class

1. Use the ClassWizard to create a new class and derive it from "generic CWnd".

2. Embed a `CBitmap` variable in this new class.

```
CBitmap m_bitmap;
```

3. In the constructor of this new class, load a bitmap into this member variable that you want to display in the MDI Client area.

```
CMDIClientWnd::CMDIClientWnd()
{
    // get bitmap and info
    m_bitmap.LoadBitmap( IDB_WZD_MDIBITMAP );
    m_bitmap.GetObject( sizeof( BITMAP ), &m_bm );
}
```

4. Use the ClassWizard to add a `WM_PAINT` message handler to this class. There, you will draw this bitmap to the client window.

```
void CMDIClientWnd::OnPaint()
{
    CPaintDC dc( this );    // device context for painting

    CDC dcMem;
    dcMem.CreateCompatibleDC( &dc );
    dcMem.SelectObject( m_bitmap );
```

```
// stretch bitmap
CRect rect;
GetClientRect(&rect);
dc.StretchBlt( rect.left, rect.top, rect.right, rect.bottom,
    &dcMem, 0, 0, m_bm.bmWidth-1, m_bm.bmHeight-1, SRCCOPY );
```

5. For a complete listing of this new MDI Client Window Class (such as it is), please refer to the MDI Client Window Class for Example 19 (page 277).

Now that we have a new MDI Window class, we need to subclass the real MDI Client window so that our new class intercepts the WM_PAINT message.

Implement the New MDI Client Window Class

1. Embed our new MDI Client Window Class in CMainFrame.

```
public:
    CMDIClientWnd m_wndMDIClient;
```

2. Then, in the OnCreate() member function of CMainFrame, subclass the current MDI Client Window with our new window class.

```
int CMainFrame::OnCreate( LPCREATESTRUCT lpCreateStruct )
{
    :    :    :

    // subclass our CMDIClientWnd class to MDIClient window
    if( !m_wndMDIClient.SubclassWindow( m_hWndMDIClient ) )
    {
        TRACE( " Failed to subclass MDI client window\n" );
        return -1;
    }
}
```

When the main window is resized, only the new portions of the MDI Client window are invalidated for painting—which creates a mess of our stretched bitmap. To fix this, we will invalidate the entire client area whenever the main window is resized.

Example 19   Drawing to the MDI Client View   **277**

3. Use the ClassWizard to add a WM_SIZE message handler and add the following to it.

```
void CMainFrame::OnSize( UINT nType, int cx, int cy )
{
    CMDIFrameWnd::OnSize( nType, cx, cy );

    m_wndMDIClient.Invalidate();
}
```

Notes

- This example paints a large "W" in our MDI Client window. A more aesthetically pleasing bitmap might be one that contains a repeated smaller image, such as a company logo which you would draw over and over again in rows and columns until the area was filled. You would, of course, need to add a lot more code to your WM_PAINT handler to tile this logo into place.
- Now that you have control of the client area of an MDI application, the sky's the limit. You can add controls here, list statistics, whatever.

CD Notes

- When executing the project on the accompanying CD, you will notice a large "W" filling the MDI Client area. Resizing the main window also causes this "W" to enlarge and shrink, accordingly.

MDI Client Window Class for Example 19

```
#if !defined MDICLIENTWND_H
#define MDICLIENTWND_H


/////////////////////////////////////////////////////////////////////////////
// CMDIClientWnd window

class CMDIClientWnd : public CWnd
{
// Construction
public:
    CMDIClientWnd();
```

```
// Attributes
public:

// Operations
public:

// Overrides
    // ClassWizard generated virtual function overrides
    // {{AFX_VIRTUAL( CMDIClientWnd )
    // }}AFX_VIRTUAL

// Implementation
public:
    virtual ~CMDIClientWnd();

    // Generated message map functions
protected:
    // {{AFX_MSG( CMDIClientWnd )
    afx_msg void OnPaint();
    // }}AFX_MSG
    DECLARE_MESSAGE_MAP()
private:
    BITMAP m_bm;
    CBitmap m_bitmap;
};

/////////////////////////////////////////////////////////////////////////

#endif
// MDIClientWnd.cpp : implementation file
//

#include "stdafx.h"
#include "wzd.h"
#include "MDIClientWnd.h"

#ifdef _DEBUG
#define new DEBUG_NEW
#undef THIS_FILE
```

Example 19   Drawing to the MDI Client View   **279**

```
static char THIS_FILE[] = __FILE__;
#endif


/////////////////////////////////////////////////////////////////////////////
// CMDIClientWnd

CMDIClientWnd::CMDIClientWnd()
{
    // get bitmap and info
    m_bitmap.LoadBitmap( IDB_WZD_MDIBITMAP );
    m_bitmap.GetObject( sizeof( BITMAP ), &m_bm );
}


CMDIClientWnd::~CMDIClientWnd()
{
}



BEGIN_MESSAGE_MAP( CMDIClientWnd, CWnd )
    // {{AFX_MSG_MAP( CMDIClientWnd )
    ON_WM_PAINT()
    // }}AFX_MSG_MAP
END_MESSAGE_MAP()



/////////////////////////////////////////////////////////////////////////////
// CMDIClientWnd message handlers

void CMDIClientWnd::OnPaint()
{
    CPaintDC dc( this );    // device context for painting

    CDC dcMem;
    dcMem.CreateCompatibleDC( &dc );
    dcMem.SelectObject( m_bitmap );

    // stretch bitmap
    CRect rect;
```

II

6

```
GetClientRect( &rect );
dc.StretchBlt( rect.left, rect.top, rect.right, rect.bottom,
    &dcMem, 0, 0, m_bm.bmWidth-1, m_bm.bmHeight-1, SRCCOPY );
}
```

Example 20   Dragging and Dropping Files into Your View

Objective

You would like to be able to open files that have been dropped into your view.

Strategy

You probably know by now that you can drag a file from the Windows Explorer or other file utility into the Developer Studio and the Studio will magically open that file. To add this same functionality to your own application, we first use CWnd::DragAcceptFiles() to tell the system it's okay to drop files on your application's window. Then, we add a WM_DROPFILES message handler to our CMainFrame class to get the file name that's being dropped and open the appropriate file. In this example, we are modifying an MDI application to open a new document and view for every file dropped on us.

Steps

Modify the Main Frame Class

1. Allow your application to accept dropped files by calling the following from your CMainFrame's OnCreate() message handler.

```
// allow files to be dropped on main window or any child window
DragAcceptFiles();
```

Example 20 Dragging and Dropping Files into Your View **281**

2. Use the ClassWizard to add a `WM_DROPFILES` message handler to your `CMainFrame` class where you will use `CWinApp::OpenDocumentFile()` to create a new Document/View for this file.

```
void CMainFrame::OnDropFiles( HDROP hDropInfo )
{

    // get filename stored in hDropInfo and use app to open it
    TCHAR szFileName[_MAX_PATH];
    ::DragQueryFile( hDropInfo, 0, szFileName, _MAX_PATH );
    ::DragFinish( hDropInfo );
    AfxGetApp() -> OpenDocumentFile( szFileName );


    CMDIFrameWnd::OnDropFiles( hDropInfo );

}
```

Notes

- This message handler will also work in an SDI application, first destroying the current Document/View before opening the new file.

CD Notes

- When executing the project on the accompanying CD, also open a copy of the Windows Explorer. Then, drag a file from the Windows Explorer into this application and notice that a new Document/View is open.

Chapter 7

Dialog Boxes and Bars

Dialog boxes and bars allow you to prompt your user with several types of controls, such as buttons and list boxes. Dialog bars are a hybrid between toolbars and dialog boxes. A dialog bar can dock to the side of your application but still have all the controls available to a dialog box.

MFC and the Developer Studio will automatically create dialog boxes and classes for you. In this chapter, we look at ways to manually modify your dialog boxes to be more useful to your user.

Example 21   Dynamically Changing Your Dialog Box Size We   will allow our user to click a button that will cause our dialog box to expand to expose more controls.

Example 22   Customizing Data Exchange and Validation We   will look at how to create our own data types that can be exchanged between our dialog class's member variables and controls.

Example 23   Overriding the Common File Dialog We will look at how to customize the File Dialog box. This dialog is available to all applications and saves you the headache of writing your own.

Example 24   Overriding the Common Color Dialog   This example is similar to the previous example, except we will look at the Color Dialog box.

Example 25   Getting a Directory Name   We will look at how to use an arcane API to get just a directory name and not a whole file path.

Example 26   Using a Child Dialog Box Like a Common Control   We will use a dialog box just like a common control in another dialog box.
Example 27   Using a Child Property Sheet Like a Common Control
   We will use a property sheet just like a common control in a dialog box.

Example 21   Dynamically Changing Your Dialog Box Size

Objective

You would like to enlarge a dialog box to expose more of its controls when the user clicks on a "More" button, as seen in Figure 7.1.

Figure 7.1   Dynamically Changing Your Dialog Box Size

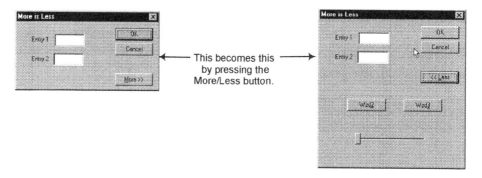

This becomes this by pressing the More/Less button.

Strategy

We will create a dialog box resource and dialog box class as usual with a "More >>" button in it. However, in the `OnInitDialog()` function of our dialog class, we will use `CWnd::MoveWindow()` to shrink our dialog window down to the "More >>" button. From then on, when our user clicks on the

Example 21   Dynamically Changing Your Dialog Box Size   **285**

"More>>" button, we will use the MoveWindow() function to toggle the size of our dialog window from large to small to large.

Steps

Toggle the Size of a Dialog Box

1. Create a dialog template and dialog class as usual with the Dialog Editor and ClassWizard. Put a "More >>" button in the middle and any extra controls to one side of it.

2. Use the ClassWizard to override its OnInitDialog() function. In OnInitDialog(), you should start by saving the current size of the dialog box. This will be its size at its largest. Then, find the location of the "More >>" button and shrink the window to this button.

```
BOOL CWzdDialog::OnInitDialog()
{
    CDialog::OnInitDialog();

    // save our full size
    GetWindowRect( &m_rectFull );
    m_rectHalf = m_rectFull;

    // calculate our half size based on bottom of "More" button
    CRect rect;
    m_ctrlMoreButton.GetWindowRect( &rect );
    m_rectHalf.bottom = rect.bottom+10;      // + 10 for cosmetics

    // toggle window size
    ToggleSize();

    return TRUE;    // return TRUE unless you set the focus to a control
                    // EXCEPTION: OCX Property Pages should return FALSE
}
```

Since our "More>>" button will also be toggling the size of the dialog window, we put the actual logic to shrink the window size in the Toggle-Size() function, which we will call in the following code fragment.

3. Use the ClassWizard to add a message handler for the "More >>" button that will also call the `ToggleSize()` function.

```
void CWzdDialog::OnMoreButton()
{
    ToggleSize();
}
```

4. The `ToggleSize()` function changes the name on the button to "<< Less" and then uses `CWnd::SetWindowPos()` to change the size of the window, as seen here.

```
void CWzdDialog::ToggleSize()
{
    CRect rect;
    CString str;
    if ( m_bToggleSize )
    {
        str = "<< &Less";
        rect = m_rectFull;
    }
    else
    {
        str = "&More >>";
        rect = m_rectHalf;
    }
    SetWindowPos( NULL,0,0,rect.Width(),rect.Height(),
        SWP_NOZORDER|SWP_NOMOVE );
    m_ctrlMoreButton.SetWindowText( str );
    m_bToggleSize = !m_bToggleSize;
}
```

To see a complete listing of this dialog class, please see the Dialog Class for Example 21 (page 287).

Notes

- If you prefer something more ornate than "More >>" and "<< Less" for your button, set it's style to BS_BITMAP with the Dialog Editor and then use CButton::SetBitmap() to change the bitmap.

Example 21   Dynamically Changing Your Dialog Box Size   **287**

CD Notes

- When executing the project on the accompanying CD, click the "Test" and then "Wzd" menu commands to open a dialog box. Then, click on the "More>>" button to open the dialog further.

Dialog Class for Example 21

```
#if !defined( AFX_WZDDIALOG_H__1EC8499A_C589_11D1_9B5C_00AA003D8695__INCLUDED_ )
#define AFX_WZDDIALOG_H__1EC8499A_C589_11D1_9B5C_00AA003D8695__INCLUDED_


#if _MSC_VER >= 1000
#pragma once
#endif    // _MSC_VER >= 1000
// WzdDialog.h : header file
//


/////////////////////////////////////////////////////////////////////////
// CWzdDialog dialog

class CWzdDialog : public CDialog
{
// Construction
public:
    CWzdDialog( CWnd* pParent = NULL );   // standard constructor

// Dialog Data
    // {{AFX_DATA( CWzdDialog )
    enum { IDD = IDD_WZD_DIALOG };
    CButton m_ctrlMoreButton;
    // }}AFX_DATA


// Overrides
    // ClassWizard generated virtual function overrides
    // {{AFX_VIRTUAL( CWzdDialog )
protected:
    virtual void DoDataExchange( CDataExchange* pDX );    // DDX/DDV support
    // }}AFX_VIRTUAL
```

II

7

```
// Implementation
protected:

    // Generated message map functions
    // {{AFX_MSG( CWzdDialog )
    virtual BOOL OnInitDialog();
    afx_msg void OnMoreButton();
    // }}AFX_MSG
    DECLARE_MESSAGE_MAP()
private:
    BOOL m_bToggleSize;
    CRect m_rectFull;
    CRect m_rectHalf;
    void ToggleSize() ;
};


// {{AFX_INSERT_LOCATION}}
// Microsoft Developer Studio will insert additional declarations immediately
//      before the previous line.

#endif
    // !defined( AFX_WZDDIALOG_H__1EC8499A_C589_11D1_9B5C_00AA003D8695__INCLUDED_ )
// WzdDialog.cpp : implementation file
//

#include "stdafx.h"
#include "wzd.h"
#include "WzdDialog.h"

#ifdef _DEBUG
#define new DEBUG_NEW
#undef THIS_FILE
static char THIS_FILE[] = __FILE__;
#endif


/////////////////////////////////////////////////////////////////////////////
// CWzdDialog dialog
```

Example 21   Dynamically Changing Your Dialog Box Size   **289**

```
CWzdDialog::CWzdDialog( CWnd* pParent /* = NULL*/ )
    : CDialog( CWzdDialog::IDD, pParent )
{

    // {{AFX_DATA_INIT( CWzdDialog )
    // NOTE: the ClassWizard will add member initialization here
    // }}AFX_DATA_INIT
    m_bToggleSize = FALSE;
}



void CWzdDialog::DoDataExchange( CDataExchange* pDX )
{
    CDialog::DoDataExchange( pDX );
    // {{AFX_DATA_MAP( CWzdDialog )
    DDX_Control( pDX, IDC_MORE_BUTTON, m_ctrlMoreButton );
    // }}AFX_DATA_MAP
}



BEGIN_MESSAGE_MAP( CWzdDialog, CDialog )
    // {{AFX_MSG_MAP( CWzdDialog )
    ON_BN_CLICKED( IDC_MORE_BUTTON, OnMoreButton )
    // }}AFX_MSG_MAP
END_MESSAGE_MAP()

/////////////////////////////////////////////////////////////////////////////
// CWzdDialog message handlers

BOOL CWzdDialog::OnInitDialog()
{
    CDialog::OnInitDialog();

    // save our full size
    GetWindowRect( &m_rectFull );
    m_rectHalf = m_rectFull;

    // calculate our half size based on bottom of "More" button
    CRect rect;
    m_ctrlMoreButton.GetWindowRect( &rect );
    m_rectHalf.bottom = rect.bottom+10;     // + 10 for cosmetics
```

II

7

```
    // toggle window size
    ToggleSize();

    return TRUE;    // return TRUE unless you set the focus to a control
                    // EXCEPTION: OCX Property Pages should return FALSE
}

void CWzdDialog::OnMoreButton()
{
    ToggleSize();
}

void CWzdDialog::ToggleSize()
{
    CRect rect;
    CString str;
    if ( m_bToggleSize )
    {
        str = "<< &Less";
        rect = m_rectFull;
    }
    else
    {
        str = "&More >>";
        rect = m_rectHalf;
    }
    SetWindowPos( NULL,0,0,rect.Width(),rect.Height(),SWP_NOZORDER|SWP_NOMOVE );
    m_ctrlMoreButton.SetWindowText( str );
    m_bToggleSize = !m_bToggleSize;
}
```

Example 22   Customizing Data Exchange and Validation

Objective

You would like to create a new data exchange and validation function to support dynamic data exchange in your dialog boxes for a new data type.

Example 22   Customizing Data Exchange and Validation   **291**

Strategy

Dynamic data exchange allows us to automatically exchange data between a dialog box control and the member variable of a dialog class. Dynamic data validation allows a user's input to be automatically checked for syntax errors. While several data formats are already supported by MFC for exchange and validation (such as for int and double), we can also support our own custom types (eg., COctalString) by supplying our own static conversion functions using the appropriate naming and calling conventions (e.g., DDX_something() or DDV_something()). In this example, we will be exchanging and validating binary data strings. The binary string we will use here will be maintained by a custom CString class we create in Example 64.

Steps

Create a Custom Data Exchange Function

1. Create a static routine with the following outline. For a complete working example, please see the Binary Data Exchange and Validation Functions for Example 22 (page 294).

```
void AFXAPI DDX_Xxxx( CDataExchange* pDX, int nIDC, CXxxx& value )
{
    CWzdString str;
    HWND hWndCtrl = pDX -> PrepareEditCtrl( nIDC );

    // getting data from control
    if ( pDX -> m_bSaveAndValidate )
    {
        // hWndCtrl == window handle of the control
        int nLen = ::GetWindowTextLength( hWndCtrl );
        ::GetWindowText( hWndCtrl, str.GetBufferSetLength( nLen ),
            nLen+1 );
        str.ReleaseBuffer();

        // temporary str now has text data, now put into member variable


    }
    // putting data into control window
```

```
    else
    {
        // copy data from member variable into str

        // sets window text only if it's different
        AfxSetWindowText( hWndCtrl, str );
    }
}
```

Create a Custom Data Validation Function

1. Create a static routine with the following outline. For a complete working example, please see the Binary Data Exchange and Validation Functions for Example 22 (page 294).

```
void AFXAPI DDV_MaxChars( CDataExchange* pDX, CByteArray const& value,
    int nBytes )
{
    // if retrieving from control, make check of value
    if ( pDX -> m_bSaveAndValidate )
    {
        if ( error )
        {
            CString str;
            str.Format( "Maximum characters you can enter is %d!",
                nBytes );
            AfxMessageBox( str, MB_ICONEXCLAMATION );
            pDX -> Fail();
        }
    }
    // else if sending to control, setup any control function
    //     that will allow it to perform it's own check
    else if ( pDX -> m_hWndLastControl !=
        NULL && pDX -> m_bEditLastControl )
    {
        ///
    }
}
```

Example 22   Customizing Data Exchange and Validation   **293**

Implement the New Data Exchange and Validation Functions

1. Include the definitions for your new exchange functions to your dialog class.

```
#include "WzdXchng.h"
```

2. Insert these functions as needed into your dialog class's DoDataExchange() routine, but make sure you insert them below the {{}} brackets so that the ClassWizard can't get at them.

```
void CWzdDialog::DoDataExchange( CDataExchange* pDX )
{
    CDialog::DoDataExchange( pDX );
    // {{AFX_DATA_MAP( CWzdDialog )
    // }}AFX_DATA_MAP
    DDX_Text( pDX, IDC_WZD_EDIT, m_WzdArray );      <<<<
    DDV_MaxChars( pDX, m_WzdArray, 7 );             <<<<
}
```

Make sure that the data validation function is called immediately after the data exchange function it checks. The m_hWndLastControl member variable of CDataExchange is initialized by the data exchange function. You can, of course, reinitialize this value in your data validation routine, but why waste CPU cycles?

Notes

- Rather than create a separate data exchange or validation function, you might instead modify the DoDataExchange() routine directly. Simply add any additions past the {{}} brackets that belong to the ClassWizard. Make any additions similar to the previous functions and remember that when m_bSaveAndValidate is TRUE, you should be transferring from the control to a dialog member variable.

```
void CWzdDialog::DoDataExchange( CDataExchange* pDX )
{
    CDialog::DoDataExchange( pDX );
    // {{AFX_DATA_MAP( CWzdDialog )
       :    :    :
    // }}AFX_DATA_MAP
```

```
    if ( pDX -> m_bSaveAndValidate )
    {
        // get from control
    }
    else
    {
        // store to control
    }
}
```

Of course, the draw back to this approach is that you can't now easily transfer this new logic to another dialog class.

CD Notes

- When executing the project on the accompanying CD, set a breakpoint on OnTestWzd() in WzdView.cpp. Then, click on "Test" and "Wzd", enter a number into the dialog box, and watch as it automatically converts that number into a byte array.

Binary Data Exchange and Validation Functions for Example 22

```
#ifndef WZDXCHNG_H
#define WZDXCHNG_H


void AFXAPI DDX_Text( CDataExchange* pDX, int nIDC, CByteArray& value );
void AFXAPI DDV_MaxChars( CDataExchange* pDX, CByteArray const& value,
    int nBytes );


#endif
// WzdXchng
//

#include "stdafx.h"
#include "WzdXchng.h"
#include "WzdString.h"
#include <afxpriv.h>
```

Example 22   Customizing Data Exchange and Validation   **295**

```
/////////////////////////////////////////////////////////////////////////////
// CWzdXchng


void AFXAPI DDX_Text( CDataExchange* pDX, int nIDC, CByteArray& value )
{
    CWzdString str;
    HWND hWndCtrl = pDX -> PrepareEditCtrl( nIDC );
    if ( pDX -> m_bSaveAndValidate )
    {
        int nLen = ::GetWindowTextLength( hWndCtrl );
        ::GetWindowText( hWndCtrl, str.GetBufferSetLength( nLen ), nLen+1 );
        str.ReleaseBuffer();
        int size = str.GetLength();
        if ( size%2 )
        {
            AfxMessageBox( "Please enter even number of digits." );
            pDX -> Fail();                    // throws exception
        }
        size /= 2;
        value.SetSize( size );
        str.GetBinary( value.GetData(),size );
    }
    else
    {
        str.PutBinary( value.GetData(),value.GetSize() );
        AfxSetWindowText( hWndCtrl, str );   // sets window text only if
                                             //    it's different

    }
}

void AFXAPI DDV_MaxChars( CDataExchange* pDX, CByteArray const& value,
    int nBytes )
{
    if ( pDX -> m_bSaveAndValidate && value.GetSize() > nBytes )
    {
        CString str;
        str.Format( "Maximum characters you can enter is %d!",nBytes );
        AfxMessageBox( str, MB_ICONEXCLAMATION );
```

```
        str.Empty();    // will not return after exception
        pDX -> Fail();
    }
    else if ( pDX -> m_hWndLastControl != NULL && pDX -> m_bEditLastControl )
    {
        // set edit box not to allow more then these bytes
        // note that this function depends on being called
        //      right after the DDX function
        ::SendMessage( pDX -> m_hWndLastControl, EM_LIMITTEXT, nBytes*2, 0 );
    }
}
```

Example 23   Overriding the Common File Dialog

Objective

You would like to add your own features and look to the Common File Dialog Box, as seen in Figure 7.2.

Figure 7.2   Customized File Dialog Box

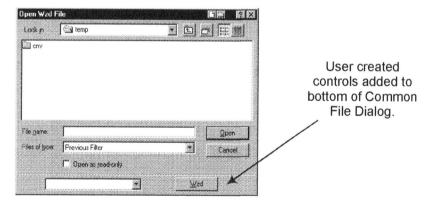

User created controls added to bottom of Common File Dialog.

Strategy

We will derive our own class from the Common File Dialog Class, CFileDialog, and add our own controls to this dialog. We will also customize this dialog using arguments in its constructor and by setting up its OPENFILENAME structure.

Example 23   Overriding the Common File Dialog   **297**

Steps

Create a New Command File Dialog Class

Note: If you don't intend to add your own controls to the File Dialog Box, you don't need to create a new class. You can, therefore, skip this section.

II

7

1. Use the ClassWizard to create a new class derived from `CFileDialog`.
2. Use the Dialog Editor to create a small dialog box template containing just the controls you intend to add to the Common File Dialog Box. The Common Dialog will dynamically create the rest of the standard controls you see in its box, so don't worry about them. You just need to create any new controls. Your template should have a child window style with no frame.
3. Use the ClassWizard to add message handlers for your new controls. To access the controls that are added automatically by the File Dialog, you can use `GetDlgItem()` with one of the following ID's. Which ID goes with which control can change in future releases of Windows, however.

```
psh1,psh2,....,psh12      // push buttons
chx1-12                   // checkboxes
rad1-12                   // radio buttons
grp1-12                   // group boxes
stc1-12                   // statics
lst1-12                   // listboxes/views
cmb1-12                   // comboboxes
edt1-12                   // edit boxes
scr1-12                   // scrollbars
```

Initialize a Common File Dialog Class

1. To use the CFileDialog class or your own derivation, first construct it, as follows.

```
CFileDialog dlg(
    TRUE,                           // TRUE = create a File Open dialog,
                                    // FALSE = create a File Save As
                                    //     dialog
    _T(".log"),                     // default filename extension
    "",                             // initial filename in edit box
                                    //     functionality flags
    OFN_ALLOWMULTISELECT|           // allow multiple files
                                    //     to be selected
    // OFN_CREATEPROMPT |           // if File Save As, prompts user
                                    //     if they want to create
                                    //     non-existant file
    // OFN_OVERWRITEPROMPT |        // if File Save As--prompts user to
                                    //     ask if they want to overwrite
                                    //     an existing file
    // OFN_ENABLESIZING |           // if Windows NT 5.0 or Win 98,
                                    //     causes box to be resizable
                                    //     by user
    // OFN_EXTENSIONDIFFERENT|      // allows user to enter a filename
                                    //     with a different extension
                                    //     from the default
    // OFN_FILEMUSTEXIST            // file must exist
    // OFN_NOLONGNAMES |            // causes dialog to use short
                                    //     filenames (8.3)
    // OFN_PATHMUSTEXIST |          // user can only type valid paths
                                    //     and filenames
    // OFN_NOVALIDATE |             // the returned filname can have
                                    //     invalid characters
                                    //     appearence flags
    // OFN_HIDEREADONLY |           // hides read-only check box
    // OFN_NONETWORKBUTTON |        // hides Network button
    // OFN_READONLY |               // initially check Read Only
                                    //     check box
    // OFN_SHOWHELP |               // Help button appears--when clicked
```

Example 23   Overriding the Common File Dialog   **299**

```
                              //      the hook procedure gets a
                              //      CDN_HELP message
    // custom template flags
    OFN_ENABLETEMPLATE |      // you will be supplying your own
                              //      custom dialog box template

    0,
    "Accounting Files ( *.log;*.txt )|*.log;*.txt|All Files
        ( *.* )|*.*||",       // file filter
        NULL);                // parent window
```

2. If you will be adding your own controls to the dialog, you must also define the dialog box template that File Dialog should open with the following.

```
dlg.m_ofn.lpTemplateName = MAKEINTRESOURCE( IDD_WZD_FILEOPEN );
```

You must also set the OFN_ENABLETEMPLATE flag when constructing the class.

3. To cause the Common File Dialog to initially use a particular directory, use

```
// set an initial directory
char lpszInitDir[] = {"c:\\temp"};
dlg.m_ofn.lpstrInitialDir = lpszInitDir;
```

4. To give the File Dialog a name other than "Open File" or "Save File As", use

```
//set the dialog's title
char lpszTitle[] = {"Open Wzd File"};
dlg.m_ofn.lpstrTitle = lpszTitle;
```

5. To set up a file filter that will be retained after the File Dialog closes, first set up a static character string.

```
static char lpstrCustomFilter[255] = {"Previous Filter\0*.log\0"};
```

Then, initialize the File Dialog with its address, as follows.

```
// retain the customer's last file filter selection
dlg.m_ofn.lpstrCustomFilter = lpstrCustomFilter;
dlg.m_ofn.nMaxCustFilter = 255;
```

II

7

Create a Common File Dialog Box and Retrieve Values

1. Use the `CFileDialog::DoModal()` function to open this file dialog box. You should then check to see that the user clicked "IDOK" and, if so, get the values from the box.

```
if ( dlg.DoModal() == IDOK )
{
```

2. To retrieve the name of the file that the user selected, you can use one of the following.

```
CString path = dlg.GetPathName();      // ex: c:\temp\temp.tmp
CString file = dlg.GetFileName();      // ex: temp.tmp
CString title = dlg.GetFileTitle();    // ex: temp
CString ext = dlg.GetFileExt();        // ex: tmp
```

3. To determine which file filter the user picked, you can use.

```
int nFilterIndex = dlg.m_ofn.nFilterIndex;
```

The filter is returned as an index into the list of file filters you provided.

4. To see if the user clicked the "Read Only" checkbox, you can use

```
BOOL bReadOnly = dlg.GetReadOnlyPref();
```

5. If the user was able to pick multiple files (you set `OFN_ALLOWMULTISELECT`), you can scroll through them using the following.

```
    for (POSITION pos = dlg.GetStartPosition();pos;)
    {
        CString pathx = dlg.GetNextPathName(pos);
            // ex: c:\temp\temp.tmp
    }
}
```

Note that when multiple files can be selected, you must parse the file-name yourself to get the file title, extension, etc. You can't use `dlg.GetPathName()`, etc.

Notes

- This example represents a way to modify the Explorer-styled File Open dialog. Using this style, not only can your user open the file they want,

Example 24   Overriding the Common Color Dialog   **301**

but they can also perform the same file maintenance capabilities available with the Windows Explorer (e.g., creating new directories, deleting old files, etc). Another older version of the File Open dialog is still available to you. The advantage of the older style is that every dialog control is easily available to you because you get to edit the actual dialog box template. This template is called FILEOPEN.DLG and can be found under \VC\INCLUDE. If you would like to work with the older File Open dialog, please refer to the next example. That example shows how to override the Common Color Dialog, which uses another complete template from \VC\INCLUDE. Of course, you should use FILEOPEN.DLG, instead. You must also reset the OFN_EXPLORER flag after constructing the File Dialog class, with the following, because this flag is automatically set when constructing the CFileDialog class.

```
dlg.m_ofn.flags& = ~OFN_EXPLORER;
```

- Rather than bend over backwards to use the "Help" button provided by CFileDialog, you might be better served to add and process your own "Help" button.

CD Notes

- When executing the project on the accompanying CD, click on the "Test" and then "Wzd" menu commands to open a Common File Dialog Box. You will notice that this dialog is similar to most file dialogs you encounter in Windows, with the addition of two new controls at the bottom.

Example 24   Overriding the Common Color Dialog

Objective

You would like to add your own features and look to the Common Color Dialog Box, as seen in Figure 7.3.

Figure 7.3   A Customized Color Dialog Box

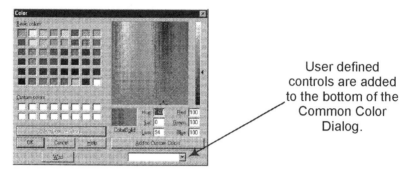

User defined controls are added to the bottom of the Common Color Dialog.

Strategy

We will derive our own class from the Common Dialog Class, `CColorDialog`, and add our own controls to this dialog. We will also customize this dialog using arguments in its constructor and by setting up its `CHOOSECOLOR` structure.

Steps

Create a New Common Color Dialog Class

Note: If you don't intend to add your own controls to the Color Dialog, you don't need to create a special Common Color Dialog Class. You can, therefore, skip this section.

---

1. Use the ClassWizard to create a new class derived from `CColorDialog`.

2. You must now cut and paste the standard color dialog template into your application's resources. The standard template can be found in the `COLOR.DLG` file in your `\VC\INCLUDE` directory. Open it with the Text Editor and paste it into the dialog section of your `.rc` file.

3. You must also add the `COLORDLG.H` resource include file for this dialog to your resource file. You can do this by clicking on your Developer Studio's "View" and "Resource Includes" menu items to open the "Resource Includes" dialog box, then enter "`COLORDLG.H`".

Example 24   Overriding the Common Color Dialog   **303**

4. You can now use the Dialog Editor to edit this dialog box. You can move controls, delete them, and add to them—just don't change the IDs of the existing controls. The Color Dialog box looks for these exact IDs.

Initialize the Common Color Dialog Class

1. To use the `CColorDialog` dialog class or your own derivation, first construct it, as follows.

```
CWzdColorDialog dlg(
    rgb,                        // initial color selection
    // CC_FULLOPEN|             // intially opens both sides
                                //      of color dialog
    // CC_PREVENTFULLOPEN|      // prevent user from opening customize
                                //      side of color dialog
    CC_SHOWHELP|                // creates help button
    // CC_SOLIDCOLOR|           // display only non-dithered colors
    CC_ENABLETEMPLATE|          // use a custom dialog template
    0,
    NULL                        // parent window
);
```

2. If you are using your own dialog template, you must also define it here in the `ColorDialog`'s `CHOOSECOLOR` structure.

```
dlg.m_cc.lpTemplateName = MAKEINTRESOURCE( IDD_WZD_CHOOSECOLOR );
```

You must also set the `CC_ENABLETEMPLATE` flag when constructing the class.

3. To save your user's customized colors between invocations of `CColorDialog`, you must supply the pointer to a static array of 16 `COLORREF` values.

```
COLORREF lpCustColors[16];
    :    :    :
    dlg.m_cc.lpCustColors = lpCustColors;
```

II

7

Create the Common Color Dialog Class

To open the Common Color Dialog Box, use the `CColorDialog::DoModal()` function. Then, check for a return value of `IDOK` to determine if your should change the color.

```
if ( dlg.DoModal() == IDOK )
{
    rgb = dlg.GetColor();
}
```

Notes

- There are also Common Dialog Boxes for fonts, finding and replacing text, opening files, and for setting up and printing files. Opening files was demonstrated in the last example.

- The Color Dialog is customarily used to select a diffused color. You can, however, use it to pick nondiffused colors by creating it with the `CC_SOLIDCOLOR` style. Unfortunately, the selection of colors becomes very sparse on a system with only 256 colors. To offer your user a much broader range of nondiffused colors on a system with 256 colors, please refer to Example 31.

CD Notes

- When executing the project on the accompanying CD, click on the "Test" and then "Wzd" menu commands to open a common color dialog box. You will notice that this dialog is similar to most color dialogs you encounter in Windows, with the addition of two new controls at the bottom.

Example 25   Getting a Directory Name

Objective

You would like to create a tree control that displays your system's disk directory, as seen in Figure 7.4.

Example 25   Getting a Directory Name   **305**

Figure 7.4   A Directory Prompt Dialog Box

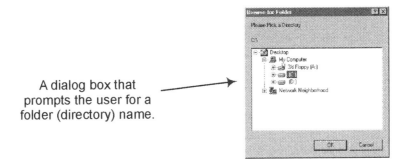

A dialog box that prompts the user for a folder (directory) name.

II

Strategy

You might think that the Common File Dialog will handle all of your file prompting needs, until one day you suddenly realize you have an application that needs to prompt the user for just a file directory. Then, you realize there's no easy way to use the Common File Dialog to prompt your user for just directories. You could make controls invisible and jerry-rig the edit control to return when the user clicks on a directory, but it can be a mess and it doesn't match the look-and-feel of other Microsoft applications when they prompt you for a directory. So what do you use? The little-documented ::SHBrowseForFolder() API call.

Since ::SHBrowseForFolder() requires some special handling, we will encapsulate its functionality in our very own Directory class.

Steps

Create a Directory Class

1. Use the ClassWizard to create a new class derived from nothing. The API we will be using requires no functionality from any of our MFC classes.

7

2. Add a `GetDirectory()` function to this class, which will start by creating the `BROWSEINFO` structure required by `::SHBrowseForFolders()`.

```
CString CWzdDirDlg::GetDirectory( CWnd *pParent,LPCSTR lpszRoot,
    LPCSTR lpszTitle )
{
    CString str;
    BROWSEINFO bi;
    bi.hwndOwner = pParent -> m_hWnd;          // owner of created dialog box
    bi.pidlRoot = 0;                           // unused
    bi.pszDisplayName = 0;                     // buffer to receive name
                                               //     displayed by folder
                                               //     (not a valid path)

    bi.lpszTitle = lpszTitle;                  // title is "Browse for
                                               //     Folder", this is
                                               //     an instruction

    bi.lpfn = BrowseCallbackProc;              // callback routine called
                                               //     when dialog has been
                                               //     initialized

    bi.lParam = 0;                             // passed to callback routine
    bi.ulFlags =
        BIF_RETURNONLYFSDIRS |                 // only allow user to select
                                               //     a directory

        BIF_STATUSTEXT |                       // create status text field
                                               //     we will be writing to
                                               //     in callback

        // BIF_BROWSEFORCOMPUTER|              // only allow user to select
                                               //     a computer

        // BIF_BROWSEFORPRINTER |              // only allow user to select
                                               //     a printer

        // BIF_BROWSEINCLUDEFILES|             // displays files too which
                                               //     user can pick

        // BIF_DONTGOBELOWDOMAIN|              // when user is exploring the
                                               //     "Entire Network" they
                                               //     are not allowed into
                                               //     any domain

        0;
    m_sRootDir = lpszRoot;                     // save for callback routine
```

Example 25   Getting a Directory Name   **307**

3. Next, you call `::SHBrowseForFolder()` to prompt the user for a directory name. You then use `::SHGetPathFromIDList()` to get that name.

```
LPITEMIDLIST lpItemId = ::SHBrowseForFolder( &bi );
if ( lpItemId )
{
    LPTSTR szBuf = str.GetBuffer( MAX_PATH );
    ::SHGetPathFromIDList( lpItemId, szBuf );
    ::GlobalFree( lpItemId );
    str.ReleaseBuffer();
}


return str;
}
```

4. As seen in Step 1, we specified a callback routine. A callback routine is necessary only to further modify the behavior of `::SHBrowseForFolder()`. If we had specified a `NULL`, no callback routine would have been required. However, in order to implement an initial directory on which to open our prompt, a callback routine is required! Therefore, add the following callback function to change the initial directory of our prompt.

```
int CALLBACK BrowseCallbackProc( HWND hwnd,UINT msg,LPARAM lp,
    LPARAM pData )
{

    TCHAR buf[MAX_PATH];

    switch( msg )
    {
        // when dialog is first initialized, change directory
        //      to one chosen above
        case BFFM_INITIALIZED:
            strcpy( buf,CWzdDirDlg::m_sRootDir );
            ::SendMessage( hwnd,BFFM_SETSELECTION,TRUE,( LPARAM) buf );
            break;
```

5. Also, if you picked the BIF_STATUSTEXT style in the BROWSEINFO structure, you can set the status using this callback routine every time the user selects a new folder.

```
        case BFFM_SELCHANGED:
            if ( ::SHGetPathFromIDList( ( LPITEMIDLIST ) lp ,buf ) )
                SendMessage( hwnd,BFFM_SETSTATUSTEXT,0,( LPARAM )buf );
            break;

    }
    return 0;
}
```

6. A complete listing of this class can be found in the Directory Class for Example 25 (page 309).

Use This New Directory Class

1. Prompting the user for a directory is now a simple matter of instantiating our directory class and calling its GetDirectory() function.

```
// get current working directory
char buf[MAX_PATH];
_getcwd( buf,MAX_PATH );

CWzdDirDlg dlg;
CString dir = dlg.GetDirectory(
    this,       // parent window
    buf,        // root directory
    "Title"     // optional title
    );
```

Notes

- You might have noticed a pidlRoot member of the BROWSEINFO structure. You'd think it could be used to specify an initial directory. Instead, it doesn't seem to serve any purpose—other than to provide you with hours of fun trying to make it work.

Example 25   Getting a Directory Name   **309**

CD Notes

- When executing the project on the accompanying CD, click on the "Test" and then "Wzd" menu commands to open a directory prompt opened to the current working directory.

Directory Class for Example 25

```
#if !defined( AFX_WZDDRDLG_H__D52656D3_8A25_11D2_9C53_00AA003D8695__INCLUDED_ )
#define AFX_WZDDRDLG_H__D52656D3_8A25_11D2_9C53_00AA003D8695__INCLUDED_


#if _MSC_VER >= 1000
#pragma once
#endif     // _MSC_VER >= 1000


class CWzdDirDlg
{
public:
    CWzdDirDlg();
    virtual ~CWzdDirDlg();

    CString GetDirectory( CWnd *pParent = NULL,
        LPCSTR lpszRoot = "c:\\",LPCSTR lpszTitle = "Please Pick a Directory" );


    static CString m_sRootDir;


};
int CALLBACK BrowseCallbackProc( HWND hwnd,UINT uMsg,LPARAM lp, LPARAM pData );


#endif
    // !defined( AFX_WZDDRDLG_H__D52656D3_8A25_11D2_9C53_00AA003D8695__INCLUDED_ )
// WzdDrDlg.cpp: implementation of the CWzdDirDlg class.
//
/////////////////////////////////////////////////////////////////////////////


#include "stdafx.h"
#include "WzdDrDlg.h"
#include "Shlobj.h"


CString CWzdDirDlg::m_sRootDir;
```

II

7

```
///////////////////////////////////////////////////////////////////////
// Construction/Destruction
///////////////////////////////////////////////////////////////////////


CWzdDirDlg::CWzdDirDlg()
{


}


CWzdDirDlg::~CWzdDirDlg()
{


}


CString CWzdDirDlg::GetDirectory( CWnd *pParent,LPCSTR lpszRoot,LPCSTR lpszTitle )
{
    CString str;
    BROWSEINFO bi;
    bi.hwndOwner = pParent -> m_hWnd;        // owner of created dialog box
    bi.pidlRoot = 0;                         // unused
    bi.pszDisplayName = 0;                   // buffer to receive name displayed by
                                             //     folder (not a valid path)
    bi.lpszTitle = lpszTitle;                // title is "Browse for Folder".
                                             //     this is an instruction
    bi.lpfn = BrowseCallbackProc;            // callback routine called when dialog
                                             //     has been initialized
    bi.lParam = 0;                           // passed to callback routine
    bi.ulFlags =
        BIF_RETURNONLYFSDIRS |               // only allow user to select
                                             //     a directory
        BIF_STATUSTEXT |                     // create status text field we will be
                                             //     writing to in callback
        // BIF_BROWSEFORCOMPUTER|            // only allow user to select a computer
        // BIF_BROWSEFORPRINTER |            // only allow user to select a printer
        // BIF_BROWSEINCLUDEFILES|           // displays files too which user
                                             //     can pick
        // BIF_DONTGOBELOWDOMAIN|            // when user is exploring the "Entire
                                             //     Network" they are not allowed
                                             //     into any domain
        0;
    m_sRootDir = lpszRoot;
```

Example 25   Getting a Directory Name   **311**

```
    LPITEMIDLIST lpItemId = ::SHBrowseForFolder( &bi );
    if ( lpItemId )
    {
        LPTSTR szBuf = str.GetBuffer( MAX_PATH );
        ::SHGetPathFromIDList( lpItemId, szBuf );
        ::GlobalFree( lpItemId );
        str.ReleaseBuffer();
    }

    return str;
}

int CALLBACK BrowseCallbackProc( HWND hwnd,UINT msg,LPARAM lp, LPARAM pData )
{
    TCHAR buf[MAX_PATH];

    switch(msg)
    {
        // when dialog is first initialized, change directory to one chosen above
        case BFFM_INITIALIZED:
            strcpy( buf,CWzdDirDlg::m_sRootDir );
            ::SendMessage( hwnd,BFFM_SETSELECTION,TRUE,( LPARAM )buf );
            break;

        // if you picked BIF_STATUSTEXT above, you can fill status here
        case BFFM_SELCHANGED:
        if ( ::SHGetPathFromIDList( ( LPITEMIDLIST ) lp ,buf ) )
            SendMessage( hwnd,BFFM_SETSTATUSTEXT,0,( LPARAM )buf );
        break;
    }
    return 0;
}
```

II

7

Example 26   Using a Child Dialog Box Like a Common Control

Objective

You would like to use a dialog box as if it were a common control in another dialog box, as seen in Figure 7.5.

Figure 7.5   A Child Dialog Box Within Another Dialog Box

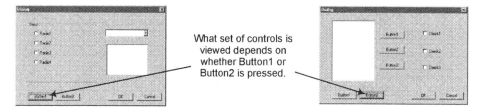

What set of controls is viewed depends on whether Button1 or Button2 is pressed.

Strategy

Considering that a dialog box is more typically used all by itself as a popup window, it's hard to remember that it can also be used as a child window of some other parent window—even another dialog box. We will, therefore, simply create a dialog box as a child window of another dialog box.

Steps

Add a Child Dialog Box to a Popup Dialog Box

1. Use the Dialog Editor to create a dialog template for each child dialog box you desire. Give them each a property of Child and no border.
2. Use the ClassWizard to create a dialog class for each of these templates.
3. Embed each of these child dialog classes in a parent dialog class.

```
CWzdDlg1 m_wzd1;
CWzdDlg2 m_wzd2;
    :    :
```

Example 26   Using a Child Dialog Box Like a Common Control   **313**

4. Use the Dialog Editor to add a static control to the parent dialog's template with the size and position at which you want the child dialogs to appear. This will act as a place holder.

5. Use the ClassWizard to add a `WM_INITDIALOG` message handler to the parent dialog and get the rectangle of this static control.

```
CRect rect;
m_ctrlStatic.GetWindowRect( &rect );
ScreenToClient( &rect );
```

6. In the same handler, create the child dialog boxes and position them all over the same static control, making sure only one is visible.

```
m_dlg1.Create( IDD_DIALOG2,this );
m_dlg1.ShowWindow( SW_SHOW );
m_dlg1.MoveWindow( &rect );

m_dlg2.Create( IDD_DIALOG3,this );
m_dlg2.ShowWindow( SW_HIDE );
m_dlg2.MoveWindow( &rect );
```

7. Based on some other control, such as a button or tab control, change which child dialog is visible.

8. Process commands from these child dialogs in their own dialog classes.

Notes

- You could use the tab control to control which dialog box is visible. However, you would be better served to use the `CPropertySheet` and `CPropertyPage` classes, instead. These classes provide other functionality you would be reinventing by using this example. In fact, you should only use this example only if some, but not all, of the controls in a dialog box will be conditionally changed.

- The AppWizard actually uses this approach to display its pages, rather than a Property Sheet with the wizard option.

CD Notes

- When executing the project on the accompanying CD, click on the "Test" and then "Wzd" menu commands to open a dialog box. Then,

II

7

click on either "Button 1" or "Button 2" to display a particular set of controls.

Example 27   Using a Child Property Sheet Like a Common Control

Objective

You would like to use a property sheet just like a common control in a dialog box, as seen in Figure 7.6.

Figure 7.6   A Property Sheet Used as a Dialog Control

This property sheet has been added just like any other control to this dialog box.

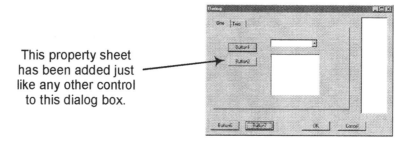

Strategy

Although the Dialog Editor allows you to add a tab control to a dialog template, the control it creates will be devoid of any functionality, other than to tell you what tab had been pressed. It's then up to you to display a child dialog box, as seen in the previous example. Rather than following this approach, we will use a Property Sheet control with all the other time-saving functionality it affords. We will create this Property Sheet using `CPropertySheet`. We will create the Property Sheet modeless and then position it over our dialog box.

Example 27   Using a Child Property Sheet Like a Common Control   **315**

Steps

Add a Property Sheet to a Dialog Box

1. Use the Dialog Editor to create a dialog template for each property sheet page you desire. Give each template a property of Child and no border.
2. Use the ClassWizard to create a property page class for each template you created, each derived from `CPropertyPage`
3. Use the ClassWizard to create your own `CPropertySheet` class. Then add a `WM_INITDIALOG` message handler where you will insert the following:

   ```
   ModifyStyleEx(0,WS_EX_CONTROLPARENT);
   ```

   Note: This prevents an infinite loop from occuring in the `CDialog` class when it attempts to locate the next control to tab to when using this dialog.
4. Embed this class in the dialog class you plan to add this property sheet to.
5. Use the Dialog Editor to add a static control to the dialog box's template with the size and position at which you want the property sheet to appear.
6. Use the ClassWizard to add a `WM_INITDIALOG` message handler to the dialog class and in that handler get the size and position of this static control.

   ```
   CRect rect;
   m_ctrlStatic.GetWindowRect( &rect );
   ScreenToClient( &rect );
   ```

7. In the same `OnInitDialog` function, create and add each page to the property sheet.

   ```
   m_pFirstPage = new CFirstPage;
   m_pSecondPage = new CSecondPage;
   m_sheet.AddPage( m_pFirstPage );
   m_sheet.AddPage( m_pSecondPage );
   ```

8. Finally in this same funtion, create the property sheet and position it over the static control.

   ```
   m_sheet.Create( this,WS_VISIBLE|WS_CHILD );
   m_sheet.MoveWindow( &rect );
   ```

Notes

- The modeless form of a Property Sheet wizard doesn't have a close or minimize button in the title bar. A simple way to add these buttons would be to create a dialog box with close and minimize buttons and embed a Property Sheet control in it with the Wizard style using this example.

CD Notes

- When executing the project on the accompanying CD, click on the "Test" and then "Wzd" menu to open a dialog box that contains, among the other controls, a Property Sheet control.

Chapter 8

Control Windows

Control windows are the buttons, list boxes, and scroll bars that allow a user to interact with your application. When a dialog box is created, the control windows you defined using the Dialog Editor are created for you. However, sometimes a control just doesn't look the way you want it to or it doesn't appear where you want it to appear.

Example 28   Drawing Your Own Control   We will look at how to draw our own control, while still making use of the other overhead associated with a control.

Example 29   Adding a Button to a Window Caption   We will add a button to the caption of a window.

Example 30   Adding a Hot Key Control   We will add a hot key to our application.

Example 28 Drawing Your Own Control

Objective

You would like to draw a control window yourself. In this example, we draw a list box so that the selected line is indented, as seen in Figure 8.1.

Figure 8.1 A Self-Drawn Listbox

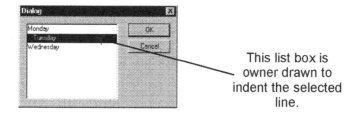

This list box is owner drawn to indent the selected line.

Strategy

First, we create our control with an owner-drawn window style. Then, we create our own control class deriving it from the MFC control class. We override the DrawItem() member function and draw the control ourselves using MFC's CDC class and its drawing functions. In this example, we will be drawing a list box control window.

Steps

Create a New List Box Class

1. Use the ClassWizard to create a new class derived from CListBox.

Draw the List Box Control

1. Use the ClassWizard to override the DrawItem() member function of this class.

Example 28   Drawing Your Own Control   **319**

2. We start `DrawItem()` by wrapping the device context handle we get from the `DRAWITEMSTRUCT` structure in a `CDC` class object. We also wrap the size of our drawing area in a `CRect` class object.

```
void CWzdListBox::DrawItem( LPDRAWITEMSTRUCT lpDIS )
{
    // get our device context and rectangle to draw to
    CDC dc;
    dc.Attach( lpDIS -> hDC );
    CRect rect( lpDIS -> rcItem );
```

How we proceed from here depends on how we want to draw this control. The sky's the limit, as long as we stay within `rect`. For this example, we will simply use `CDC::DrawText()` to draw the items in our list box as usual. However, if an item is selected by the mouse, we will draw that item using the system's highlight colors. We will also indent this line by 10 pixels.

3. Determine and set the foreground and background colors of this line based on whether or not it's selected.

```
// if our item is selected, then set colors accordingly
//      and fill in background
COLORREF bk = dc.GetBkColor();
COLORREF fg = dc.GetTextColor();
if ( lpDIS -> itemState & ODS_SELECTED )
{
    bk = ::GetSysColor( COLOR_HIGHLIGHT );
    fg = ::GetSysColor( COLOR_HIGHLIGHTTEXT );
}
dc.SetTextColor( fg );
```

4. Now, draw the background for this list item.

```
CBrush brush( bk );
dc.FillRect( &rect, &brush );
```

5. Next, draw this list item's text using `CDC::DrawText()`. If this line is selected, we indent the start of the text by 10 pixels.

```
if ( lpDIS -> itemState & ODS_SELECTED )
    rect.left += 10;
int nBkMode = dc.SetBkMode( TRANSPARENT );

CString str;
GetText( lpDIS -> itemID,str );
dc.DrawText( str, &rect, DT_LEFT|DT_VCENTER );
```

Notice that we get the text to draw from the list box control itself. Normally, a list box doesn't maintain this list for an owner-drawn control. However, since we will be setting the "Has Strings" property for this control in the Dialog Editor, the listbox will continue to maintain this list internally. This allows us to continue to be able to use this control class's `AddString()` function.

6. The last thing we do in `DrawItem()` is cleanup.

```
dc.SetTextColor( fg );
dc.SetBkMode( nBkMode );
dc.Detach();
```

7. For a complete listing of this list box control class, please refer to the List Box Class for Example 28 (page 322).

Use the New List Box Class

1. To make use of this new control class, first use the Dialog Editor to add a list box control to a dialog template. Make sure to set the "Owner-Drawn Fixed" and "Has Strings" properties.

2. If you haven't already done so, use the ClassWizard to create a dialog class for this template.

3. Use the ClassWizard to add a control member variable to your dialog class. Then, substitute your new class name for the `CListBox` class name that the ClassWizard adds automatically. You should also pull this definition out of the {{}} brackets so that the ClassWizard can't mess with it. This list box will now be subclassed by your new list box class the next time that `UpdateData(FALSE)` is called.

Example 28   Drawing Your Own Control   **321**

Notes

- Along with list box control windows, the other controls that can be owner-drawn include: combo boxes, extended combo boxes, list controls, buttons, header controls, and tab controls. Both list boxes and combo boxes allow you to also set the size of each entry by setting the "Owner-Drawn Variable" style using the Dialog Editor.

- When using the "Owner-Drawn Variable" style, you must also override the `MeasureItem()` member function of your new control class and return the size you want each entry to be. Please see Example 13 for an example of overriding the `MeasureItem()` function.

- If you have also set the "Sort" style for this control using the Dialog Editor, you must also override the `SortItem()` member function of your new control class.

- Actually, owner-drawn is somewhat of a misnomer here. Technically, this is a self-drawn control. Owner-drawn originally meant that the owner of a control, such as a dialog box, would draw that control. However, in this example, the message sent by the control to tell the owner that it's time to draw (`WM_DRAWITEM`) is being reflected back to the control's MFC class, where it's processed. So, technically speaking, this control is being drawn by the MFC class that owns the control. However, since these two are considered one entity in the MFC world, the control should be thought of as drawing itself, or self-drawn. For more on message reflection, please see Chapter 1.

- Why bother with the owner-drawn style of a control when you could just write a `WM_PAINT` message handler to draw the control? Because you want to synchronize your drawing with the other functionality of a control. For example, an owner-drawn combo box will tell you when to draw its list box. More importantly, it will tell you where to draw the list box so that mouse clicks on that part of the screen will relate to the proper entry of that control. Even an owner-drawn button control tells you when to draw a depressed or a nondepressed button to coincide with the command message it generates when clicked.

CD Notes

- When executing the project on the accompanying CD, click on the "Test" and then "Wzd" menu commands to open a dialog box with a list box. Notice that clicking on any line in the list box causes the text in that line to be indented.

List Box Class for Example 28

```
#if !defined( AFX_WZDLISTBOX_H__41500055_E450_11D1_9B7D_00AA003D8695__INCLUDED_ )
#define AFX_WZDLISTBOX_H__41500055_E450_11D1_9B7D_00AA003D8695__INCLUDED_

#if _MSC_VER >= 1000
#pragma once
#endif    // _MSC_VER >= 1000
// WzdListBox.h : header file
//

/////////////////////////////////////////////////////////////////////////////
// CWzdListBox window

class CWzdListBox : public CListBox
{
// Construction
public:
    CWzdListBox();

// Attributes
public:

// Operations
public:

// Overrides
    // ClassWizard generated virtual function overrides
    // {{AFX_VIRTUAL( CWzdListBox )
public:
    virtual void DrawItem( LPDRAWITEMSTRUCT lpDrawItemStruct );
    // }}AFX_VIRTUAL

// Implementation
public:
    virtual ~CWzdListBox();

// Generated message map functions
protected:
    // {{AFX_MSG( CWzdListBox )
    // }}AFX_MSG
```

Example 28   Drawing Your Own Control   **323**

```
    DECLARE_MESSAGE_MAP()
};


/////////////////////////////////////////////////////////////////////////////


// {{AFX_INSERT_LOCATION}}
// Microsoft Developer Studio will insert additional declarations immediately
//      before the previous line.

#endif
    // !defined( AFX_WZDLISTBOX_H__41500055_E450_11D1_9B7D_00AA003D8695__INCLUDED_ )
// WzdListBox.cpp : implementation file
//

#include "stdafx.h"
#include "wzd.h"
#include "WzdListBox.h"

#ifdef _DEBUG
#define new DEBUG_NEW
#undef THIS_FILE
static char THIS_FILE[] = __FILE__;
#endif

/////////////////////////////////////////////////////////////////////////////
// CWzdListBox

CWzdListBox::CWzdListBox()
{
}

CWzdListBox::~CWzdListBox()
{
}


BEGIN_MESSAGE_MAP( CWzdListBox, CListBox )
    // {{AFX_MSG_MAP( CWzdListBox )
    // }}AFX_MSG_MAP
END_MESSAGE_MAP()
```

```cpp
/////////////////////////////////////////////////////////////////////////////
// CWzdListBox message handlers

void CWzdListBox::DrawItem( LPDRAWITEMSTRUCT lpDIS )
{
    // get our device context and rectangle to draw to
    CDC dc;
    dc.Attach( lpDIS -> hDC );
    CRect rect( lpDIS -> rcItem );

    // if our item is selected, then set colors accordingly
    //     and fill in background
    COLORREF bk = dc.GetBkColor();
    COLORREF fg = dc.GetTextColor();
    if ( lpDIS -> itemState & ODS_SELECTED )
    {
        bk = ::GetSysColor( COLOR_HIGHLIGHT );
        fg = ::GetSysColor( COLOR_HIGHLIGHTTEXT );
    }
    dc.SetTextColor( fg );
    CBrush brush( bk );
    dc.FillRect( &rect, &brush );

    // draw text
    if ( lpDIS -> itemState & ODS_SELECTED )
        rect.left += 10;
    int nBkMode = dc.SetBkMode( TRANSPARENT );

    CString str;
    GetText( lpDIS -> itemID,str) ;
    dc.DrawText(str, &rect, DT_LEFT|DT_VCENTER);

    // cleanup
    dc.SetTextColor( fg );
    dc.SetBkMode( nBkMode );
    dc.Detach();
}
```

Example 29   Adding a Button to a Window Caption   **325**

Example 29   Adding a Button to a Window Caption

Objective

You would like to put a button in the caption bar of a window, as seen in Figure 8.2.

Figure 8.2   A Button in the Caption Bar

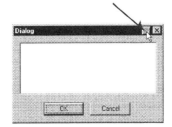

Put a button in the caption of your dialog box.

Strategy

Technically speaking, the button we are about to create is not a control window at all. Instead, we will be creating a "pseudo" button, one which we will draw ourselves and for which we will process mouse messages. The system draws a window caption in response to two messages (WM_NCPAINT and WM_ACTIVATE) and we will draw our button using each of these messages and the drawing capabilities of the CDC class. We will also process another message (WM_NCLBUTTONDOWN) to determine whether the user has clicked on our "button".

Steps

Set Up the Bitmaps for Your Caption Button

1. Use the Bitmap Editor to create two bitmaps: one for the button when it's up and one for when its down. You accomplish this effect by playing

with the highlight and shadow colors in the border of the bitmap. Use gray for the background color—you will be substituting this color when loading these bitmaps. For a normal-sized button to be used in a regular window caption, make your bitmap is 16 by 16 pixels. For the smaller button size required for a tool window, make your bitmap 12 by 12 pixels.

2. In the constructor of your dialog class, load these two bitmaps into embedded CBitmap member variables. Use the CWzdBitmap provided on the CD with this example so that you can substitute the gray background color for the current system color for a button. Also, initialize the state of the button as unpressed.

```
CWzdDialog::CWzdDialog( CWnd* pParent /* = NULL*/ )
    : CDialog( CWzdDialog::IDD, pParent )
{

    // {{AFX_DATA_INIT( CWzdDialog )
    // }}AFX_DATA_INIT
    m_bPressed = FALSE;
    m_bitmapPressed.LoadBitmapEx( IDB_PRESSED_BITMAP,TRUE );
    m_bitmapUnpressed.LoadBitmapEx( IDB_UNPRESSED_BITMAP,TRUE );

}
```

Note: This bitmap class is an example in my previous book, *Visual C++ MFC Programming by Example*.

Draw the Caption Button

1. Use the ClassWizard to add a WM_NCPAINT and WM_ACTIVATE message handler to this class. From both of these message handlers, call the DrawButton() function, which will draw your button. We will write the DrawButton() function next.

```
void CWzdDialog::OnNcPaint()
{

    // draw caption first
    CDialog::OnNcPaint();

    // then draw button on top
```

Example 29   Adding a Button to a Window Caption   **327**

```
    DrawButton();
}

void CWzdDialog::OnActivate( UINT nState, CWnd* pWndOther,
    BOOL bMinimized )
{
    CDialog::OnActivate( nState, pWndOther, bMinimized );

    DrawButton();
}
```

2. Create the DrawButton() function. It should determine which button bit-map to draw and then draw it using StretchBlt(), as seen here.

```
void CWzdDialog::DrawButton()
{
    // if window isn't visible or is minimized, skip
    if ( !IsWindowVisible() || IsIconic() )
        return;

    // get appropriate bitmap
    CDC memDC;
    CDC* pDC = GetWindowDC();
    memDC.CreateCompatibleDC( pDC );
    memDC.SelectObject( m_bPressed ? &m_bitmapPressed :
        &m_bitmapUnpressed );

    // get button rect and convert into non-client area coordinates
    CRect rect, rectWnd;
    GetButtonRect(rect);
    GetWindowRect(rectWnd);
    rect.OffsetRect( -rectWnd.left, -rectWnd.top );

    // draw it
    pDC -> StretchBlt( rect.left, rect.top, rect.Width(),
        rect.Height(), &memDC, 0, 0, m_bitmapPressed.m_Width,
        m_bitmapPressed.m_Height, SRCCOPY );
```

```
    memDC.DeleteDC();
    ReleaseDC( pDC );
}
```

Notice that we get the device context from `GetWindowDC()` so that our context will draw to the entire window, including the nonclient area where our caption is. We use `StretchBlt()` because we won't be sure on which resolution our dialog box will be drawn. We use another local helper function, called `GetButtonRect()`, to determine the dimensions of our button, which we will create next.

3. Create the `GetButtonRect()` function to determine the dimensions of your button. We will be relying extensively on the system to tell us the appropriate dimensions for our button. Use the `GetSystemMetrics()` function to get those dimensions.

```
void CWzdDialog::GetButtonRect( CRect& rect )
{
    GetWindowRect(&rect);

    // for small caption use SM_CYDLGFRAME
    rect.top += GetSystemMetrics(SM_CYFRAME)+1;

    // for small caption use SM_CYSMSIZE
    rect.bottom = rect.top + GetSystemMetrics( SM_CYSIZE )-4;

    // for small caption use SM_CXDLGFRAME
    rect.left = rect.right - GetSystemMetrics( SM_CXFRAME ) -
        ( 2 * // set to number of buttons already in caption + 1
        GetSystemMetrics( SM_CXSIZE ) )-1;
        // for small caption use SM_CXSMSIZE

    // for small caption use SM_CXSMSIZE
    rect.right = rect.left + GetSystemMetrics(SM_CXSIZE)-3;
}
```

Example 29   Adding a Button to a Window Caption   **329**

Allow the User to Click the Button

1. To determine if the user has clicked on our "button", use the ClassWizard to add a `WM_NCLBUTTONDOWN` message handler to this class. There, we will check to see if the user has clicked in the caption and, if so, if they have clicked where our button is. Notice that we use the `GetButtonRect()` function again here.

```
void CWzdDialog::OnNcLButtonDown( UINT nHitTest, CPoint point )
{
    if ( nHitTest == HTCAPTION )
    {
        // see if in area we reserved for button
        CRect rect;
        GetButtonRect( rect );
        if ( rect.PtInRect( point ) )
        {
            m_bPressed = !m_bPressed;
            DrawButton();
        }
    }
    CDialog::OnNcLButtonDown( nHitTest, point );
}
```

II

8

For a complete listing of this dialog class, please refer to the Dialog Class for Example 29 (page 330).

Notes

- Not only do we need to process the `WM_NCPAINT` message to draw our button, we also need to process the `WM_ACTIVATE` message. The `WM_ACTIVATE` message causes the system to redraw the color of the caption whenever the window is activated (the window caption goes from a dull color to a bright color when it's active).

- Make sure you draw your "button" control after the system has drawn the caption. All we are doing here is drawing a bitmap on top of what the system draws.

- This example demonstrates a push button that remains pushed. To toggle the button as the user clicks it, you should always draw the "down" button bitmap when processing the WM_LBUTTONDOWN message and then add a new WM_NCLBUTTONUP message handler to draw the "up" button bitmap.

- This same approach can be applied to any type of control as long as you don't mind supplying the logic.

CD Notes

- When executing the project on the accompanying CD, click on the "Test" and then "Wzd" menu commands to open a dialog box that has a new button in the caption. Clicking on that button causes it to remain depressed or undepressed.

Dialog Class for Example 29

```
#if !defined( AFX_WZDDIALOG_H__A76FC804_D3BD_11D1_9B67_00AA003D8695__INCLUDED_ )
#define AFX_WZDDIALOG_H__A76FC804_D3BD_11D1_9B67_00AA003D8695__INCLUDED_

#if _MSC_VER >= 1000
#pragma once
#endif    // _MSC_VER >= 1000
// WzdDialog.h : header file
//

#include "WzdBitmap.h"


/////////////////////////////////////////////////////////////////////////////
// CWzdDialog dialog

class CWzdDialog : public CDialog
{
// Construction
public:
    CWzdDialog( CWnd* pParent = NULL );    // standard constructor

// Dialog Data
    // {{AFX_DATA( CWzdDialog )
    enum { IDD = IDD_WZD_DIALOG };
    // NOTE: the ClassWizard will add data members here
    // }}AFX_DATA
```

Example 29   Adding a Button to a Window Caption   **331**

```
// Overrides
    // ClassWizard generated virtual function overrides
    // {{AFX_VIRTUAL( CWzdDialog )
protected:
    virtual void DoDataExchange( CDataExchange* pDX );    // DDX/DDV support
    // }}AFX_VIRTUAL


// Implementation
protected:

    // Generated message map functions
    // {{AFX_MSG( CWzdDialog )
    afx_msg void OnNcPaint();
    afx_msg void OnNcLButtonDown( UINT nHitTest, CPoint point );
    afx_msg void OnActivate( UINT nState, CWnd* pWndOther, BOOL bMinimized );
    // }}AFX_MSG
    DECLARE_MESSAGE_MAP()
private:
    BOOL m_bPressed;
    CWzdBitmap m_bitmapPressed;
    CWzdBitmap m_bitmapUnpressed;

    void DrawButton();
    void GetButtonRect( CRect &rect );
};


// {{AFX_INSERT_LOCATION}}
// Microsoft Developer Studio will insert additional declarations immediately
//    before the previous line.

#endif
    // !defined( AFX_WZDDIALOG_H__A76FC804_D3BD_11D1_9B67_00AA003D8695__INCLUDED_ )
// WzdDialog.cpp : implementation file
//

#include "stdafx.h"
#include "wzd.h"
#include "WzdDialog.h"
```

```
#ifdef _DEBUG
#define new DEBUG_NEW
#undef THIS_FILE
static char THIS_FILE[] = __FILE__;
#endif


#define HTBUTTON 32
/////////////////////////////////////////////////////////////////////////////
// CWzdDialog dialog


CWzdDialog::CWzdDialog( CWnd* pParent /* = NULL*/ )
    : CDialog( CWzdDialog::IDD, pParent )
{
    // {{AFX_DATA_INIT( CWzdDialog)
    // NOTE: the ClassWizard will add member initialization here
    // }}AFX_DATA_INIT
    m_bPressed = FALSE;
    m_bitmapPressed.LoadBitmapEx( IDB_PRESSED_BITMAP,TRUE );
    m_bitmapUnpressed.LoadBitmapEx( IDB_UNPRESSED_BITMAP,TRUE );
}


void CWzdDialog::DoDataExchange( CDataExchange* pDX )
{
    CDialog::DoDataExchange( pDX );
    // {{AFX_DATA_MAP( CWzdDialog )
    // NOTE: the ClassWizard will add DDX and DDV calls here
    // }}AFX_DATA_MAP
}


BEGIN_MESSAGE_MAP( CWzdDialog, CDialog )
    // {{AFX_MSG_MAP( CWzdDialog )
    ON_WM_NCPAINT()
    ON_WM_NCLBUTTONDOWN()
    ON_WM_ACTIVATE()
    // }}AFX_MSG_MAP
END_MESSAGE_MAP()
```

Example 29   Adding a Button to a Window Caption   **333**

```
///////////////////////////////////////////////////////////////////////
// CWzdDialog message handlers


void CWzdDialog::OnNcPaint()
{
    // draw caption first
    CDialog::OnNcPaint();

    // then draw button on top
    DrawButton();
}

// caption is also redrawn with WM_ACTIVATE message for color change
void CWzdDialog::OnActivate( UINT nState, CWnd* pWndOther, BOOL bMinimized )
{
    CDialog::OnActivate( nState, pWndOther, bMinimized );

    DrawButton();
}

void CWzdDialog::OnNcLButtonDown( UINT nHitTest, CPoint point )
{
    if ( nHitTest == HTCAPTION )
    {
        // see if in area we reserved for button
        CRect rect;
        GetButtonRect( rect );
        if ( rect.PtInRect( point ) )
        {
            m_bPressed = !m_bPressed;
            DrawButton();
        }
    }
    CDialog::OnNcLButtonDown( nHitTest, point );
}

void CWzdDialog::GetButtonRect( CRect& rect )
{
```

```
    GetWindowRect( &rect );

    // for small caption use SM_CYDLGFRAME
    rect.top += GetSystemMetrics( SM_CYFRAME )+1;

    // for small caption use SM_CYSMSIZE
    rect.bottom = rect.top + GetSystemMetrics( SM_CYSIZE )-4;

    // for small caption use SM_CXDLGFRAME
    rect.left = rect.right - GetSystemMetrics( SM_CXFRAME ) -
        // set to number of buttons already in caption + 1
        ( 2 *
        // for small caption use SM_CXSMSIZE
        GetSystemMetrics( SM_CXSIZE ) )-1;

    // for small caption use SM_CXSMSIZE
    rect.right = rect.left + GetSystemMetrics( SM_CXSIZE )-3;
}


void CWzdDialog::DrawButton()
{
    // if window isn't visible or is minimized, skip
    if ( !IsWindowVisible() || IsIconic() )
        return;

    // get appropriate bitmap
    CDC memDC;
    CDC* pDC = GetWindowDC();
    memDC.CreateCompatibleDC( pDC );
    memDC.SelectObject( m_bPressed ? &m_bitmapPressed : &m_bitmapUnpressed );

    // get button rect and convert into non-client area coordinates
    CRect rect, rectWnd;
    GetButtonRect( rect );
    GetWindowRect( rectWnd );
    rect.OffsetRect( -rectWnd.left, -rectWnd.top );

    // draw it
    pDC -> StretchBlt( rect.left, rect.top, rect.Width(), rect.Height(), &memDC,
        0, 0, m_bitmapPressed.m_Width, m_bitmapPressed.m_Height, SRCCOPY );
```

Example 30   Adding a Hot Key Control   **335**

```
    memDC.DeleteDC();
    ReleaseDC( pDC );
}
```

Example 30   Adding a Hot Key Control

Objective

You would like to add a hot key to your application.

Strategy

Hot keys are a little-known ability for your user to communicate with your application when it isn't the active desktop application. When the user presses a hot key, which is usually a combination of Shift, Control, Alt, and alphanumeric keys, your application can either be made the active application or it can receive a WM_HOTKEY message to process.

The Dialog Editor allows us to place a hot key control in a dialog template, but this really isn't a hot key control. Instead, it's more of a hot key edit box that converts any combination of keys (Shift/alt/control) into their text equivalent (e.g., pressing the Shift and F keys when this control has focus causes "Shift+F" to appear in the control). The output of this control, however, can be used directly in the Windows API function that registers a hot key.

To register a hot key with the system so that our application is made active when a hot key combination is pressed, we will use WM_SETHOTKEY. To cause a WM_HOTKEY message to be sent to our application when a hot key combination is pressed, we will use ::RegisterHotKey().

Steps

Prompt the User for a Hot Key

1. Use the Dialog Editor to add a hot key edit box control to your dialog template.
2. Use the ClassWizard to add a hot key control member variable to this dialog template's dialog class.

3. Use the following to get the hot key key codes from the hot key control.

```
WORD m_wVkCode;
WORD m_wModifier;


m_ctrlHotKey.GetHotKey( m_wVkCode,m_wModifier );
```

Register the Hot Key to Activate Your Application

1. Send a WM_SETHOTKEY message to the main window with the hot key values returned previously.

```
AfxGetMainWnd() -> SendMessage(WM_SETHOTKEY,
    ( WPARAM )MAKEWORD( m_wVkCode,m_wModifier ) );
```

Register the Hot Key to Send WM_HOTKEY

1. Using the output of the hot key edit box control with RegisterHotKey() requires that you first translate the key modifier returned by GetHotKey() into a value that RegisterHotKey() can use.

```
// must translate hot key control's returned modifier to
//      RegisterHotKey format
UINT mod = 0;
if ( m_wModifier&HOTKEYF_ALT ) mod| = MOD_ALT;
if ( m_wModifier&HOTKEYF_CONTROL ) mod| = MOD_CONTROL;
if ( m_wModifier&HOTKEYF_SHIFT ) mod| = MOD_SHIFT;
if ( m_wModifier&HOTKEYF_EXT ) mod| = MOD_WIN;
m_wModifier = mod;
```

2. Register the hot key.

```
::RegisterHotKey(
    AfxGetMainWnd() -> m_hWnd,     // window to receive hot-key
                                   //      notification
    1234,                          // identifier of hot key
    m_wModifier,                   // key-modifier flags
    m_wVkCode                      // virtual-key code
);
```

Example 30   Adding a Hot Key Control   **337**

3. Process `WM_HOTKEY` messages in the appropriate nonchild window (Main Frame and Child Frame windows are typically the only eligible windows) by manually adding

```
// to the message map
ON_MESSAGE( WM_HOTKEY,OnHotKey )

LRESULT CMainFrame::OnHotKey( WPARAM wParam,LPARAM lParam )
{
    wParam;     // id specified in RegisterHotKey()
    // (in this example: 1234)
    lParam;     // virtual key code and modifiers of the hotkey
    return 0L;
}
```

Notes

- You must be careful what you allow your user to enter as a hot key, because that key combination will take precedence over any shortcut keys any other application might have. For example, if your user specified Alt+S for a hot key, even though this combination might mean save for the application with which they're currently working, it will instead either activate your application or send it a `WM_HOTKEY` message.

- The difference between hot keys and accelerator keys is that hot keys will interact with your application even when it's not active.

CD Notes

- When executing the project on the accompanying CD, click on the "Test" and then "Wzd" menu commands to open a dialog box that allows a hot key combination to be entered. Click on "Make Active Application" and "OK". Then, activate another application and reenter your hot key combination. The first application should again be activated. Click on "Send `WM_HOTKEY` Message" and place a break point on the `OnHotKey()` function in `CMainFrame`. Then, press the hot key combination and watch as the application breaks at this point.

Chapter 9

Drawing

Bitmaps and icons allow you to add color and style to your application. Because all Windows interfaces are essentially alike, logos and splash screens are really your only way to distinguish the look of your application from someone else's. Obviously, drawing is also important for creating your own controls and displaying figures in a CAD application. You can use the examples in this chapter to give your application some distinguishing characteristics.

Example 31   Using Nondiffused Colors   We will look at how to draw in nondiffused colors. The sad story about diffused colors can be found in Chapter 1. However, the conclusion is that for graphic applications, your application will need to draw in nondiffused colors.

Example 32   Stretching a Bitmap   We will stretch an existing bitmap to fill a particular spot.

Example 33   Capturing the Screen   We will look at loading a bitmap with the image from an area of the screen.

Example 34 Outputting a DIB Bitmap File We will output a bitmap to a device independent bitmap (DIB) file.

Example 31 Using Nondiffused Colors

Objective

You would like to draw in a nondiffused color. You would also like the user to be able to set and save their own nondiffused color preferences, as seen in Figure 9.1.

Figure 9.1 Nondiffused Color Options

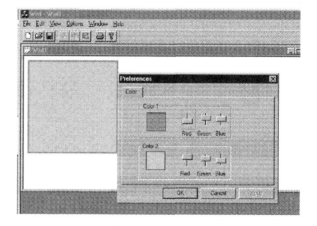

Strategy

To draw in nondiffused colors, our application will need to create, maintain and use its own palette. To do this, we will use MFC's CPalette class. To prompt our user for color preferences, we will create a new property page for our preferences. This property page will have a slider bar for each primary color (i.e., red, green, and blue), which will allow our user to pick a color by dragging a bar to the selected intensity. We will then use the AnimatePalette() member function of CPalette to dynamically change the selected color in the view. Finally, we will save our user's color selection using GetProfileBinary() and WriteProfileBinary().

Example 31   Using Nondiffused Colors   **341**

Steps

Create an Application Palette

1. Define an array of colors in your `CMainFrame` class and set them to some default colors in that class's constructor.

```
COLORREF m_rgbColors[NUM_COLORS];
:    :    :
CMainFrame::CMainFrame()
{
    m_rgbColors[COLOR1_COLOR] = RGB( 200,20,150 );
    m_rgbColors[COLOR2_COLOR] = RGB( 0,200,100 );
}
```

2. In `CMainFrame`, add a function that will create a palette from these colors. Create a `CPalette` object and then create the palette with `CPalette`'s `CreatePalette()` function. `CPalette::CreatePalette()` takes one argument, a `LOGPALETTE` structure containing all of the colors we want to add to the palette, which we initialize here.

```
void CMainFrame::CreatePalette()
{
    LOGPALETTE *lp = ( LOGPALETTE * )new BYTE[sizeof( LOGPALETTE ) +
        ( NUM_COLORS * sizeof( PALETTEENTRY ) )];
    lp -> palVersion = 0x300;
    lp -> palNumEntries = NUM_COLORS;
    for ( int i = 0; i<NUM_COLORS; i++ )
    {
        lp -> palPalEntry[i].peRed = GetRValue( m_rgbColors[i] );
        lp -> palPalEntry[i].peGreen = GetGValue( m_rgbColors[i] );
        lp -> palPalEntry[i].peBlue = GetBValue( m_rgbColors[i] );
        // reserve for animation
        lp -> palPalEntry[i].peFlags = PC_RESERVED;

    }
    if ( m_pPalette ) delete m_pPalette;
```

II

9

```
    m_pPalette = new CPalette;
    m_pPalette -> CreatePalette( lp );
    delete []lp;
}
```

3. Then, call this new `CMainFrame::CreatePalette()` function from the `OnCre-ate()` function of `CMainFrame` just after we load any program options from the system registry. As we will see, these program options will soon contain color options, too.

Draw with Your Application Palette

1. To draw using your new palette, you must first select the palette into your device context and then realize the colors in the palette. Realizing colors simply means Windows attempts to stick all of your application palette colors into the system palette, where it can be used.

```
// select palette into device context and realize colors
CPalette *pOPalette = pDC -> SelectPalette(
    ( ( CMainFrame* )AfxGetMainWnd() ) -> GetPalette(),FALSE );
pDC -> RealizePalette();
```

Notice we also added a wrapper function to `CMainFrame` called `GetPal-ette()` to allow other classes to access the application palette.

1. Whenever a nondiffused color is required for drawing, use the `PALETTEIN-DEX()` macro to select the color.

```
CPen pen( PS_SOLID,3,PALETTEINDEX( COLOR1_COLOR ) );
CPen *pOPen = pDC -> SelectObject( &pen );
CBrush brush( PALETTEINDEX( COLOR2_COLOR ) );
CBrush *pOBrush = pDC -> SelectObject( &brush );
```

The values `COLOR1_COLOR` and `COLOR2_COLOR` are simple indexes into our application palette.

2. When this pen and brush are used to draw lines and figures, the appropriate nondiffused color will be used.

Example 31   Using Nondiffused Colors   **343**

3. After you are finished drawing, reselect the old pen, brush, and palette.

```
pDC -> SelectObject( pOPen );
pDC -> SelectObject( pOBrush );
pDC -> SelectPalette( pOPalette,FALSE );
```

Create a Color Options Property Page

1. Use the Developer Studio to create a new dialog template. Give this template a style of child with a thin frame. The caption for this template will be what appears in the tab for this property page.

2. Use the Dialog Editor to add a picture control and three slider bars, as seen in Figure 9.1. You might also simply use the template on the CD accompanying this book.

3. Use the ClassWizard to create a new property page class for this template derived from CPropertyPage.

4. Add this property page to your application's preferences.

```
void CMainFrame::OnOptionsPreferences()
{
    CPropertySheet sheet( _T( "Preferences" ),this );
    m_pColorPage = new CColorPage;


    sheet.AddPage( m_pColorPage );


    m_pColorPage -> m_rgbColors[COLOR1_COLOR] =
        m_rgbColors[COLOR1_COLOR];
    m_pColorPage -> m_rgbColors[COLOR2_COLOR] =
        m_rgbColors[COLOR2_COLOR];


    sheet.DoModal();


    delete m_pColorPage;
}
```

We will skip over the details of how to pass your application's current color values to this page and how the slider bars change these values—please refer to the Main Frame Class for Example 31 (page 346) and the Colors Property Page for Example 31 (page 352). Suffice it to say that the slider bars create a new RGB value for our color, which we now need to put into the system palette.

5. Store the new RGB value in the system palette by using the `AnimatePalette()` function of `CPalette`, as seen here.

```
// select our palette into a device context
CDC *pDC = GetDC();
CMainFrame *pFrame = ( CMainFrame * )AfxGetMainWnd();
CPalette *pPalette = pFrame -> GetPalette();
CPalette *pOPalette = pDC -> SelectPalette( pPalette,FALSE );


// add new color to palette using animation
PALETTEENTRY  pentry;
pentry.peRed = GetRValue( m_rgbColors[i] );
pentry.peGreen = GetGValue( m_rgbColors[i] );
pentry.peBlue = GetBValue( m_rgbColors[i] );
pentry.peFlags = PC_RESERVED;
pPalette -> AnimatePalette( i,1,&pentry );


// reselect old palette
pDC -> SelectPalette( pOPalette,FALSE );
```

Notice that even though the `AnimatePalette()` function does not use a device context for an argument, *the palette `AnimatePalette()` works with must itself be selected into a device context* for it to work. In other words, `PPalette` above had to be selected into `pDC` in order for `AnimatePalette()` to work.

Save and Restore Color Preferences

1. To save your application's colors for the next time, you will need to add the following code to whatever function you currently use to save your user's preferences.

```
UINT size = sizeof( m_rgbColors );
AfxGetApp() -> WriteProfileBinary( "Settings","Colors",
    ( BYTE* )m_rgbColors,size);
```

Example 31   Using Nondiffused Colors   **345**

2. To restore your user's color preferences, you will need to add the following code to whatever function you currently use to restore your user's preferences.

```
BYTE *p;
UINT size;
if ( AfxGetApp() -> GetProfileBinary( "Settings","Colors", &p,&size ) )


{
    memcpy( m_rgbColors,p,size );
    delete []p;
}
```

Notes

- MFC application colors default to diffused colors, which means that several primary colors are clustered together to give the appearance of that color. For some applications, especially a CAD application, this is unacceptable because diffused colors tend to be fuzzy. For more on the difference between diffused and nondiffused colors, please refer to Chapter 1.

- You will notice when you run the example on the CD that not only does AnimatePalette() change the color of the rectangles in the Color Property Page, but it will also change the color in your view. This is because you are modifying the system palette directly. Each pixel in your view points to a color in the system palette—so if you change the color in the system palette, the pixel color changes, too.

- You might notice that the nondiffused colors in your application will become distorted when another application becomes active. This is because the other application has taken over the system palette and tossed out your colors. You can eliminate some of this distortion by handling two messages in CMainFrame: WM_UPDATEPALETTE and WM_ONQUERYNEWPALETTE. The recommended way of handling these functions is to use UpdateColor() to refresh your colors using the new system palette. However, UpdateColor() tends to be slow and make more color mistakes than it fixes. Most graphic applications simply ignore what their colors do when they're in the background or just tell their application to redraw itself using the new system palette by calling RedrawWindow() from the WM_QUERYNEWPALETTE message handler. This function causes the main window and all of its children windows, especially the views, to redraw themselves.

II

9

CD Notes

- When executing the project on the accompanying CD, you will notice the view is filled with a two-colored square. Click on the "Options" and then the "Preferences" menu commands to open the preferences property sheet. Drag any of the slider bars and notice that the color of the square changes incrementally to another nondiffused color. If you then exit and re-enter the application, you will also notice that the application remembers the last color you set.

Main Frame Class for Example 31

```cpp
// MainFrm.h : interface of the CMainFrame class
//
/////////////////////////////////////////////////////////////////////////////

#if !defined( AFX_MAINFRM_H__CA9038EA_B0DF_11D1_A18C_DCB3C85EBD34__INCLUDED_ )
#define AFX_MAINFRM_H__CA9038EA_B0DF_11D1_A18C_DCB3C85EBD34__INCLUDED_

#if _MSC_VER >= 1000
#pragma once
#endif    // _MSC_VER >= 1000

#include "colorpage.h"

class CMainFrame : public CMDIFrameWnd
{
    DECLARE_DYNAMIC(CMainFrame)
public:
    CMainFrame();

// Attributes
public:

// Operations
public:
    void LoadOptions();
    void SaveOptions();

    void SetColorRef( int id,COLORREF rgb ){m_rgbColors[id] = rgb;};
    CPalette *GetPalette(){return m_pPalette;};
```

Example 31   Using Nondiffused Colors   **347**

```
// Overrides
    // ClassWizard generated virtual function overrides
    // {{AFX_VIRTUAL( CMainFrame )
    virtual BOOL PreCreateWindow( CREATESTRUCT& cs );
    // }}AFX_VIRTUAL

// Implementation
public:
    virtual ~CMainFrame();
#ifdef _DEBUG
    virtual void AssertValid() const;
    virtual void Dump( CDumpContext& dc ) const;
#endif

protected:    // control bar embedded members
    CStatusBar    m_wndStatusBar;
    CToolBar      m_wndToolBar;

// Generated message map functions
protected:
    // {{AFX_MSG( CMainFrame )
    afx_msg int OnCreate( LPCREATESTRUCT lpCreateStruct );
    afx_msg void OnClose();
    afx_msg void OnOptionsPreferences();
    // }}AFX_MSG
    DECLARE_MESSAGE_MAP()
private:
    COLORREF m_rgbColors[NUM_COLORS];
    CColorPage *m_pColorPage;
    CPalette *m_pPalette;

    void CreatePalette();

};


/////////////////////////////////////////////////////////////////////////////

// {{AFX_INSERT_LOCATION}}
```

II

9

```
// Microsoft Developer Studio will insert additional declarations immediately
//     before the previous line.

#endif
    // !defined( AFX_MAINFRM_H__CA9038EA_B0DF_11D1_A18C_DCB3C85EBD34__INCLUDED_ )
// MainFrm.cpp : implementation of the CMainFrame class
//

#include "stdafx.h"
#include "Wzd.h"
#include "WzdProject.h"

#include "MainFrm.h"

#include <afxpriv.h>

#ifdef _DEBUG
#define new DEBUG_NEW
#undef THIS_FILE
static char THIS_FILE[] = __FILE__;
#endif


/////////////////////////////////////////////////////////////////////////////
// CMainFrame

IMPLEMENT_DYNAMIC( CMainFrame, CMDIFrameWnd )

BEGIN_MESSAGE_MAP( CMainFrame, CMDIFrameWnd )
    // {{AFX_MSG_MAP( CMainFrame )
    ON_WM_CREATE()
    ON_WM_CLOSE()
    ON_COMMAND( ID_OPTIONS_PREFERENCES, OnOptionsPreferences )
    // }}AFX_MSG_MAP
END_MESSAGE_MAP()

static UINT indicators[] =
{
    ID_SEPARATOR,          // status line indicator
    ID_INDICATOR_CAPS,
    ID_INDICATOR_NUM,
```

Example 31   Using Nondiffused Colors   **349**

```
    ID_INDICATOR_SCRL,
};


///////////////////////////////////////////////////////////////////////////
// CMainFrame construction/destruction

CMainFrame::CMainFrame()
{
    m_rgbColors[COLOR1_COLOR] = RGB( 200,20,150 );
    m_rgbColors[COLOR2_COLOR] = RGB( 0,200,100 );
    m_pPalette = NULL;
}


CMainFrame::~CMainFrame()
{
    m_pPalette -> DeleteObject();
    delete m_pPalette;
    m_pPalette = NULL;
}

int CMainFrame::OnCreate( LPCREATESTRUCT lpCreateStruct )
{
    if ( CMDIFrameWnd::OnCreate( lpCreateStruct ) == -1 )
        return -1;

    LoadOptions();
    CreatePalette();

    if ( !m_wndToolBar.Create( this ) ||
        !m_wndToolBar.LoadToolBar( IDR_MAINFRAME ) )
    {
        TRACE0( "Failed to create toolbar\n" );
        return -1;    // fail to create
    }

    if ( !m_wndStatusBar.Create( this ) ||
        !m_wndStatusBar.SetIndicators( indicators,
            sizeof( indicators )/sizeof( UINT ) ) )
    {
        TRACE0( "Failed to create status bar\n" );
```

II

9

```
            return -1;    // fail to create
    }

    // TODO: Remove this if you don't want tool tips or a resizeable toolbar
    m_wndToolBar.SetBarStyle( m_wndToolBar.GetBarStyle() |
        CBRS_TOOLTIPS | CBRS_FLYBY | CBRS_SIZE_DYNAMIC ):

    // TODO: Delete these three lines if you don't want the toolbar to
    //     be dockable
    m_wndToolBar.EnableDocking( CBRS_ALIGN_ANY ):
    EnableDocking( CBRS_ALIGN_ANY ):
    DockControlBar( &m_wndToolBar ):

    return 0:
}


BOOL CMainFrame::PreCreateWindow( CREATESTRUCT& cs )
{
    // TODO: Modify the Window class or styles here by modifying
    //     the CREATESTRUCT cs

    return CMDIFrameWnd::PreCreateWindow( cs ):
}


///////////////////////////////////////////////////////////////////////////
// CMainFrame diagnostics

#ifdef _DEBUG
void CMainFrame::AssertValid() const
{
    CMDIFrameWnd::AssertValid():
}

void CMainFrame::Dump( CDumpContext& dc ) const
{
    CMDIFrameWnd::Dump( dc ):
}

#endif    //_DEBUG
```

Example 31   Using Nondiffused Colors   **351**

```
//////////////////////////////////////////////////////////////////////////////
// CMainFrame message handlers

void CMainFrame::OnClose()
{
    SaveOptions();

    CMDIFrameWnd::OnClose();
}


void CMainFrame::LoadOptions()
{
    BYTE *p;
    UINT size;
    if ( AfxGetApp() -> GetProfileBinary( SETTINGS_KEY,COLORS_KEY,&p,&size ) )
    {
    memcpy( m_rgbColors,p,size );
    delete []p;
    }
}


void CMainFrame::SaveOptions()
{
    UINT size = sizeof( m_rgbColors );
    AfxGetApp() ->
        WriteProfileBinary( SETTINGS_KEY,COLORS_KEY,( BYTE* )m_rgbColors,size );
}


void CMainFrame::OnOptionsPreferences()
{
    CPropertySheet sheet( _T( "Preferences" ),this );
    m_pColorPage = new CColorPage;

    sheet.AddPage( m_pColorPage );

    m_pColorPage -> m_rgbColors[COLOR1_COLOR] = m_rgbColors[COLOR1_COLOR];
    m_pColorPage -> m_rgbColors[COLOR2_COLOR] = m_rgbColors[COLOR2_COLOR];

    sheet.DoModal();
```

II

9

```
        delete m_pColorPage;
}


void CMainFrame::CreatePalette()
{

    LOGPALETTE *lp = ( LOGPALETTE * )calloc( 1, sizeof( LOGPALETTE ) +
        ( NUM_COLORS * sizeof( PALETTEENTRY ) ) );
    lp -> palVersion = 0x300;
    lp -> palNumEntries = NUM_COLORS;
    for ( int i = 0; i < NUM_COLORS; i++ )
    {
        lp -> palPalEntry[i].peRed = GetRValue( m_rgbColors[i] );
        lp -> palPalEntry[i].peGreen = GetGValue( m_rgbColors[i] );
        lp -> palPalEntry[i].peBlue = GetBValue( m_rgbColors[i] );
        // reserve for animation
        lp -> palPalEntry[i].peFlags = PC_RESERVED;


    }
    if ( m_pPalette ) delete m_pPalette;
    m_pPalette = new CPalette;
    m_pPalette -> CreatePalette( lp );
    free( lp );
}
```

Colors Property Page for Example 31

```
#if !defined COLORPAGE_H
#define COLORPAGE_H


// ColorPage.h : header file
//


#include "WzdProject.h"


/////////////////////////////////////////////////////////////////////////////
// CColorPage dialog


class CColorPage : public CPropertyPage
{

    DECLARE_DYNCREATE( CColorPage )
```

Example 31   Using Nondiffused Colors   **353**

```
// Construction
public:
    CColorPage();
    ~CColorPage();

// Dialog Data
    // {{AFX_DATA( CColorPage )
    enum { IDD = IDD_COLOR_PAGE };
    CStatic m_ctrlColor1Display;
    CStatic m_ctrlColor2Display;
    CSliderCtrl m_ctrlBlueSlider1;
    CSliderCtrl m_ctrlGrnSlider1;
    CSliderCtrl m_ctrlRedSlider1;
    CSliderCtrl m_ctrlBlueSlider2;
    CSliderCtrl m_ctrlGrnSlider2;
    CSliderCtrl m_ctrlRedSlider2;
    // }}AFX_DATA

    COLORREF m_rgbColors[NUM_COLORS];

// Overrides
    // ClassWizard generate virtual function overrides
    // {{AFX_VIRTUAL( CColorPage )
protected:
    virtual void DoDataExchange(CDataExchange* pDX);    // DDX/DDV support
    // }}AFX_VIRTUAL

// Implementation
protected:
    // Generated message map functions
    // {{AFX_MSG( CColorPage )
    virtual BOOL OnInitDialog();
    afx_msg void OnVScroll( UINT nSBCode, UINT nPos, CScrollBar* pScrollBar );
    afx_msg void OnPaint();
    //}}AFX_MSG
    DECLARE_MESSAGE_MAP()
private:
    void DrawColorRects();
```

II

9

```
};

#endif
// ColorPage.cpp : implementation file
//

#include "stdafx.h"
#include "wzd.h"
#include "ColorPage.h"
#include "WzdProject.h"
#include "MainFrm.h"

#ifdef _DEBUG
#define new DEBUG_NEW
#undef THIS_FILE
static char THIS_FILE[] = __FILE__;
#endif

/////////////////////////////////////////////////////////////////////////////
// CColorPage property page

IMPLEMENT_DYNCREATE( CColorPage, CPropertyPage )

CColorPage::CColorPage() : CPropertyPage( CColorPage::IDD )
{
    // {{AFX_DATA_INIT( CColorPage )
    // }}AFX_DATA_INIT
}

CColorPage::~CColorPage()
{
}

void CColorPage::DoDataExchange( CDataExchange* pDX )
{
    CPropertyPage::DoDataExchange( pDX );
    // {{AFX_DATA_MAP( CColorPage )
    DDX_Control( pDX, IDC_COLOR1_DISPLAY, m_ctrlColor1Display );
    DDX_Control( pDX, IDC_COLOR2_DISPLAY, m_ctrlColor2Display );
    DDX_Control( pDX, IDC_BLUE_SLIDER1, m_ctrlBlueSlider1 );
```

Example 31 Using Nondiffused Colors **355**

```
   DDX_Control( pDX, IDC_GRN_SLIDER1, m_ctrlGrnSlider1 );
   DDX_Control( pDX, IDC_RED_SLIDER1, m_ctrlRedSlider1 );
   DDX_Control( pDX, IDC_BLUE_SLIDER2, m_ctrlBlueSlider2 );
   DDX_Control( pDX, IDC_GRN_SLIDER2, m_ctrlGrnSlider2 );
   DDX_Control( pDX, IDC_RED_SLIDER2, m_ctrlRedSlider2 );
   // }}AFX_DATA_MAP
}

BEGIN_MESSAGE_MAP( CColorPage, CPropertyPage )
   // {{AFX_MSG_MAP( CColorPage )
   ON_WM_VSCROLL()
   ON_WM_PAINT()
   // }}AFX_MSG_MAP
END_MESSAGE_MAP()

/////////////////////////////////////////////////////////////////////////////
// CColorPage message handlers

BOOL CColorPage::OnInitDialog()
{
   CPropertyPage::OnInitDialog();

   // setup slider bars
   m_ctrlBlueSlider1.SetTicFreq( 15 );
   m_ctrlBlueSlider1.SetRange( 0,255,TRUE );
   m_ctrlBlueSlider1.SetPos( 255-GetBValue( m_rgbColors[COLOR1_COLOR] ) );
   m_ctrlGrnSlider1.SetTicFreq( 15 );
   m_ctrlGrnSlider1.SetRange( 0,255,TRUE );
   m_ctrlGrnSlider1.SetPos( 255-GetGValue( m_rgbColors[COLOR1_COLOR] ) );
   m_ctrlRedSlider1.SetTicFreq( 15 );
   m_ctrlRedSlider1.SetRange( 0,255,TRUE );
   m_ctrlRedSlider1.SetPos( 255-GetRValue( m_rgbColors[COLOR1_COLOR] ) );
   m_ctrlBlueSlider2.SetTicFreq( 15 );
   m_ctrlBlueSlider2.SetRange( 0,255,TRUE );
   m_ctrlBlueSlider2.SetPos( 255-GetBValue( m_rgbColors[COLOR2_COLOR] ) );
   m_ctrlGrnSlider2.SetTicFreq( 15 );
   m_ctrlGrnSlider2.SetRange( 0,255,TRUE );
   m_ctrlGrnSlider2.SetPos( 255-GetGValue( m_rgbColors[COLOR2_COLOR] ) );
   m_ctrlRedSlider2.SetTicFreq( 15 );
   m_ctrlRedSlider2.SetRange( 0,255,TRUE );
```

II

9

```
        m_ctrlRedSlider2.SetPos( 255-GetRValue( m_rgbColors[COLOR2_COLOR] ) );


    return TRUE;    // return TRUE unless you set the focus to a control
                    //      EXCEPTION: OCX Property Pages should return FALSE
}


void CColorPage::OnPaint()
{
    CPaintDC dc( this );     // device context for painting


    DrawColorRects();
}


void CColorPage::OnVScroll( UINT nSBCode, UINT nPos, CScrollBar* pScrollBar )
{
    int i;
    int color = 0;
    UINT id = pScrollBar -> GetDlgCtrlID();
    switch ( nSBCode )
    {
        case SB_TOP:
            color = 255;
            break;
        case SB_BOTTOM:
            color = 0;
            break;
        case SB_LINEDOWN:
        case SB_LINEUP:
        case SB_PAGEDOWN:
        case SB_PAGEUP:
            switch ( id )
            {
                case IDC_RED_SLIDER1:
                    color = m_ctrlRedSlider1.GetPos();
                    break;
                case IDC_BLUE_SLIDER1:
                    color = m_ctrlBlueSlider1.GetPos();
                    break;
                case IDC_GRN_SLIDER1:
```

Example 31   Using Nondiffused Colors   **357**

```
                    color = m_ctrlGrnSlider1.GetPos();
                    break;
                case IDC_RED_SLIDER2:
                    color = m_ctrlRedSlider2.GetPos();
                    break;
                case IDC_BLUE_SLIDER2:
                    color = m_ctrlBlueSlider2.GetPos();
                    break;
                case IDC_GRN_SLIDER2:
                    color = m_ctrlGrnSlider2.GetPos();
                    break;
            }
        break;
    case SB_THUMBPOSITION:
    case SB_THUMBTRACK:
        color = nPos;
        break;
    case SB_ENDSCROLL:
        break;
}

if ( nSBCode! = SB_ENDSCROLL )
{
    color = 255-color;

    switch ( id )
    {
        case IDC_RED_SLIDER1:
        case IDC_RED_SLIDER2:
            i = COLOR1_COLOR;
            if ( id == IDC_RED_SLIDER2 ) i = COLOR2_COLOR;
                m_rgbColors[i] = RGB( color,GetGValue( m_rgbColors[i] ),
                    GetBValue( m_rgbColors[i] ) );
            break;
        case IDC_BLUE_SLIDER1:
        case IDC_BLUE_SLIDER2:
            i = COLOR1_COLOR;
            if ( id == IDC_BLUE_SLIDER2 ) i = COLOR2_COLOR;
                m_rgbColors[i] = RGB( GetRValue( m_rgbColors[i] ),
                    GetGValue( m_rgbColors[i] ),color );
```

II

9

```
                        break;
                case IDC_GRN_SLIDER1:
                case IDC_GRN_SLIDER2:
                    i = COLOR1_COLOR;
                    if ( id == IDC_GRN_SLIDER2 ) i = COLOR2_COLOR;
                        m_rgbColors[i] = RGB(GetRValue( m_rgbColors[i] ),color,
                            GetBValue( m_rgbColors[i] ) );
                    break;
            }

        // select palette into a device context
        // (NOTE: you MUST select a palette into a device context for
        //     AnimatePalette() to work!)
        CDC *pDC = GetDC();
        CMainFrame *pFrame = ( CMainFrame * )AfxGetMainWnd();
        CPalette *pPalette = pFrame -> GetPalette();
        CPalette *pOPalette = pDC -> SelectPalette( pPalette,FALSE );

        // add new color to palette using animation
        PALETTEENTRY  pentry;
        pentry.peRed = GetRValue( m_rgbColors[i] );
        pentry.peGreen = GetGValue( m_rgbColors[i] );
        pentry.peBlue = GetBValue( m_rgbColors[i] );
        pentry.peFlags = PC_RESERVED;
        pPalette -> AnimatePalette( i,1,&pentry );

        // reselect old palette
        pDC -> SelectPalette( pOPalette,FALSE );

        // change color in CMainFrame
        pFrame -> SetColorRef( i,m_rgbColors[i] );
    }

    CPropertyPage::OnVScroll( nSBCode, nPos, pScrollBar );
}

void CColorPage::DrawColorRects()
{
    // create a solid brush using our palette
    CDC *pDC = GetDC();
    CMainFrame *pFrame = ( CMainFrame * )AfxGetMainWnd();
```

Example 32   Stretching a Bitmap   **359**

```
CPalette *pPalette = pFrame -> GetPalette();
CPalette *pOPalette = pDC -> SelectPalette( pPalette,FALSE );
CBrush brush1( PALETTEINDEX( COLOR1_COLOR ) );
CBrush brush2( PALETTEINDEX( COLOR2_COLOR ) );

// draw the rectangle
CRect rect;
m_ctrlColor1Display.GetWindowRect( &rect );
ScreenToClient( &rect );
rect.DeflateRect( 2,2,2,2 );
CBrush *pOBrush = pDC -> SelectObject( &brush1 );
pDC -> Rectangle( &rect );

m_ctrlColor2Display.GetWindowRect( &rect );
ScreenToClient( &rect );
rect.DeflateRect( 2,2,2,2 );
pDC -> SelectObject( &brush2 );
pDC -> Rectangle( &rect );

// unselect everything
pDC -> SelectObject( pOBrush );
pDC -> SelectPalette( pOPalette,FALSE );
ReleaseDC( pDC );
}
```

Example 32   Stretching a Bitmap

Objective

You would like to stretch a bitmap to fit a particular area on the screen, as seen in Figure 9.2.

Figure 9.2 Stretching Bitmaps

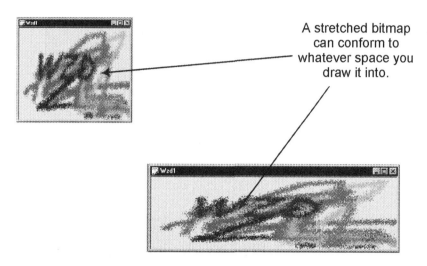

A stretched bitmap can conform to whatever space you draw it into.

Strategy

We will be using the CDC::SetStretchMode() function to set the stretching mode and the CDC::StretchBlt() function to do the actual stretching. We will encapsulate this functionality into our own bitmap class.

Steps

Create a New Bitmap Class

1. Use the ClassWizard to create a new bitmap class derived from CBitmap. Add a new member function to this class called Stretch().

Create a Stretched Bitmap Object

1. Give Stretch() three arguments. The first two arguments will be the width and height of the stretched bitmap to create. The last argument is the stretching mode.

```
CBitmap *CWzdBitmap::Stretch( int nWidth, int nHeight, int nMode )
{
```

Example 32   Stretching a Bitmap   **361**

2. Create a device context. Since we don't want to have to pass a device context to this function, get your device context from the desktop.

```
CDC dcTo,dcFrom,dcScreen;
dcScreen.Attach( ::GetDC( NULL ) );
```

3. Then, create two memory device contexts from this screen context, one for the unstretched bitmap and one for the stretched bitmap.

```
// create "from" device context and select the loaded bitmap into it
dcFrom.CreateCompatibleDC( &dcScreen );
dcFrom.SelectObject( this );


// create a "to" device context select a memory bitmap into it
dcTo.CreateCompatibleDC( &dcScreen );
```

4. Next, create an empty bitmap in memory with the width and height of the bitmap we want to create and select it into the "To" device context.

```
CBitmap *pBitmap = new CBitmap;
pBitmap -> CreateCompatibleBitmap( &dcScreen, nWidth, nHeight );
dcTo.SelectObject( pBitmap );
```

5. We assume in Stretch() that a bitmap object has already been loaded into this bitmap class using LoadBitmap() or through some other means. We now get the size of that bitmap.

```
// get original bitmap size
BITMAP bmInfo;
GetObject( sizeof( bmInfo ),&bmInfo );
```

6. Set the stretch mode and then do the stretch. Notice we are stretching any bitmap loaded into this class to the empty bitmap object we just created.

```
// set the stretching mode
dcTo.SetStretchBltMode( nMode );


// stretch loaded bitmap into memory bitmap
dcTo.StretchBlt( 0, 0, nWidth, nHeight,
    &dcFrom, 0, 0, bmInfo.bmWidth, bmInfo.bmHeight, SRCCOPY );
```

II

9

7. The last thing we do in Stretch() is clean up our device contexts and return a pointer to this new, stretched bitmap class object. It's up to the caller to delete this bitmap object when they're done with it:

```
// delete and release device contexts
dcTo.DeleteDC();
dcFrom.DeleteDC();
::ReleaseDC( NULL, dcScreen.Detach() );

// it's up to the caller to delete this new bitmap
return pBitmap;

}
```

To see the new bitmap class in its entirety, please refer to the Bitmap Class for Example 32 (page 364).

Use the New Bitmap Class

1. To use this new bitmap class, we start by loading a bitmap into it.

```
CWzdBitmap m_bitmap;
:    :    :
m_bitmap.LoadBitmap( IDB_WZD_BITMAP );
```

2. Then, we create the stretched version of this bitmap with

```
// get a bitmap stretched to size of client area of view
CRect rect;
GetClientRect( &rect );
CBitmap *pBitmap = m_bitmap.Stretch( rect.Width(),rect.Height(),
    COLORONCOLOR );
    // also HALFTONE -- slower but attempts to average colors
    // BLACKONWHITE  -- monotone--scacrifices white for black pixels
    // WHITEONBLACK  -- monotone--sacrifices black for white pixels

    // if you use HALFTONE, must also call the following next to
    //     realign the brush
::SetBrushOrgEx( pDC -> m_hDC, 0,0, NULL );
```

Example 32   Stretching a Bitmap   **363**

3. You can then draw this stretched bitmap to the screen, as usual.

```
// get device context to select bitmap into
CDC dcComp;
dcComp.CreateCompatibleDC( pDC );
dcComp.SelectObject( pBitmap );

// draw bitmap
pDC -> BitBlt( 0,0,rect.Width(),rect.Height(), &dcComp, 0,0,SRCCOPY );

delete pBitmap;
```

II

Notes

- The CDC::StretchBlt() function works especially well on large bitmaps where it has hundreds of pixels to work with. At smaller resolutions, such as the 16 by 16 pixel size of a standard toolbar button, any compromises the function has to make are readily apparent. Lines blur and images distort to levels unacceptable for a professional appearance. At this resolution, you are better served to avoid using StretchBlt() and to use individual bitmaps manually drawn for different resolutions, instead. For bitmaps that are neither large nor small, the stretching mode becomes important. When expanding or contracting an image using the HALFTONE mode, the StretchBlt() function creates color averages. If there's a choice between putting one color or the other in a pixel, the function averages the color values and puts that value there instead. This can be slower, but the affect can be more acceptable than the COLORON-COLOR mode, which doesn't average. For monotone bitmaps (with black and white pixels only), you can pick either BLACKONWHITE or WHITEONBLACK for a stretching mode. When the function is in BLACKONWHITE mode and it has to choose between putting a black pixel in the new image or a white pixel, it always chooses black. The WHITEONBLACK mode has the opposite effect.

- It's always preferable to stretch a reference bitmap rather than to continually stretch the bitmap that results from a stretch, especially when stretching several times. In other words, don't stretch bitmap A to create B, then stretch B to create C. Instead, stretch A to get B and C. Each trip through StretchBlt() causes some distortion, which will be cumulative if you keep using the same bitmap each time.

9

CD Notes

- When executing the project on the accompanying CD, you will notice that the view is filled with a bitmap that changes size as you squeeze or enlarge the view.

Bitmap Class for Example 32

```
#ifndef WZDBITMAP_H
#define WZDBITMAP_H


class CWzdBitmap : public CBitmap
{
public:
    DECLARE_DYNAMIC( CWzdBitmap )

// Constructors
    CWzdBitmap();


    CBitmap *Stretch( int nWidth, int nHeight, int nMode );

// Implementation
public:
    virtual ~CWzdBitmap();

// Attributes
// Operations


};
#endif
// WzdBitmap.cpp : implementation of the CWzdBitmap class
//


#include "stdafx.h"
#include "WzdBitmap.h"
#include "resource.h"


/////////////////////////////////////////////////////////////////////////////
// CWzdBitmap
```

Example 32   Stretching a Bitmap   **365**

```
IMPLEMENT_DYNAMIC( CWzdBitmap, CBitmap )

CWzdBitmap::CWzdBitmap()
{
}

CWzdBitmap::~CWzdBitmap()
{
}

CBitmap *CWzdBitmap::Stretch( int nWidth, int nHeight, int nMode )
{
    CDC dcTo,dcFrom,dcScreen;
    dcScreen.Attach( ::GetDC( NULL ) );

    // create "from" device context and select the loaded bitmap into it
    dcFrom.CreateCompatibleDC( &dcScreen );
    dcFrom.SelectObject( this );

    // create a "to" device context select a memory bitmap into it
    dcTo.CreateCompatibleDC( &dcScreen );
    CBitmap *pBitmap = new CBitmap;
    pBitmap -> CreateCompatibleBitmap( &dcScreen, nWidth, nHeight );
    dcTo.SelectObject( pBitmap );

    // get original bitmap size
    BITMAP bmInfo;
    GetObject( sizeof( bmInfo ),&bmInfo );

    // set the stretching mode
    dcTo.SetStretchBltMode( nMode );

    // stretch loaded bitmap into memory bitmap
    dcTo.StretchBlt( 0, 0, nWidth, nHeight,
        &dcFrom, 0, 0, bmInfo.bmWidth, bmInfo.bmHeight, SRCCOPY );

    // delete and release device contexts
    dcTo.DeleteDC();
    dcFrom.DeleteDC();
    ::ReleaseDC( NULL, dcScreen.Detach() );
```

II

9

```
        // it's up to the caller to delete this new bitmap
        return pBitmap;

}
```

Example 33   Capturing the Screen

Objective

You would like to capture an area of the screen in a bitmap.

Strategy

You probably already knew that you could draw a bitmap on the screen with BitBlt(). But did you know you could reverse the process and BitBlt() an area of the screen back to a bitmap? What you get, however, is nothing without the current system palette since a bitmap is not much more than an array of pointers into a color table. So, we also need to capture the current system palette. Since all of this functionality relates nicely to a bitmap object, we will be encapsulating it in our own bitmap class.

Steps

Capture the Screen

1. Use the ClassWizard to create a new class derived from CBitmap.
2. Use the Text Editor to add a new function to this class called Capture(), which takes a RECT structure as its calling argument. This RECT will contain the screen coordinates of the area of the screen to capture. Start this function by deleting the last palette and bitmap object stored in this class.

```
void CWzdBitmap::Capture( CRect &rect )
{
    // cleanup old captures
    if (m_pPalette)
```

Example 33   Capturing the Screen   **367**

```
{
    DeleteObject();
    delete m_pPalette;
}
```

3. Next, since the caller has just supplied us with the dimension that this bitmap will finally have, let's store them for posterity.

```
// save width and height
m_nWidth = rect.Width();
m_nHeight = rect.Height();
```

4. Create a device context for the screen. Normally, we would be using this context to write to. This time, however, we will be reading from it.

```
CDC dcScreen;
dcScreen.CreateDC( "DISPLAY", NULL, NULL, NULL );
```

5. Create an empty bitmap object and select it into a memory device context.

```
CDC dcMem;
dcMem.CreateCompatibleDC( &dcScreen );
CreateCompatibleBitmap( &dcScreen, m_nWidth, m_nHeight );
dcMem.SelectObject( this );
```

6. Now just BitBlt() from the screen to our empty bitmap.

```
dcMem.BitBlt( 0,0,m_nWidth,m_nHeight, &dcScreen, rect.left, rect.top,
    SRCCOPY );
```

Capture the System Palette

As mentioned previously, this bitmap is worthless without the current system palette. We need this palette so that we can draw this bitmap again and expect the same colors. We will be using CPalette's CreatePalette() member function to create this palette, whose sole calling argument is a LOGPALETTE structure that we need to fill with the current system palette.

1. Start by creating an empty LOGPALETTE that will be big enough to contain all the colors in the screen.

```
int nColors = ( 1 << ( dcScreen.GetDeviceCaps( BITSPIXEL ) *
    dcScreen.GetDeviceCaps( PLANES ) ) );
LOGPALETTE *pLogPal = ( LOGPALETTE * )new BYTE[
    sizeof(LOGPALETTE) + ( nColors * sizeof( PALETTEENTRY ) )];
```

2. Now, initialize the header with version and number of colors.

```
pLogPal -> palVersion    = 0x300;
pLogPal -> palNumEntries = nColors;
```

3. Next, we capture the current system colors with ::GetSystemPaletteEntries().

```
::GetSystemPaletteEntries( dcScreen.m_hDC, 0, nColors,
    ( LPPALETTEENTRY )( pLogPal -> palPalEntry ) );
```

4. Now, create the palette and cleanup.

```
m_pPalette = new CPalette;
m_pPalette -> CreatePalette( pLogPal );

// clean up
delete []pLogPal;
dcMem.DeleteDC();
dcScreen.DeleteDC();
```

5. To see a complete listing of this new class, please refer to the Bitmap Class for Example 34 (page 377).

Use This New Bitmap Class

1. In this example, we will capture the entire desktop to our bitmap.

```
CWnd *pWnd = GetDesktopWindow();
pWnd -> GetWindowRect( &rect );

CWzdBitmap bitmap;
bitmap.Capture( rect );
```

Example 33   Capturing the Screen   **369**

2. You can now display this bitmap using the following:

```
// select bitmap palette
CPalette *pOldPal =
    pDC -> SelectPalette( m_bitmap.GetPalette(), FALSE );
pDC -> RealizePalette();

// get device context to select bitmap into
CDC dcComp;
dcComp.CreateCompatibleDC( pDC );
dcComp.SelectObject( &m_bitmap );

// draw bitmap
pDC -> BitBlt( 0,0,m_bitmap.m_nWidth,m_bitmap.m_nHeight,
    &dcComp, 0,0,SRCCOPY );

// reselect old palette
pDC -> SelectPalette( pOldPal,FALSE );
```

Notes

- We use this example to print our application's view in Example 18. First, the view is captured using this example. Then, it's converted into a Device Independent Bitmap (DIB). Finally, it's BitBlted to the printer.

CD Notes

- When executing the project on the accompanying CD, click on the Test/Wzd menu items. The current desktop will appear in the view.

Bitmap Class for Example 33

```
#ifndef WZDBITMAP_H
#define WZDBITMAP_H

class CWzdBitmap : public CBitmap
{
public:
    DECLARE_DYNAMIC( CWzdBitmap )
```

```
// Constructors
    CWzdBitmap();


    void Capture( CRect &rect );
    CPalette *GetPalette(){return m_pPalette;};

// Implementation
public:
    virtual ~CWzdBitmap();

// Attributes
    int m_nWidth;
    int m_nHeight;

// Operations
private:
    CPalette *m_pPalette;
};
#endif
// WzdBitmap.cpp : implementation of the CWzdBitmap class
//

#include "stdafx.h"
#include "WzdBtmap.h"


/////////////////////////////////////////////////////////////////////////////
// CWzdBitmap

IMPLEMENT_DYNAMIC( CWzdBitmap, CBitmap )


CWzdBitmap::CWzdBitmap()
{
    m_pPalette = NULL;
}


CWzdBitmap::~CWzdBitmap()
{
    if ( m_pPalette )
    {
```

Example 33   Capturing the Screen   **371**

```
        delete m_pPalette;
    }
}


void CWzdBitmap::Capture( CRect &rect )
{
    // cleanup old captures
    if ( m_pPalette )
    {
        DeleteObject();
        delete m_pPalette;
    }

    // save width and height
    m_nWidth = rect.Width();
    m_nHeight = rect.Height();

    /////////////////////////////////////////////
    // copy screen image into a bitmap object
    /////////////////////////////////////////////

    // create a device context that accesses the whole screen
    CDC dcScreen;
    dcScreen.CreateDC( "DISPLAY", NULL, NULL, NULL );

    // create an empty bitmap in memory
    CDC dcMem;
    dcMem.CreateCompatibleDC( &dcScreen );
    CreateCompatibleBitmap( &dcScreen, m_nWidth, m_nHeight );
    dcMem.SelectObject( this );

    // copy screen into empty bitmap
    dcMem.BitBlt(0,0,m_nWidth,m_nHeight,&dcScreen,rect.left,rect.top,SRCCOPY);

    // this bitmap is worthless without the current system palette, so...

    /////////////////////////////////////////////
    // save system palette in this bitmap's palette
    /////////////////////////////////////////////
```

II

9

```
// create an empty logical palette that's big enough to hold all the colors
int nColors = ( 1 << ( dcScreen.GetDeviceCaps( BITSPIXEL ) *
    dcScreen.GetDeviceCaps( PLANES ) ) );
LOGPALETTE *pLogPal = ( LOGPALETTE * )new BYTE[
    sizeof( LOGPALETTE ) + ( nColors * sizeof( PALETTEENTRY ) )];

// initialize this empty palette's header
pLogPal -> palVersion    = 0x300;
pLogPal -> palNumEntries = nColors;

// load this empty palette with the system palette's colors
::GetSystemPaletteEntries( dcScreen.m_hDC, 0, nColors,
        ( LPPALETTEENTRY )( pLogPal -> palPalEntry ) );

// create the palette with this logical palette
m_pPalette = new CPalette;
m_pPalette -> CreatePalette( pLogPal );

// clean up
delete []pLogPal;
dcMem.DeleteDC();
dcScreen.DeleteDC();
}
```

Example 34   Outputting a DIB Bitmap File

Objective

You would like to output a bitmap to a DIB file.

Strategy

All bitmaps in a file are device independent, which is to say they contain their own color table. When a bitmap is loaded into an application for use, it's broken up into a header, color table, and picture data (pointers into the color table). The color table is put into a palette that must be selected into a device context before the bitmap is drawn. The header is converted into a bitmap header.

Here, we will reverse this process and create a DIB from its bitmap header, palette, and picture data. The ::GetDIBits() API will create the final

Example 34   Outputting a DIB Bitmap File   **373**

product. Then, we'll just write this object out to disk like any other binary file. We will also encapsulate all of this functionality in a new bitmap class.

Steps

Create a DIB Object

1. Use the ClassWizard to create a new class derived from CBitmap.

2. Add a new function to this class called CreateDIB(). This function will create a DIB from the bitmap currently loaded in this bitmap class (e.g., with LoadBitmap()). Start this function by creating an empty DIB header, which you will fill with information from the bitmap's header.

```
HANDLE CWzdBitmap::CreateDIB( int *pbmData )
{
    BITMAPINFOHEADER bi;
    memset( &bi, 0, sizeof( bi ) );
    bi.biSize = sizeof( BITMAPINFOHEADER );
    bi.biPlanes = 1;
    bi.biCompression = BI_RGB;
```

3. Get and store dimensions of this bitmap from the bitmap header:

```
    BITMAP bm;
    GetObject(sizeof( bm ),( LPSTR )&bm );
    bi.biWidth = bm.bmWidth;
    bi.biHeight = bm.bmHeight;
```

4. Get the number of bits required per pixel. This is the size of each picture data pointer. The bigger the data pointer, the more colors that can be defined in the color table, but the larger the bitmap.

```
    int bits = bm.bmPlanes * bm.bmBitsPixel;
    if ( bits <= 1 )
        bi.biBitCount = 1;
    else if ( bits <= 4 )
        bi.biBitCount = 4;
```

```
else if ( bits <= 8 )
    bi.biBitCount = 8;
else
    bi.biBitCount = 24;
```

We will now use `::GetDIBits()` to fill in the color table and picture data. Unfortunately, we first have to create a memory area large enough to hold this bitmap. We first have to find out how big both our color table and our picture data will be.

5. Calculate the size of the color table based on the number of possible colors times the size of one color definition (four bytes).

```
int biColorSize = 0;
if ( bi.biBitCount! = 24 ) biColorSize = ( 1 << bi.biBitCount );
biColorSize* = sizeof( RGBQUAD );
```

6. Calculate the picture data size based on the size of each color pointer times the dimensions of the bitmap (width times height). `::GetDIBits()` will be outputting each row on a DWORD alignment, so you must also make sure this size accounts for that.

```
bi.biSizeImage = ( DWORD )bm.bmWidth * bi.biBitCount;
    // bits per row
bi.biSizeImage = ( ( ( bi.biSizeImage ) + 31 ) / 32 ) * 4;
    // DWORD aligned
bi.biSizeImage* = bm.bmHeight;    // bytes required for whole bitmap
```

7. Now allocate an area of global memory big enough to hold the header, color table, and picture data and copy the header we've created so far into it.

```
HANDLE hDIB = ::GlobalAlloc( GHND, bi.biSize + biColorSize +
    bi.biSizeImage );
LPBITMAPINFOHEADER lpbi =
    ( LPBITMAPINFOHEADER )::GlobalLock( hDIB );
*lpbi = bi;
```

Example 34   Outputting a DIB Bitmap File   **375**

8. One last step before you can use `::GetDIBits()` is to create a device context with this bitmap's pallete selected into it.

```
CDC dc;
dc.Attach( ::GetDC( NULL ) );
CPalette *pPal = dc.SelectPalette( m_pPalette,FALSE );
dc.RealizePalette();
```

9. Now that you've finally created all the ingredients `::GetDIBits()` needs, let it do its job.

```
::GetDIBits( dc.m_hDC, ( HBITMAP )m_hObject, 0,
    ( UINT )bi.biHeight, ( LPSTR )lpbi + ( WORD )lpbi ->
    biSize + biColorSize, ( LPBITMAPINFO )lpbi, DIB_RGB_COLORS );
```

10. Now let's cleanup and return the DIB's handle.

```
::GlobalUnlock( hDIB );
dc.SelectPalette( pPal,FALSE );
dc.RealizePalette();

// return handle to the DIB
return hDIB;
}
```

Save This DIB Object to Disk

We will now add another function to this class, which will take as an argument the name of a file to which we will save this bitmap. We will essentially be just copying the DIB created in the `CreateDIB()` function to a disk file, but first there's yet one more header we need to create: the bitmap file header.

1. Start by adding a new function to this class, which will start by using `CreateDIB()` to get a DIB handle to the current bitmap object.

```
void CWzdBitmap::SaveBitmapEx( CString sFile )
{
    // create a DIB bitmap
    int bmData;
    HANDLE hDIB = CreateDIB( &bmData );
```

2. Next, lock the handle to get a memory pointer to the DIB.

```
LPBYTE lpBitmap = ( LPBYTE )::GlobalLock( hDIB );
int bmSize = ::GlobalSize( hDIB );
```

Note: A handle is, in fact, a pointer to a pointer. This allows the object it eventually points to in memory to move around. Why is the object moving around? So that the system can compress memory whenever it's filled with holes—holes that are created by other applications deallocating memory. Because your handle doesn't directly point to this memory (in fact, it points to a table of memory pointers), all the system has to do is update the pointer table. However, when you lock a handle, you're telling the system to stop moving the object around and to return a real pointer to the object through which you can access your object like any other memory object. To allow the system to continue to compress out unallocated memory you should never lock memory for any significant length of time.

---

3. Create an empty bitmap file header and fill it with the particulars of this bitmap.

```
BITMAPFILEHEADER bmfh;
bmfh.bfType = 'MB';      // (actually 'BM' for bitmap)
bmfh.bfSize = sizeof(BITMAPFILEHEADER)+bmSize;
bmfh.bfReserved1 = 0;
bmfh.bfReserved2 = 0;
bmfh.bfOffBits = bmData;
```

4. Use CFile to create a binary file and write out the header, followed by the DIB.

```
CFile file;
file.Open( sFile, CFile::modeCreate|CFile::modeWrite );
file.Write( &bmfh,sizeof( BITMAPFILEHEADER ) );
file.Write( lpBitmap,bmSize );
file.Close();
```

Example 34   Outputting a DIB Bitmap File   **377**

5. Now, just cleanup.

```
    ::GlobalUnlock( hDIB );
    ::GlobalFree( hDIB );
}
```

Use This New Bitmap Function

1. After you've created a bitmap, either by capturing the screen or drawing to it, you can save it to disk with

```
bitmap.SaveBitmapEx(
    "dib.bmp"    // file name and path to save to
    );
```

Notes

- We use the CreateDIB() portion of this example in Example 18 to convert a bitmap to a DIB before we print it. In fact, you need to convert any bitmap to a DIB before your printer will accept it. In theory, you should only have to realize your bitmap's palette into a printer device context and then BitBlt() the bitmap over, but BitBlt() just can't seem to hack the conversion from a screen palette to a printer palette.

CD Notes

When executing the project on the accompanying CD, click on the Test/Wzd menu items. The current desktop will be captured and written to a file called dib.bmp.

Bitmap Class for Example 34

```
#ifndef WZDBITMAP_H
#define WZDBITMAP_H

class CWzdBitmap : public CBitmap
{
public:
    DECLARE_DYNAMIC( CWzdBitmap )
```

II

9

```
// Constructors
    CWzdBitmap();


    void SaveBitmapEx( CString sFile );
    HANDLE CreateDIB( int *pbmData = NULL );

// Implementation
public:
    virtual ~CWzdBitmap();

// Attributes
    int m_nWidth;
    int m_nHeight;

// Operations
private:
    CPalette *m_pPalette;
};
#endif
// WzdBitmap.cpp : implementation of the CWzdBitmap class
//


#include "stdafx.h"
#include "WzdBtmap.h"


//////////////////////////////////////////////////////////////////////////
// CWzdBitmap

IMPLEMENT_DYNAMIC( CWzdBitmap, CBitmap )


CWzdBitmap::CWzdBitmap()
{
    m_pPalette = NULL;
}


CWzdBitmap::~CWzdBitmap()
{
    if ( m_pPalette )
    {
```

Example 34   Outputting a DIB Bitmap File   **379**

```
        delete m_pPalette;
    }
}


void CWzdBitmap::SaveBitmapEx( CString sFile )
{
    // create a DIB bitmap
    int bmData;
    HANDLE hDIB = CreateDIB( &bmData );

    // get a memory pointer to it
    LPBYTE lpBitmap = ( LPBYTE )::GlobalLock( hDIB );
    int bmSize = ::GlobalSize( hDIB );

    // create file
    CFile file;
    file.Open( sFile, CFile::modeCreate|CFile::modeWrite );

    // write the bitmap header
    BITMAPFILEHEADER bmfh;
    bmfh.bfType = 'MB';    // (actually 'BM' for bitmap)
    bmfh.bfSize = sizeof( BITMAPFILEHEADER ) + bmSize;
    bmfh.bfReserved1 = 0;
    bmfh.bfReserved2 = 0;
    bmfh.bfOffBits = bmData;
    file.Write( &bmfh,sizeof( BITMAPFILEHEADER ) );

    // write the bitmap body
    file.Write( lpBitmap,bmSize );

    // cleanup
    file.Close();
    ::GlobalUnlock( hDIB );
    ::GlobalFree( hDIB );
}


HANDLE CWzdBitmap::CreateDIB(int *pbmData)
{
```

II

9

```
/////////////////////////////////////////////
// create DIB header from our BITMAP header
/////////////////////////////////////////////

BITMAPINFOHEADER bi;
memset( &bi, 0, sizeof( bi ) );
bi.biSize = sizeof( BITMAPINFOHEADER );
bi.biPlanes = 1;
bi.biCompression = BI_RGB;

// get and store dimensions of bitmap
BITMAP bm;
GetObject( sizeof( bm ),( LPSTR )&bm );
bi.biWidth = bm.bmWidth;
bi.biHeight = bm.bmHeight;

// get number of bits required per pixel
int bits = bm.bmPlanes * bm.bmBitsPixel;
if ( bits <= 1 )
    bi.biBitCount = 1;
else if ( bits <= 4 )
    bi.biBitCount = 4;
else if ( bits <= 8 )
    bi.biBitCount = 8;
else
    bi.biBitCount = 24;


// calculate color table size
int biColorSize = 0;
if ( bi.biBitCount! = 24 ) biColorSize = ( 1 << bi.biBitCount );
biColorSize* = sizeof( RGBQUAD );

// calculate picture data size
bi.biSizeImage = ( DWORD )bm.bmWidth * bi.biBitCount;     // bits per row
bi.biSizeImage = ( ( ( bi.biSizeImage ) + 31 ) / 32 ) * 4;
    // DWORD aligned
bi.biSizeImage* = bm.bmHeight;     // bytes required for whole bitmap

// return size to caler in case they want to save to file
```

Example 34   Outputting a DIB Bitmap File   **381**

```
if ( pbmData )
*pbmData = bi.biSize + biColorSize;

/////////////////////////////////////////////
// get DIB color table and picture data
/////////////////////////////////////////////

// allocate a hunk of memory to hold header, color table and picture data
HANDLE hDIB = ::GlobalAlloc( GHND, bi.biSize + biColorSize +
    bi.biSizeImage );

// get a memory pointer to this hunk by locking it
LPBITMAPINFOHEADER lpbi = ( LPBITMAPINFOHEADER )::GlobalLock( hDIB );

// copy our header structure into hunk
*lpbi = bi;

// get a device context and select our bitmap's palette into it
CDC dc;
dc.Attach( ::GetDC( NULL ) );
CPalette *pPal = dc.SelectPalette( m_pPalette,FALSE );
dc.RealizePalette();

// load our memory hunk with the color table and picture data
::GetDIBits( dc.m_hDC, ( HBITMAP )m_hObject, 0, ( UINT )bi.biHeight,
    ( LPSTR )lpbi + ( WORD )lpbi -> biSize + biColorSize,
    ( LPBITMAPINFO )lpbi, DIB_RGB_COLORS );

// clean up
::GlobalUnlock( hDIB );
dc.SelectPalette( pPal,FALSE );
dc.RealizePalette();

// return handle to the DIB
return hDIB;
}
```

II

9

10

Chapter 10

Help

On-line help can drastically reduce the learning curve required for your application. Rather than trying to find the appropriate command in a manual, your user can directly interrogate the control itself. Rather than depending on a printed index to find the subject, your user can ask your application to look through the whole manual at once. The cost of printing a manual also plummets.

On-line help, however, is only useful if your user has a general idea of what your application does. It provides no more help to a complete novice than a Finnish dictionary can train someone to speak Finnish. For those times that you need to offer the basics, only the overview possible in a manual or tutorial can help.

Example 35 Adding a Help Menu Item We will add the standard "Contents" and "Search" menu items to your application's "Help" menu.

Example 36 Adding Context Sensitive Help We will add point and click help to your application.

Example 37 Adding Bubble Help We will add point and hover bubble help to your application.

The Three Types of On-Line Help

There are three types of on-line help.

Menu Help is the help your user can retrieve through your "Help" menu. Although the AppWizard only adds an "About" command to this menu, you can also add "Index" and "Contents" commands to allow your user to seamlessly open your own help files into Window's WinHelp application. Please see Example 35 for how to add menu help to your application.

Context Sensitive Help allows your user to summon help based on the menu item or dialog control with which they're currently interacting. In effect, context sensitive help allows your user to turn to the exact page in your on-line documentation that describes that control or menu item. Under the category of context help there are two types.

- **F1 Help** allows your user to press the F1 key to get help on the currently selected menu item or active dialog box.
- **What's This Help** allows your user to be more selective by actually using the mouse to click on the control or screen area they want help on. When in "what's this" mode, the normal mouse cursor turns into a question mark arrow.

Examples of both of these types of context sensitive help may be found in Example 36.

Bubble Help allows your user to interrogate the purpose of a control or area of the view just by allowing the mouse cursor to linger over that spot. A small window then opens and describes the area. Bubble help isn't as descriptive as Context Help, but it's faster. For an example of bubble help, please see Example 37.

Example 35   Adding a Help Menu Item

Objective

You would like to add the standard "Contents" and "Search" menu items to your application's "Help" menu, as seen in Figure 10.1.

Example 35   Adding a Help Menu Item   **385**

Figure 10.1   Menu Help

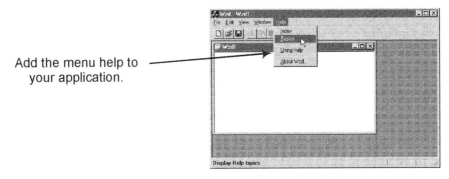

Add the menu help to
your application.

II

Strategy

Actually, most of the functionality required to provide menu help is already
contained in several MFC classes. Therefore, all we will be doing in this
example is selectively "waking up" that functionality.

Steps

Enable Menu Help

10

1. Use the Menu Editor to add the following commands to your Help
   menu.

   Name: &Index
   ID: ID_HELP_INDEX
   Comment: "Display Help Index\nHelp Index"

   Name: &Topics
   ID: ID_HELP_FINDER
   Comment: "Display Help topics\nHelp Topics"

   Separator

   Name: &Using Help
   ID: ID_HELP_USING
   Comment: "Display instructions about how to use help\nHelp"

Separator

Note: This last help command, "Using Help", is a freebee provided by Microsoft's WinHelp that allows your user to find out how to use WinHelp.

---

2. Use the Text Editor to add the following message macros to your `CMain-Frame`'s message map below the ClassWizard's `{{}}` markers.

```
ON_COMMAND( ID_HELP_INDEX, CMDIFrameWnd::OnHelpIndex )
ON_COMMAND( ID_HELP_USING, CMDIFrameWnd::OnHelpUsing )
ON_COMMAND( ID_HELP_FINDER, CMDIFrameWnd::OnHelpFinder )
```

3. That's it! Your MFC classes will now take care of the job of summoning Window's WinHelp application and passing your application's help files (`.hlp` and `.cnt`) to it when your user clicks on the menu help. MFC, by default, assumes these files are in your application's executable directory and uses your application's name. If, however, these files will be located someplace else or called something else, add the following to your application class's `InitInstance()` function.

```
m_pszHelpFilePath = _tcsdup( _T(
    "d:\\somedir\\myhelp.hlp"    // the directory and name of
                                 //     your help file
    ) );
```

4. Please refer to the Appendix C for information on how to create your own `.hlp` and `.cnt` help files.

Notes

- The AppWizard will automatically add these menu items, along with some context sensitive help functionality, if you select the "Context Help" option in AppWizard's Step 4. Sample `.hlp` and `.cnt` files are also added. However, in most companies, the job of creating the `.hlp` and `.cnt` files usually falls to the tech writers and product managers, using some sort of help authoring system like RoboHELP™. Therefore, it's better to

Example 36   Adding Context Sensitive Help   **387**

just manually add these menu items yourself than to add all that dead weight to your project.

CD Notes

- When executing the project on the accompanying CD, you will notice that the help menu has been expanded to include three new help commands.

Example 36   Adding Context Sensitive Help

Objective

You would like to add point and click help to your application, as seen in Figure 10.2.

Figure 10.2   Context Sensitive Help

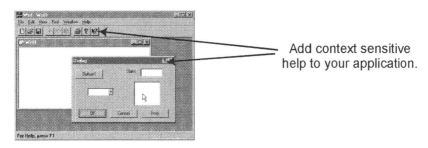

Add context sensitive help to your application.

Strategy

We will again be waking up some functionality in MFC to provide our application with context sensitive help. This example will add both F1 Help and What's This Help to your application. Your user will be able to access F1 Help by pressing F1 with a menu item selected or a dialog box open or by pressing the "Help" button in a dialog box. Your user will be able to access What's This Help by pressing Shift+F1 or clicking on the menu item "What's this" or a toolbar button or a question mark button in the caption bar of your dialog boxes. What's This Help causes the mouse cursor to appear as a question mark arrow until the user clicks on an item.

Steps

Add Context Help to the Main Application

1. Use the Text Editor to add the following to the message map in your `CMainFrame` class, below the ClassWizard's {{}} markers.

```
ON_COMMAND( ID_HELP, CMDIFrameWnd::OnHelp )
ON_COMMAND( ID_CONTEXT_HELP, CMDIFrameWnd::OnContextHelp )
ON_COMMAND( ID_DEFAULT_HELP, CMDIFrameWnd::OnHelpFinder )
```

2. If your help files won't be installed in the same directory as your application's executable, add the following to your application class's `InitInstance()` function.

```
m_pszHelpFilePath = _tcsdup( _T(
    "d:\\somedir\\myhelp.hlp"    // the directory and name of
                                 //    your help file
    ) );
```

3. Use the Menu Editor to add a "What's this" command to your Help menu.

   Name: &What's this
   ID: ID_CONTEXT_HELP
   Comment: "Display help for clicked on buttons, menus and windows\nHelp"

4. You can also optionally add a "What's this" button to your toolbar using the Toolbar Editor. Make the ID equal `ID_CONTEXT_HELP`. You can get a bitmap for this button (i.e., an arrow with question mark) by using the AppWizard to create a new dummy project and specify "Context help" in Step 4. Then, cut the "What's this" bitmap from this new project's toolbar and paste it into your current project.

5. Implement the F1 and Shift+F1 keys as accelerator keys using the Accelerator Table Editor with the following IDs.

VK_F1	ID_CONTEXT_HELP	VIRTKEY,SHIFT
VK_F1	ID_HELP	VIRTKEY

6. Use the String Table Editor to add two new idle messages to your string table: one for normal idle, informing your user they can press F1 to get

Example 36   Adding Context Sensitive Help   **389**

help; and the other for "What's this" idle, informing your user that they can click on an item for help.

(a) Change `AFX_IDS_IDLEMESSAGE`'s message to "For Help, press F1".

(b) Add an `AFX_IDS_HELPMODEMESSAGE` message as "Select an object on which to get Help".

Add Context Help to Dialog Boxes

1. Use the Dialog Editor and open the property box for the dialog template itself to specify the "Context help" style. This puts a "?" button in the caption bar of the dialog when it's created. Do this for every dialog template you want to provide with What's This Help.

2. You can also add a "Help" button to each template. Give this button an ID of `ID_HELP`.

3. Edit the properties of each control in a dialog template and choose the "Help ID" option. This adds a help ID to that control, which is used to reference this control when WinHelp is called. This ID will also be added automatically to a `resource.hm` file by the Dialog Editor. If the `resource.hm` file doesn't already exist, the editor will create one for you. You or your tech writer will be using this `.hm` file later, when writing your help files. Please refer to Example C2 in Appendix C (page 622) for more on this subject.

NOTE: The help ID is automatically generated by the Dialog Editor and is based on your control's IDC number. In other words, if your control's ID is

```
#define IDC_BUTTON1                        1000
```

your help ID will be

```
#define HIDC_BUTTON1                       0x808203e8
```

where `3e8` at the end of the ID is the hex equivalent of the 1000 in `IDC_BUTTON1`. If you were to create another control in another dialog box with the exact same ID number of 1000, the help system will invoke the same help message as this one. Therefore, it's important to keep track of your controls' ID numbers and perhaps assign them manually so that the help for two controls don't interfere with each other.

Supporting the context help in a dialog box requires the handling of two messages to that dialog box: WM_HELPINFO and WM_HELPHITTEST. The first is sent by the system when your user has clicked on an item in your dialog box when in "What's this" mode. The second message is sent for the same reason, except only if your dialog is not currently active, as can be the case with modeless dialogs or dialog bars. In both cases, you will need to determine what was clicked on and, if it has a help ID, either return it or call up the help system yourself.

1. In each dialog class supporting What's This Help, add a WM_HELPINFO and WM_HELPHITTEST message handlers. (WM_HELPHITTEST must be added manually. WM_HELPINFO can be added with the Class Wizard)

```
BEGIN_MESSAGE_MAP( CWzdDialog, CDialog )
    // {{AFX_MSG_MAP( CWzdDialog )
    ON_WM_HELPINFO()
    // }}AFX_MSG_MAP
    ON_MESSAGE( WM_HELPHITTEST,OnHelpHitTest )
END_MESSAGE_MAP()
```

2. For the OnHelpInfo() handler, determine which control was selected and query its help ID. Then, use CWinApp::WinHelp() to call up the appropriate context help.

```
BOOL CWzdDialog::OnHelpInfo( HELPINFO* pHelpInfo )
{
    if ( pHelpInfo -> iContextType == HELPINFO_WINDOW )
    {
        DWORD helpId = -1;
        CWnd* pWnd = GetDlgItem( pHelpInfo -> iCtrlId );
        if ( pWnd )
        {
            helpId = pWnd -> GetWindowContextHelpId();
        }
        AfxGetApp() -> WinHelp( helpId, HELP_CONTEXTPOPUP );
    }
```

Example 36   Adding Context Sensitive Help   **391**

```
    // return CDialog::OnHelpInfo( pHelpInfo );
    return TRUE;
}
```

3. In the case of `OnHelpHitTest()`, you will have to determine what control was clicked before being able to query its help ID. This time, however, you can return the ID to MFC, which will invoke the help system.

```
LRESULT CWzdDialog::OnHelpHitTest( WPARAM wParam, LPARAM lParam )
{
    DWORD helpId = -1;
    // find the window that was clicked
    CWnd* pWnd =  ChildWindowFromPoint( CPoint( LOWORD( lParam ),
        HIWORD( lParam ) ), CWP_SKIPINVISIBLE|CWP_SKIPTRANSPARENT );
    if ( pWnd )
    {
        helpId = pWnd -> GetWindowContextHelpId();
    }
    return helpId;
}
```

4. Refer to the Appendix C for how to create an `.hlp` and `.cnt` file for your application.

Notes

- The AppWizard can also add some of this functionality for you automatically. Please see the Notes section in the previous example (page 386).

CD Notes

- When executing the project on the accompanying CD, you will notice that the help menu has been expanded to include "What's this" and that the toolbar has an additional "What's this" button. Click on the Test/Wzd menu commands to open a dialog box that has a "Help" button and a "?" button in the caption bar.

Example 37   Adding Bubble Help

Objective

You would like to add point and hover Bubble Help to your application, as seen in Figure 10.3.

Figure 10.3   Bubble Help

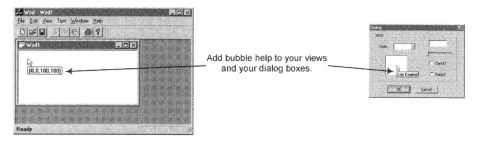

Add bubble help to your views and your dialog boxes.

Strategy

We will be using MFC's CToolTipCtrl class to provide Bubble Help. This class simply opens a small popup window when the user allows the mouse cursor to linger over a control or an area of the view. In this window will appear a short text message that you define for CToolTipCtrl. The trick to using CToolTipCtl is getting the mouse messages to it, since the WM_MOUSEMOVE messages it has to monitor are actually going to the window procedure of the window over which the mouse is positioned. For example, if the mouse is hovering over a button control, those mouse messages are going to that button control. Luckily, the tool tip control that CToolTipCtrl encapsulates provides subclassing functionality that allows you to intercept these messages. Unfortunately, the CToolTipCtrl implementation of the tool tip control, for whatever reason, doesn't make using this option easy. We will, therefore, be adding our own functions to our own CToolTipCtrl derived class to access the tool tip control directly.

Example 37   Adding Bubble Help   **393**

Steps

Derive a New `CToolTipCtrl` Class

1. Use the ClassWizard to create a new class derived from `CToolTipCtrl`.
2. Use the Text Editor to add a new function, `AddTool()`. This function will allow you to embed a single tool tip control in your dialog class, yet use it to support all of the controls in your dialog. You start by creating an empty `TOOLINFO` structure that is sent to the tool tip control.

```
BOOL CWzdToolTipCtrl::AddTool( UINT nID, LPCTSTR lpszText  )
{
    TOOLINFO ti;
    memset( &ti, 0, sizeof( TOOLINFO ) );
    ti.cbSize = sizeof( TOOLINFO );
```

3. Next, you store the text message the user passed you.

```
    ti.lpszText = ( LPSTR )lpszText;
```

4. Next, you ask the tool tip control to subclass the messages coming from the specified control ID.

```
    ti.uFlags = TTF_IDISHWND|TTF_SUBCLASS;
    ti.uId = ( UINT )GetParent() -> GetDlgItem( nID ) -> m_hWnd;
```

5. Next, you specify the owner of this tool tip control to give the tool tip window an owner.

```
    ti.hwnd = GetOwner() -> GetSafeHwnd();
```

6. Finally, you send this information to the control.

```
    return ( BOOL )SendMessage( TTM_ADDTOOL, 0, ( LPARAM )&ti );
}
```

Implement the New `CToolTipCtrl` Class in a Dialog Box

1. Embed this new class in your dialog class.

```
CWzdToolTipCtrl m_tooltips;
```

2. Use the Dialog Editor to modify the style of any static controls in this dialog to "Parent notify" if you want bubble help to appear over them.

You *do not* need to make this change for any control other than a static control because all other types of controls automatically notify their parent window.

3. Use the ClassWizard to add a `WM_INITDIALOG` message handler to this dialog class and create the bubble help window.

```
m_tooltips.Create( this );
```

Note: Even though we just created a window, it remains invisible until the tool tip control needs to display bubble help for a control. Also, note that the exact same window is being used to display bubble help for all of the controls in this dialog.

---

4. Also in the `WM_INITDIALOG` handler, specify the help message for any control for which you want Bubble Help by using the `AddTool()` function you added previously.

```
m_tooltips.AddTool(
    IDC_RADIO1,        // control id
    "Radio Control"    // bubble help message
    );
```

5. Finally, activate the tooltip.

```
m_tooltips.Activate( TRUE );
```

6. To change the help text message for a control later on, you can use

```
m_tooltips.UpdateTipText(
    "Checked Box",     // new text
    GetDlgItem(
        IDC_CHECK1     // the control's id
        )
    );
```

That does it for adding Bubble Help for the controls in a dialog box. This solution can also be used whenever you have child windows in a parent window but don't want to have to implement a separate `CToolTipCtrl` control for each child window. Next, we will create a new `CToolTipCtrl` derived class to cause Bubble Help to appear only over certain areas of another window, typically the view.

Example 37   Adding Bubble Help   **395**

Add to the New `CToolTipCtrl` **Derived Class**

1. Use the Text Editor to add a new function, `AddArea()` to your new tool tip class. Again, you start by creating an empty `TOOLINFO` structure that is sent to the tool tip control.

```
BOOL CWzdToolTipCtrl::AddArea( UINT nID, LPRECT lpRect,
    LPCTSTR lpszText )
{
    TOOLINFO ti;
    memset( &ti, 0, sizeof( TOOLINFO ) );
    ti.cbSize = sizeof( TOOLINFO );
```

2. Next, specify the text to be displayed.

```
    ti.lpszText = ( LPSTR )lpszText;
```

3. Next, tell the tool tip control which window should be monitored (the view) and which area to monitor.

```
    ti.hwnd = GetOwner() -> GetSafeHwnd();
    memcpy( &ti.rect, lpRect, sizeof( RECT ) );
```

4. Also, tell the tool tip control to subclass this window so the control can intercept its mouse messages.

```
    ti.uFlags = TTF_SUBCLASS;
```

5. Next, specify an ID you can define that will allow you to modify this tool tip later. This can be any unsigned integer number.

```
    ti.uId = nID;
```

6. Now, send this definition to the tool tip control and return.

```
    return ( BOOL )SendMessage( TTM_ADDTOOL, 0, ( LPARAM )&ti );
}
```

Implement the New `CToolTipCtrl` **Class in a View**

1. Embed this new tool tip control class in your View Class.

```
CWzdToolTipCtrl m_tooltips;
```

II

10

2. Create this new tool tip control class in the WM_CREATE message handler of your view.

```
m_tooltips.Create( this );
```

3. In this same handler, or any where else, define areas of the View in which you want to display a Bubble Help message.

```
m_tooltips.AddArea(
    1,                  // user defined id
    &rect1,             // a CRect value defining the area in view
                        //     coordinates
    "message",          // the bubble help message
    );
```

4. Before you leave this handler, activate the tool tip control with

```
m_tooltips.Activate( TRUE );
```

5. To change the help text message later on, use

```
m_tooltips.UpdateTipText(
    "new text",     // the new text message
    this,           // pointer to view class
    1               // the id defined above
    );
```

6. To change the area to which this tip refers, you can use

```
m_tooltips.SetToolRect(
    this,       // pointer to view class
    1,          // the id defined above
    &rect       // a new CRect value defining
                //     the area in view coordinates
    );
```

7. To see a complete listing of the tool tip class, please refer to Tooltip Class for Example 37 (page 397).

Notes

- Another way to approach the dialog solution would have been to add an individual tool tip control to each dialog control. However, this would involve creating your own derivation of every control class, just so you could embed a tool tip class in it.

Example 37   Adding Bubble Help   **397**

- The tool tip control displays a single line of text. For a particularly long piece of information or for certain types of information, you may want to create multiple lines of text. Unfortunately, just embedding a \n character in your text won't do it. Instead, you must add a WM_PAINT and WM_NCPAINT message handler to your derivation of the tool tip control and paint the window yourself. Get the text message to display with GetWindowText().

CD Notes

- When executing the project on the accompanying CD, you will notice that allowing the mouse cursor to hover in the upper-left corner of the view will cause a Bubble Help window to open. Then, click on the Test/Wzd menu commands to open a dialog. Allowing the cursor to hover over any cntrols in this dialog will cause a Bubble Help message to open.

- Also provided in this project are WzdTT2.cpp and .h files that are the basis of creating your own tool tip help from scratch.

Tooltip Class for Example 37

```
#if !defined( AFX_WZDTLTIP_H__1E222DA3_97EF_11D2_A18D_C32FFDBA4686__INCLUDED_ )
#define AFX_WZDTLTIP_H__1E222DA3_97EF_11D2_A18D_C32FFDBA4686__INCLUDED_

#if _MSC_VER > 1000
#pragma once
#endif    // _MSC_VER > 1000
// WzdTlTip.h : header file
//

//////////////////////////////////////////////////////////////////////////////////
// CWzdToolTipCtrl window

class CWzdToolTipCtrl : public CToolTipCtrl
{
// Construction
public:
    CWzdToolTipCtrl();
```

```
// Attributes
public:

// Operations
public:
    BOOL AddTool( UINT nID, LPCTSTR lpszText = LPSTR_TEXTCALLBACK  );
    BOOL AddTool( UINT nID, UINT nIDText  );
    BOOL AddArea( UINT nID, LPRECT lpRect,
        LPCTSTR lpszText = LPSTR_TEXTCALLBACK  );
    BOOL AddArea( UINT nID, LPRECT lpRect, UINT nIDText );

// Overrides
    // ClassWizard generated virtual function overrides
    // {{AFX_VIRTUAL( CWzdToolTipCtrl )
    // }}AFX_VIRTUAL

// Implementation
public:
    virtual ~CWzdToolTipCtrl();

    // Generated message map functions
protected:
    // {{AFX_MSG( CWzdToolTipCtrl )
    // NOTE - the ClassWizard will add and remove member functions here.
    // }}AFX_MSG

    DECLARE_MESSAGE_MAP()
};


//////////////////////////////////////////////////////////////////////////


// {{AFX_INSERT_LOCATION}}
// Microsoft Visual C++ will insert additional declarations immediately before
//    the previous line.

#endif
    // !defined( AFX_WZDTLTIP_H__1E222DA3_97EF_11D2_A18D_C32FFDBA4686__INCLUDED_ )
// WzdTlTip.cpp : implementation file
//
```

Example 37   Adding Bubble Help   **399**

```
#include "stdafx.h"
#include "wzd.h"
#include "WzdTlTip.h"


#ifdef _DEBUG
#define new DEBUG_NEW
#undef THIS_FILE
static char THIS_FILE[] = __FILE__;
#endif


/////////////////////////////////////////////////////////////////////////////
// CWzdToolTipCtrl


CWzdToolTipCtrl::CWzdToolTipCtrl()
{
}


CWzdToolTipCtrl::~CWzdToolTipCtrl()
{
}



BEGIN_MESSAGE_MAP( CWzdToolTipCtrl, CToolTipCtrl )
    // {{AFX_MSG_MAP( CWzdToolTipCtrl )
    // NOTE - the ClassWizard will add and remove mapping macros here.
    // }}AFX_MSG_MAP
END_MESSAGE_MAP()


/////////////////////////////////////////////////////////////////////////////
// CWzdToolTipCtrl message handlers

BOOL CWzdToolTipCtrl::AddTool( UINT nID, LPCTSTR lpszText )
{
    TOOLINFO ti;
    memset( &ti, 0, sizeof( TOOLINFO ) );
    ti.cbSize = sizeof( TOOLINFO );
    ti.hwnd = GetOwner() -> GetSafeHwnd();
    ti.uFlags = TTF_IDISHWND|TTF_SUBCLASS;
    ti.uId = ( UINT )GetParent() -> GetDlgItem( nID ) -> m_hWnd;
    ti.lpszText = ( LPSTR )lpszText;
```

II

10

```
        return ( BOOL )SendMessage( TTM_ADDTOOL, 0, ( LPARAM )&ti );
}


BOOL CWzdToolTipCtrl::AddTool( UINT nID, UINT nIDText )
{

    CString str;
    VERIFY( str.LoadString( nIDText ) );
    return AddTool( nID,str );
}


BOOL CWzdToolTipCtrl::AddArea( UINT nID, LPRECT lpRect, LPCTSTR lpszText )
{

    TOOLINFO ti;
    memset( &ti, 0, sizeof( TOOLINFO ) );
    ti.cbSize = sizeof( TOOLINFO );
    ti.hwnd = GetOwner() -> GetSafeHwnd();
    ti.uFlags = TTF_SUBCLASS;
    ti.uId = nID;
    memcpy( &ti.rect, lpRect, sizeof( RECT ) );
    ti.lpszText = ( LPSTR )lpszText;
    return ( BOOL )SendMessage( TTM_ADDTOOL, 0, ( LPARAM )&ti );
}


BOOL CWzdToolTipCtrl::AddArea( UINT nID, LPRECT lpRect, UINT nIDText )
{

    CString str;
    VERIFY( str.LoadString( nIDText ) );
    return AddArea( nID,lpRect,str );
}
```

Chapter 11

Plain Windows

Almost all of the windows in an MFC application are specialized—they're either a dialog box or a view window or a control window. They're all variations of the same basic, plain window. The examples in this chapter show you how to create a general-purpose window, which can be used for anything.

The user interface of a Windows application is composed entirely of individual windows with different sizes and styles. More often than not, you will use the editors and wizards in the Developer's Studio to automatically add windows to your application. (Examples of this can be found throughout this book.) Unfortunately, the wizards can only create a select set of windows, outside of which you need to create your own window manually.

In this chapter are examples of manually creating any type of window anywhere within your application's user interface.

Example 38   Creating a Generic Window   We will look at how to create a window using just MFC's generic window process.

Example 39   Creating a Window Class—The Short Version   We will look at how to create a custom Window Class using MFC's `AfxRegisterWndClass()`, which fills in a lot of the class options automatically.

Example 40   Creating a Window Class—The Long Version   We will look at how to create a Window Class using MFC's AfxRegisterClass(), which gives us total control over the creation of a Window Class.

Example 38   Creating a Generic Window

Objective

You would like to create a general-purpose window, such as one of the four shown in Figure 11.1.

Figure 11.1   Four Plain Windows

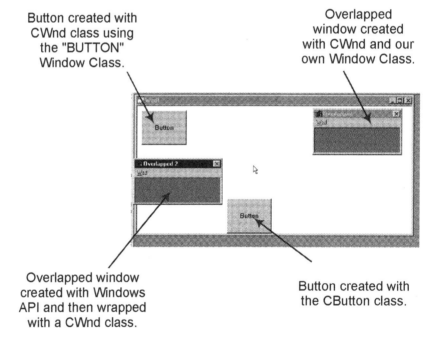

Button created with
CWnd class using
the "BUTTON"
Window Class.

Overlapped
window created
with CWnd and our
own Window Class.

Overlapped window
created with Windows
API and then wrapped
with a CWnd class.

Button created with
the CButton class.

Strategy

We will use the MFC generic window class, CWnd, to create windows. We will also look at creating a window directly using the Windows API. We will then look at a way to connect a CWnd object to an existing window.

Example 38   Creating a Generic Window   **403**

Steps

Create a Generic Window Using CWnd

1. To create a generic window using MFC, you can use

```
CWnd wnd;
HMENU hMenu = ::LoadMenu( NULL,MAKEINTRESOURCE( IDR_WZD_MENU ) );
wnd.CreateEx(
    0,                          // extended window style
    _T( "AfxWnd" ),             // MFC window class name
    "Caption",                  // window caption
    WS_CHILD|WS_VISIBLE,        // window style
    10,10,                      // x,y position
    100,75,                     // width and height
    m_hWnd,                     // parent window handle
    hMenu                       // menu handle,
                                //    or if child window, a window id

    );
```

2. The window class name we used previously is for a generic MFC window. To create a button control, you would simply substitute AfxWnd for BUTTON.

```
CWnd wndButton;
wndButton.CreateEx( 0,         // extended window style
    _T( "BUTTON" ),            // window class name
    "My Button",               // window caption
    WS_CHILD|WS_VISIBLE,       // window style
    10,10,                     // x,y position
    100,75,                    // width and height
    hWnd,                      // parent window handle
    ( HMENU )IDC_WZD_BUTTON    // in this case, a button id
    );
```

II

11

3. To create the exact same button using the MFC `CButton` class, you can use the following. The `Create()` member function of `CButton` simply does what we did above.

```
CRect rect( 200,200,300,275 );
CButton button;
button.Create(
    "Button",                    // window caption
    WS_CHILD|WS_VISIBLE,         // window style
    rect,                        // position and dimensions
    this,                        // parent window class
    IDC_WZD_BUTTON               // button id
    );
```

Create a Generic Window Using the Windows API

1. To create a window using the Windows API directly, you can use the following. In this example, we are creating an overlapped window. Please see the section "Notes" on page 405 for the difference between overlapped, popup, and child windows.

```
HWND hWnd = ::CreateWindowEx(
    WS_EX_CLIENTEDGE,            // extended window style
    "AfxWnd",                    // windows class name
    "Overlapped",                // window caption
    WS_CAPTION|WS_SYSMENU|WS_OVERLAPPED|WS_VISIBLE|WS_DLGFRAME,
                                 // window style
    220,220,200,100,             // position and dimensions
    NULL,                        // owner window handle--NULL is Desktop
    hMenu,                       // for popup and overlapped windows
    AfxGetInstanceHandle(),      // handle to application instance
    NULL                         // pointer to window-creation data
    );
```

Example 38   Creating a Generic Window   **405**

Wrap Window Objects with CWnd **Objects**

1. You can encapsule the window you just created with a CWnd class by using

```
CWnd wndWrapper;
wndWrapper.Attach( hWnd );
```

Once wrapped, the CWnd class will now destroy this window when the CWnd class is deconstructed.

2. To deconstruct a CWnd class or any of its derivations (e.g., CButton) without destroying the window to which that it's attached, you can detach it first with

```
HWND hWnd = wndWrapper.Detach();
::DestroyWindow( hWnd );
```

3. If you want a window to destroy the MFC class to which it's attached when the window is destroyed, use the ClassWizard to add a WM_NCDESTROY message handler to that class. Then, add the following code to that handler.

```
void CWzdWnd::OnNcDestroy()
{
    CWnd::OnNcDestroy();

    delete this;
}
```

Notes

- Generic windows are the basis of all the windows you see in a Windows interface. There are three general types of windows—overlapped, popup and child, which are the basis for, respectively, Main Windows, Dialog and Message Boxes, and Control windows (e.g., buttons). The Windows API usually draws a great deal of an overlapped or popup window for you in what's called the nonclient area of the window. Child windows, on the other hand, are mostly drawn by its window procedure, which is specified in a Window Class. Control windows are child windows with unique window procedures that draw unique controls. For more detail on windows, please refer to Chapter 1.

- After calling CreateEx() or CreateWindow(), an MFC window is created using a sequence of windows messages from the Windows API to that window's window procedure. For this sequence of events, please refer to Appendix A.

- Adding a WM_NCDESTROY message handler to deconstruct an MFC class is usually unnecessary. The class is usually either embedded in another class that will deconstruct it when the application terminates or the class has been allocated on the stack in a function and will deconstruct when the function returns. One case that requires you to manually destroy the class, however, is the with modeless dialog box. A modeless dialog's window can be destroyed by the user just by clicking on the close button and your class can be left without a window to talk to and no way to be destroyed. The result is memory leaks.

- The window classes we used here come supplied by Windows and MFC. To create your own windows class, please refer to the next two examples.

CD Notes

- When executing the project on the accompanying CD, you will notice the view is filled with four basic windows created in four different ways.

Example 39 Creating a Window Class—The Short Version

Objective

You would like to create a general-purpose window class to be used when creating a window.

Strategy

There are two functions provided by the MFC framework for creating and registering a windows class. In this example, we will be using the shorter version, AfxRegisterWndClass(). To get more control over the creation of a windows class, please refer to the next example.

Example 39   Creating a Window Class—The Short Version   **407**

Steps

Create a New Window Class with `AfxRegisterWndClass()`

1. To create a window class, you can use

```
lpszClass = AfxRegisterWndClass(
    // window class styles
    CS_DBLCLKS |                          // convert two mouse clicks into
                                          //      a double click to this
                                          //      window's process

    CS_HREDRAW |                          // send WM_PAINT to window
                                          //      if horizontal size changes

    CS_VREDRAW                            // send WM_PAINT to window
                                          //      if vertical size changes

    // CS_OWNDC |                         // every window created from
                                          //      this class gets its very
                                          //      own device context

    // CS_PARENTDC |                      // device context created for
                                          //      this window allows drawing
                                          //      in parent window too

    // CS_NOCLOSE |                       // disable the close command
                                          //      on the System menu

    ::LoadCursor(NULL,IDC_CROSS),         // window class cursor
                                          //      or NULL for default arrow
                                          //      cursor (this cursor is
                                          //      displayed the when mouse
                                          //      cursor is over a window
                                          //      created with this class)

    (HBRUSH)(COLOR_BACKGROUND+1),         // background color
                                          //      or NULL for no background
                                          //      erase (if NULL, window
                                          //      will not erase
                                          //      background for you)
```

II

11

```
        AfxGetApp() -> LoadIcon(IDI_WZD_ICON)    // window icon or
                                                 //      NULL for default
                                                 //      icon (icon
                                                 //      displayed in
                                                 //      window caption
                                                 //      or in minimized
                                                 //      window)
    );
```

The `AfxRegisterWndClass()` generates a new Window Class name for you. To create a window using this new Window Class, simply create it using this generated name.

Use the Window Class Created with `AfxRegisterWndClass()`

1. To use the new Window Class, put the name generated by `AfxRegister-WndClass()` into the `CWnd::CreateEx()` function.

```
CWnd wnd;
wnd.CreateEx( 0,lpszClass,"",WS_OVERLAPPEDWINDOW|WS_VISIBLE,
    100,100,200,100,NULL,NULL );
```

Create the Simplest Window Class

1. To create the simplest window class using `AfxRegisterWndClass()`, you can use

```
lpszClass = AfxRegisterWndClass( 0 );
```

A window created with this class will have an arrow cursor, default icon, and no background erasing.

Notes

- A window class name is simply a text string identifying a window class structure that's been registered in the system. The class structure maintains your window class's style, background color, and window process, which are then all used to initialize a window when it's created. For more on window classes, please refer to Chapter 1.

- The `AfxRegisterWndClass()` creates and initializes a window class structure for you. For the window procedure, it uses the generic MFC window process call `AfxWndProc`. It even creates a window class name for you

Example 40   Creating a Window Class—The Long Version   **409**

based on the arguments you pass to it. One drawback to this approach, however, is that if you call this function twice with the exact same arguments, you will be creating only one window class. Normally, this isn't a problem unless you use the discontinued `CS_CLASSDC`. Even then, it's only a problem if two windows created with this window class try to draw at the same time.

- If you would like more control over your window class, including what name it will have, please refer to the next example.

CD Notes

- When executing the project on the accompanying CD, set a breakpoint on the `OnTestWzd1()` function in `WzdView`. Then, click on the "Test" and then "Wzd1" menu commands and trace through the application as it creates two window classes and then creates two windows from them.

Example 40   Creating a Window Class—The Long Version

Objective

You would like to create a specific window class and be able to specify a window class name.

Strategy

There are two functions provided by the MFC framework for creating and registering a windows class. In this example, we will be using the longer version, `AfxRegisterClass()`, which allows the most control over creation of a windows class. To create a quick, general-purpose window class, please refer to the previous example.

Steps

Create a Window Class with AfxRegisterClass()

1. To create a window class, you must first initialize a WNDCLASS structure.

```
WNDCLASS wndclass =
{

    // window class styles
    CS_DBLCLKS |             // convert two mouse clicks into a double
                            //     click to this window's process

    CS_HREDRAW |            // send WM_PAINT to window if horizontal
                            //     size changes

    CS_VREDRAW             // send WM_PAINT to window if vertical
                            //     size changes

    // CS_GLOBALCLASS |     // class is available to all process threads

    // CS_OWNDC |           // every window created from this class
                            //     gets its very own device context

    // CS_PARENTDC |        // device context created for this window
                            //     allows drawing in parent window too

    // CS_NOCLOSE |         // disable the close command on the
                            //     System menu

    AfxWndProc,            // window process for every window created
                            //     from this class

    0,0,                    // extra window and class bytes unused
                            //     in MFC

    AfxGetInstanceHandle(),    // handle of this application's instance
    AfxGetApp() -> LoadIcon( IDI_WZD_ICON ),    // window icon or NULL
    ::LoadCursor( NULL,IDC_CROSS ),          // window class cursor or
                                            //     NULL for default
                                            //     arrow cursor

    ( HBRUSH )( COLOR_BACKGROUND + 1 ),    // background color or NULL
                                            //     for no background
                                            //     erase

    MAKEINTRESOURCE( IDR_WZD_MENU ),       // menu to be used when
```

Example 40   Creating a Window Class—The Long Version   **411**

```
                                        //      creating windows
                                        //      using this class
    "MyClassName"                       // a class name you are
                                        //      assigning this
                                        //      windows class

};
```

2. Then, you can register your window class with the system using

```
AfxRegisterClass( &wndclass );
```

Use the Window Class Created with AfxRegisterClass()

1. To use the new Window Class, put its name into the CWnd::CreateEx() function.

```
CWnd wnd;
wnd.CreateEx( 0, "MyClassName",          <<< new class name
    "",WS_OVERLAPPEDWINDOW|WS_VISIBLE,
    100,100,200,100,NULL,NULL );
```

Notes

- All windows created for use with MFC use the AfxWndProc window procedure to allow messages to MFC windows to be processed in the same way. The window procedure determines the look and feel of a window. A BUTTON window class has a special window procedure that will draw a button in response to a WM_PAINT message to its window procedure. For more on this, please refer to Chapter 1.

- The icon specified here is the one that appears in the upper-left corner of a window with a caption or in a minimized window. A default icon is supplied by the system and varies between operating systems.

- The cursor specified here will appear when the mouse moves over the client area of the window created with this class. If unspecified, a default arrow cursor is used.

- If you set the background brush to NULL, the system won't erase the background of your window. This means that when your window is created, all of the other nonclient areas of your window will still be drawn but the client area won't be erased. Whatever was below the window will still appear in your client area. To erase the background of the client area yourself, use the ClassWizard to add a WM_ERASEBKGRND message handler to your MFC class. There, you can draw a rectangle to fill in the background, or supply some other fancy background.

- The menu handle you supply here will be incorporated into every window created from this window class. You can override this default menu by simply supplying a new menu handle to CWnd::CreateEx() when creating the window. If a menu handle is omitted from the window class and CWnd::CreateEx(), your window will be created without a menu. Child windows cannot have a menu—this entry in the window class is unused. However, in CWnd::CreateEx(), it's used as a control ID.

- The class name you supply here can be any text name, including the name of an existing window class, such as BUTTON. Naming your class after an existing class, however, means that all future windows will be created from your new class. For example, if you name your window class BUTTON, all future button controls will be created using your class and not the system's. Please see the section "Subclassing a Window Procedure" on page 9 for a brief discussion of superclassing.

- The window classes your application registers will only be available to your application. If you use the CS_GLOBALCLASS class style, your window class will also be available to every thread your application creates, but no other application in the system (as was the case in Windows 3.1). When your application terminates, any windows classes it has created will be unregistered and destroyed.

Chapter 12

Specialized Applications

The examples in this chapter have little in common, other than that they are simple to create or are frequently created. MFC already comes with two elaborate examples of editors (Notepad and Wordpad), but because of that elaboration, it's hard to see how easy it is to create an unadorned one. An Explorer interface, while not exactly easy to create, is still an interface that's as popular to use as an SDI or MDI interface.

Example 41   Creating a Simple Text Editor We will create a fairly complete text editor without much effort.

Example 42   Creating a Simple RTF Editor We will create an editor with the limited ability to manipulate fonts and paragraphs like a word processor.

Example 43   Creating an Explorer Interface We will create the outsides of an Explorer-like application, with the tree control on the left side and a list control on the right side.

Example 44   Creating a Simple ODBC Database Editor   We will create a simple database editor that can manipulate an ODBC-compliant database.

Example 45   Creating a Simple DAO Database Editor   We will create a simple database editor that can manipulate a DAO-compliant database.

Example 46   Creating a Simple Wizard   We will create a simple wizard using a property sheet.

Example 41   Creating a Simple Text Editor

Objective

You would like to create a simple text editor using the edit control, as seen in Figure 12.1.

Figure 12.1   A Simple Text Editor

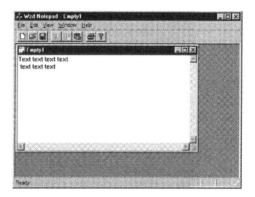

Strategy

We will use the AppWizard to make a simple text editor using MFC's CEdit-View. We will also look at a way of creating a text editor view for an existing application using the ClassWizard.

Example 41   Creating a Simple Text Editor   **415**

Steps

Create a Text Editor with the AppWizard

1. Use the AppWizard to create either an SDI or MDI application.

2. While still in the AppWizard, use the "Advanced Options" to make the "File extension" for this project *.txt.

3. In the last step of the AppWizard, you are shown an inventory of the classes that will be created for this project. Select the CXxxView class, where Xxx is your project's name. Below, you'll notice that the base class combo box has become enabled. You should now select CEditView from that combo box.

4. That's it! The AppWizard will now create the appropriate View Class, which will now support cut and paste, and the Document Class will be modified to load and save text documents. The Common File Dialog will be used to prompt your user for file names to load and save using the *.txt extension.

Create a Text Editor View with the ClassWizard

1. In an existing application, use the ClassWizard to create a new view class derived from CEditView.

2. Use the ClassWizard to create a new document class derived from CDocument.

3. In the new document class, add the following to the Serialize() function to allow text documents to be loaded and saved.

```
void CWzdDoc::Serialize(CArchive& ar)
{
    POSITION pos = GetFirstViewPosition();
    ( ( CWzdEditView * )GetNextView( pos ) ) -> SerializeRaw( ar );
}
```

4. If you are creating an SDI application, you will need to blank out your View Class when a new document is loaded. To do this, add the following code to your Document Class's `OnNewDocument()` function.

```
BOOL CWzdDoc::OnNewDocument()
{
    if ( !CDocument::OnNewDocument() )
        return FALSE;

    POSITION pos = GetFirstViewPosition();         <<< add
    ( ( CWzdEditView * )GetNextView( pos ) ) ->
        SetWindowText(NULL);                        <<< add

    return TRUE;
}
```

5. You can now create a new document template class in your Application Class using these two new classes.

Notes

- The `CEditView` class is a hybrid of the `CView` class and a `CEdit` control class. The `CEditView` class also includes a member function, `SerializeRaw()` that makes loading and saving text documents easy.

- Several excellent examples of a text editor already come with your MFC VC++ package. However, because they all include several example bells and whistles, the resulting code tends to obscure just how easy it is to create a text editor using the built-in features of MFC.

CD Notes

- When executing the project on the accompanying CD, you will notice that it has all of the functionality of a simple text editor.

Example 42   Creating a Simple RTF Editor

Objective

You would like to create a simple rich text editor using the rich edit control, as seen in Figure 12.2.

Example 42   Creating a Simple RTF Editor   **417**

Figure 12.2   A Simple Rich Text Editor

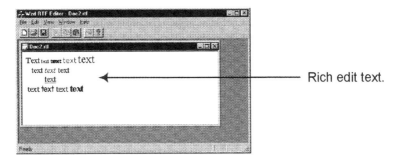

Rich edit text.

II

Strategy

We will use the AppWizard to create a simple Rich Text Format (RTF) text editor using MFC's `CRichEditView` and `CRichEditDoc` classes. We will also look at how to use the ClassWizard to add an RTF text editor view to an existing application.

Steps

Create a Text Editor with the AppWizard

1. Use the AppWizard to create either an SDI or MDI application.
2. While still in the AppWizard, use the "Advanced Options" to make the "File extension" for this project `*.rtf`.
3. In the last step of the AppWizard, you are shown an inventory of the classes that will be created for this project. Select the `CXxxView` class, where `Xxx` is your project's name. Below, you'll notice that the base class combo box has become enabled. You should now select `CRichEditView` from that combo box.
4. That's it! The AppWizard will now create the appropriate View Class, which will now support cut and paste, and the Document Class will be modified to load and save rich text documents. The Common File Dialog will be used to prompt your user for file names to load and save using the `*.rtf` extension.

12

Create a Rich Text Editor View with the ClassWizard

1. Use the ClassWizard to create a new view class derived from `CRichEditView`.

2. There is currently no way to automatically create a Document Class derived from `CRichEditDoc`, so use the ClassWizard to create a new class derived from `CDocument`, instead. Then, manually edit the resulting `.h` and `.cpp` files to substitute all occurrences of `CDocument` for `CRichEditDoc`.

3. Manually add the following override to your new Document Class. This is a helper function for your `CRichEditDoc` that returns a class you will create in the next step and allows your document to contain COM objects, as well as RTF text.

```
CRichEditCntrItem* CWzdDoc::CreateClientItem( REOBJECT* pReo ) const
{
    return new CWzdCntrItem( pReo, ( CWzdDoc* )this );
}
```

4. Manually create a class from the source found in the Rich Edit Container Item Class for Example 42 (page 419).

5. You can now create a new document template class in your Application Class using these two new classes.

Notes

- The `CRichEditView` class is a hybrid of the `CView` class and a `CRichEditCtrl` control class. The `CRichEditDoc` class simplifies loading and saving RTF files. A third MFC class called `CRichEditCntrItem` helps the `CRichEditDoc` class to load and save your RTF documents.

- If you plan to use the `CRichEditCtrl` control class alone, you must call `AfxInitRichEdit()` from your Application Class's `OnInitInstance()` function. This initializes your application to use a Rich Edit Control. The `CRichEditView` class calls this function for you.

- The Rich Text Control embedded into your Rich Edit View also has built-in functionality that makes formatting fonts and paragraphs trivial. Please refer to your MFC documentation on the `CRichEditCtrl` class for how.

Example 42   Creating a Simple RTF Editor   **419**

CD Notes

When executing the project on the accompanying CD, you will notice that it can open and display .rtf files.

Rich Edit Container Item Class for Example 42

```
// WzdCntrItem.h : interface of the CWzdCntrItem class
//
/////////////////////////////////////////////////////////////////////////////

#if !defined WZDCNTRITEM
#define WZDCNTRITEM

class CWzdDoc;
class CWzdRichEditView;

class CWzdCntrItem : public CRichEditCntrItem
{
    DECLARE_SERIAL( CWzdCntrItem )

// Constructors
public:
    CWzdCntrItem( REOBJECT* pReo = NULL, CWzdDoc* pContainer = NULL );

// Attributes
public:
    CWzdDoc* GetDocument()
        { return (CWzdDoc*)COleClientItem::GetDocument(); }
    CWzdRichEditView* GetActiveView()
        { return (CWzdRichEditView*)COleClientItem::GetActiveView(); }

    // ClassWizard generated virtual function overrides
    // {{AFX_VIRTUAL(CWzdCntrItem)
public:
protected:
    // }}AFX_VIRTUAL

// Implementation
public:
```

II

12

```
#ifdef _DEBUG
    virtual void AssertValid() const;
    virtual void Dump(CDumpContext& dc) const;
#endif
};


/////////////////////////////////////////////////////////////////////////////
#endif
// WzdCntrItem.cpp : implementation of the CWzdCntrItem class
//

#include "stdafx.h"
#include "Wzd.h"

#include "WzdDoc.h"
#include "WzdRichEditView.h"
#include "WzdCntrItem.h"

#ifdef _DEBUG
#undef THIS_FILE
static char BASED_CODE THIS_FILE[] = __FILE__;
#endif


/////////////////////////////////////////////////////////////////////////////
// CWzdCntrItem implementation

IMPLEMENT_SERIAL( CWzdCntrItem, CRichEditCntrItem, 0 )

CWzdCntrItem::CWzdCntrItem( REOBJECT *pReo, CWzdDoc* pContainer )
    : CRichEditCntrItem( pReo, pContainer )
{
}


/////////////////////////////////////////////////////////////////////////////
// CWzdCntrItem diagnostics

#ifdef _DEBUG
void CWzdCntrItem::AssertValid() const
{
```

Example 43   Creating an Explorer Interface   **421**

```
    CRichEditCntrItem::AssertValid();
}

void CWzdCntrItem::Dump( CDumpContext& dc ) const
{
    CRichEditCntrItem::Dump( dc );
}
#endif
```

Example 43   Creating an Explorer Interface

Objective

You would like to create an application interface to resemble Windows Explorer, as seen in Figure 12.3.

Figure 12.3   An Explorer Interface

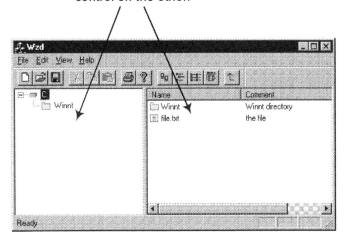

Just like Windows
Explorer(tm), this example
shows how to create an
interface with a tree control
on one side and a list
control on the other.

Strategy

Windows Explorer is an SDI application that has one document, but two views. The first view is a tree view and the second is a list control view. Both of these views, in fact, sit inside of another window called a splitter window that fills up the inside of the main window. The AppWizard that comes with v6.0 of Visual C++ will allow you to select and create this interface automatically (in Step Five). However, it doesn't fill in the logic necessary to synchronize these two views.

Therefore, in this example, we will first create this interface manually, in the event you're not using v6.0, by modifying the CMainframe class to split a tree view into two parts and put a list view into the second part. Then, for both cases, we will fill these views using standard CTreeView and CListView class calls. We will synchronize these views so that what's in the tree view relates to what's in the list view, and vice versa. To allow any level of subdirectories to be displayed, we will be making heavy use of recursion.

We will be displaying file directories in this example, but any type of hierarchical organization can be displayed using these methods.

Steps

Create the Application

1. Use the AppWizard to create an SDI application. In the last step, you are given an inventory of the classes that will be created for your project. Select the CXxxView class where Xxx is your project's name. Then, under "Base class", select "CTreeView". Click on "Finish" to complete your project.

2. Use the ClassWizard to create a new view class derived from CListView.

3. Embed a splitter window class in your CMainFrame class.

```
private:
    CSplitterWnd m_wndSplitter;
```

Example 43   Creating an Explorer Interface   **423**

4. Use the ClassWizard to add an override to your `CMainFrame` class for the `OnCreateClient()` function. There, you will start by creating a splitter window with two columns.

```
BOOL CMainFrame::OnCreateClient( LPCREATESTRUCT lpcs,
    CCreateContext* pContext )
{
    // create a split window with 1 row and 2 columns
    if ( !m_wndSplitter.CreateStatic( this, 1, 2 ) )
    {
        TRACE0( "Failed to CreateStaticSplitter\n" );
        return FALSE;
    }
}
```

5. Next, in `OnCreateClient()`, tell the splitter window to use the tree view passed to this function as our the view in the first column.

```
CRect rect;
GetClientRect( &rect );
if ( !m_wndSplitter.CreateView( 0, 0, pContext ->
    m_pNewViewClass, CSize( rect.Width()/4, 50 ), pContext ) )

{
    TRACE0( "Failed to create first pane\n" );
    return FALSE;
}
```

Notice that we are making the width of this column one-fourth the width of our application's main window.

6. Next, in `OnCreateClient()`, tell the splitter window that our second column is the list view we created with the ClassWizard.

```
if ( !m_wndSplitter.CreateView( 0, 1, RUNTIME_CLASS( CWzdListView ),
    CSize( 200,50), pContext ) )
{
    TRACE0( "Failed to create second pane\n" );
    return FALSE;
}
```

7. Finally, in `OnCreateClient()`, we tell `CMainFrame` that the list view will initially be the active view. We also return `TRUE` without calling the original

`CFrameWnd::OnCreateClient()`, which would just undo all the work we just did!

```
SetActiveView( ( CView* )m_wndSplitter.GetPane( 0,1 ) );


    return TRUE;      // CFrameWnd::OnCreateClient( lpcs, pContext );
```

Fill in the Tree View Class

1. Use the ClassWizard to add an `OnInitialUpdate()` override to your tree view class. There, you will create and load the bitmaps we will be displaying in our tree control.

```
void CWzdTreeView::OnInitialUpdate()
{

    CTreeView::OnInitialUpdate();


    m_ImageList.Create( IDB_SMALL_BITMAP, 16, 1, RGB( 0,0,0 ) );
    GetTreeCtrl().SetImageList( &m_ImageList, TVSIL_NORMAL );


}
```

The basic action of an Explorer-type interface is that a single click on an item in the tree view opens the list view to that point in the document. A double-click on the list view does the same for the tree view. We will be synchronizing these two views using our Document Class's `UpdateAllViews()`. Therefore, both views will need to override the `OnUpdate()` function and use the supplied `lHint` to determine its action, as seen next.

2. Use the ClassWizard to override the `OnUpdate()` function of the tree view class. Then, check to see if this update was intended solely to update the list view and, if so, return immediately.

```
void CWzdTreeView::OnUpdate( CView* pSender, LPARAM lHint,
    CObject* pHint )
{

    if ( lHint&LIST_VIEW_ONLY )

    {

        return;

    }
```

Example 43   Creating an Explorer Interface   **425**

3. Next, in this `OnUpdate()`, use a case statement to determine how to update the view. In the default case, we will simply use a helper function to fill in the view with the document, as seen here.

```
switch ( lHint )
{
    default:
        GetTreeCtrl().DeleteAllItems();
        AddBranch( TVI_ROOT, GetDocument() -> GetWzdList() );
        break;
```

The helper function `AddBranch()` can be found in the Tree View for Example 43 (page 433). It uses a lot of recursion to fill in the view with not only a name but also a pointer to a data object that further identifies the item. This data object contains information on the name, category (e.g., file, directory, etc.), comment, and object ID of an item. The object ID is especially important when trying to locate an item from one view inside the other view, as we will see next.

4. If this update is as a result of the user having picked something new in the list view, we will need to select the same item here in the tree view. We accomplish this by first locating the identical item in the tree view. Then we select it, like so.

```
case TREE_VIEW_SELECT:
    pInfo = GetDocument() -> GetSelection();
    if ( (hitem =
        FindTreeItem( GetTreeCtrl().GetChildItem( GetTreeCtrl().GetRootItem() ),
        pInfo -> m_nObjectID ) ) != NULL )
    {
        GetTreeCtrl().SelectItem( hitem );
    }
    break;
```

As seen here, we use the object ID of the list view item to locate the identical item in the tree view.

II

12

5. We also handle the case when the user wants to go up one tree level (from this subdirectory to the one above). Actually, in this example, the user makes this request through a toolbar button.

```
case TREE_VIEW_UP_LEVEL:
    hitem = GetTreeCtrl().GetSelectedItem();
    hitem = GetTreeCtrl().GetParentItem( hitem );
    GetTreeCtrl().SelectItem( hitem );
    break;

}
```

6. To determine when the user has selected a new item in our tree view, we use the ClassWizard to add a TVN_SELCHANGED control notification handler. There, we tell the Document Class that the selection has changed and the Document Class then uses UpdateAllViews() to notify the list view that it needs to change.

```
void CWzdTreeView::OnSelchanged( NMHDR* pNMHDR, LRESULT* pResult )
{
    NM_TREEVIEW* pNMTreeView = ( NM_TREEVIEW* )pNMHDR;

    GetDocument() -> SaveSelection( ( CWzdInfo * )pNMTreeView ->
        itemNew.lParam,LIST_VIEW_ONLY );

    *pResult = 0;
}
```

For a complete listing of the tree view class, please refer to the Tree View for Example 43 (page 433).

Fill in the List View Class

1. Use the ClassWizard to override the OnInitialUpdate() function. There, you should load up the small and large icons you will be using in the list

Example 43   Creating an Explorer Interface   **427**

view. You should also initialize the list view's columns and column names in the event the user will be using the report style.

```
void CWzdListView::OnInitialUpdate()
{
    CListView::OnInitialUpdate();

    // attach image lists to list control
    m_ImageList.Create( IDB_NORMAL_BITMAP, 32, 1, RGB( 0,0,0 ) );
    GetListCtrl().SetImageList( &m_ImageList, LVSIL_NORMAL );
    m_ImageSmallList.Create( IDB_SMALL_BITMAP, 16, 1, RGB( 0,0,0 ) );
    GetListCtrl().SetImageList( &m_ImageSmallList, LVSIL_SMALL );

    // initialize list control
    CClientDC dc( this );
    TEXTMETRIC tm;
    dc.GetTextMetrics( &tm );
    GetListCtrl().ModifyStyle( 0,LVS_REPORT|LVS_SHOWSELALWAYS );
    GetListCtrl().SendMessage( LVM_SETEXTENDEDLISTVIEWSTYLE,0,
        LVS_EX_FULLROWSELECT );
    GetListCtrl().InsertColumn( 0,"Name",LVCFMT_LEFT,
        20*tm.tmAveCharWidth,0 );
    GetListCtrl().InsertColumn( 1,"Comment",LVCFMT_LEFT,
        30*tm.tmAveCharWidth,1 );
}
```

2. Next, use the ClassWizard to override the OnUpdate() function. Again, we return if this is just a tree view update. Again, we process the update action using a case statement.

```
void CWzdListView::OnUpdate( CView* pSender, LPARAM lHint,
    CObject* pHint )
{
    if ( lHint&TREE_VIEW_ONLY )
    {
        return;
    }
}
```

II

12

```
GetListCtrl().DeleteAllItems();

switch ( lHint )
{
```

3. A list view has four styles: icon, small icon, list, and report. In this example, the user can select one of these through buttons in the toolbar. The Document Class sends the request here, where we modify this list view's control, accordingly.

```
case ICON_LIST_VIEW:
    GetListCtrl().ModifyStyle( LVS_LIST|LVS_SMALLICON|LVS_REPORT,
        LVS_ICON );
    break;

case LIST_LIST_VIEW:
    GetListCtrl().ModifyStyle( LVS_ICON|LVS_SMALLICON|LVS_REPORT,
        LVS_LIST );
    break;

case SMALL_ICON_LIST_VIEW:
    GetListCtrl().ModifyStyle( LVS_LIST|LVS_ICON|LVS_REPORT,
        LVS_SMALLICON );
    break;

case REPORT_LIST_VIEW:
    GetListCtrl().ModifyStyle( LVS_LIST|LVS_SMALLICON|LVS_ICON,
        LVS_REPORT );
    break;
}
```

4. Next, in `OnUpdate()`, fill in the list view with document data. Again, we use a helper function (`AddLine()`) and, again, it's heavily recursed. A listing of the `AddLine()` function can be found in the List View for Example 43 (page 441). First, we list all of the folders in this subdirectory. Then, we list all of the files.

```
// list all folders in selection
POSITION pos;
CWzdInfo *pSelInfo = GetDocument() -> GetSelection();
for ( pos = pSelInfo -> m_list.GetHeadPosition(); pos; )
```

Example 43   Creating an Explorer Interface   **429**

```
        {
            CWzdInfo *pInfo = pSelInfo -> m_list.GetNext( pos );
            if ( pInfo -> m_nCategory == CWzdInfo::FOLDER )
            {
                AddLine( pInfo );
            }
        }


        // list all files in selection
        for ( pos = pSelInfo -> m_list.GetHeadPosition(); pos; )
        {
            CWzdInfo *pInfo = pSelInfo -> m_list.GetNext(pos);
            if (pInfo -> m_nCategory == CWzdInfo::FILE)
            {
                AddLine(pInfo);
            }
        }
}
```

5. As mentioned previously, the user can change the tree view's selection by double-clicking on a list view item. Use the ClassWizard to add a WM_LBUTTONDBLCLK message handler to your list view class. There, look to see what was selected and, when found, tell the Document Class. The Document Class will then relay the information to the tree view class.

```
void CWzdListView::OnLButtonDblClk(UINT nFlags, CPoint point)
{
    int ndx;
    if ( GetListCtrl().GetSelectedCount() == 1 &&
        ( ndx = GetListCtrl().GetNextItem( -1, LVIS_SELECTED ) ) != -1 )
    {
        GetDocument() ->
            SaveSelection( (CWzdInfo *)GetListCtrl().GetItemData(ndx),
            TREE_VIEW_SELECT);
    }


    CListView::OnLButtonDblClk( nFlags, point );
}
```

6. This completes our list view class. For a complete listing of this class, please refer to the List View for Example 43 (page 441).

The Document Class is responsible for maintaining our data items and synchronizing our two views. The data items we used in this example can be found in the Data Class for Example 43 (page 432). The toolbar buttons that change the list view style and the button that causes the tree view to move up one level are both processed in the Document Class. As alluded to earlier, UpdateAllViews() is used to communicate with the views.

Fill in the Document Class

1. Use the Toolbar Editor to add buttons for the four different styles of a list view: large icon, small icon, list, and report.

2. Use the ClassWizard to add command handlers for these new toolbar buttons in your Document Class.

3. In those command handlers, use UpdateAllViews() to inform the list view which style to use.

```
void CWzdDoc::OnIconMode()
{
    UpdateAllViews( NULL,ICON_LIST_VIEW );
}


void CWzdDoc::OnListMode()
{
    UpdateAllViews( NULL,LIST_LIST_VIEW );
}


void CWzdDoc::OnReportMode()
{
    UpdateAllViews( NULL,REPORT_LIST_VIEW );
}


void CWzdDoc::OnSmalliconMode()
{
    UpdateAllViews( NULL,SMALL_ICON_LIST_VIEW );
}
```

Example 43   Creating an Explorer Interface   **431**

The commands we used here (e.g., SMALL_ICON_LIST_VIEW) are defined by us in our Document Class's include file.

4. Use the Toolbar Editor and ClassWizard to also add a button control, which will allow the user to go up one level, and fill it in.

```
void CWzdDoc::OnUplevel()
{

    UpdateAllViews( NULL,TREE_VIEW_UP_LEVEL );

}
```

5. Use the Text Editor to add a new function to your Document Class that can be called by either view to update which data item is currently selected by both.

```
void CWzdDoc::SaveSelection( CWzdInfo *pSelectionInfo,int nMode )
{

    m_pSelectionInfo = pSelectionInfo;
    UpdateAllViews( NULL,nMode );

}
```

That's it. Clicking in either view should cause the other view to respond by selecting the same item.

Notes

- Windows Explorer is an SDI application. However, you can certainly make this an MDI application too—but not automatically with the App-Wizard. Follow this example from the start, but use the AppWizard to create an MDI application, instead of an SDI. Then, instead of overriding CMainFrame's OnCreateClient() function, override your CChildFrame's OnCreateClient() to create a splitter window. Everything else should be essentially the same. You might also consider creating four views in your splitter window, instead of two. Then, you could show two sets of tree and list views. You will need to account for this, however, in the Document Class, since all four views will be using the same document.

- To actually fill your views with real directory information, you can use the Windows API calls ::FindFirstFile() and ::FindNextFile().

- As mentioned previously, an Explorer-type interface can be used for many applications other than maintaining file directories. I have personally seen it used to access forms in a CAD system, where the forms were files that were grouped under directories called models. Another use was

II

12

to maintain a network, where each node was a "directory" and the cards inside were the "files".

CD Notes

- When executing the project on the accompanying CD, you will notice that it has the same look and feel of the Windows Explorer, albeit non-functional.

Data Class for Example 43

```
#ifndef WZDINFO_H
#define WZDINFO_H

#include "afxtempl.h"
#include "WzdInfo.h"

class CWzdInfo : public CObject
{
public:
    CWzdInfo( CString sName,CString sComment,int nCategory,int nObjectID );
    ~CWzdInfo();

    enum {
    DEVICE,
    FOLDER,
    FILE
    };

    // misc info
    CString m_sName;
    CString m_sComment;
    int m_nCategory;
    int m_nObjectID;

    CList<CWzdInfo*,CWzdInfo*> m_list;

};
#endif
// WzdInfo.cpp : implementation of the CWzdInfo class
```

Example 43   Creating an Explorer Interface   **433**

```
//

#include "stdafx.h"
#include "WzdInfo.h"


/////////////////////////////////////////////////////////////////////////////
// CWzdInfo

CWzdInfo::CWzdInfo( CString sName,CString sComment,int nCategory,
    int nObjectID ) :
    m_sName( sName ), m_sComment( sComment ),m_nCategory( nCategory ),
    m_nObjectID( nObjectID )
{
}


CWzdInfo::~CWzdInfo()
{
    while (!m_list.IsEmpty())
    {
        delete m_list.RemoveHead();
    }
}
```

II

Tree View for Example 43

12

```
// WzdTreeView.h : interface of the CWzdTreeView class
//
/////////////////////////////////////////////////////////////////////////////

#if !defined WZDTREEVIEW_H
#define WZDTREEVIEW_H

#if _MSC_VER >= 1000
#pragma once
#endif    // _MSC_VER >= 1000

#include <afxcview.h>

class CWzdTreeView : public CTreeView
{
```

```
protected:    // create from serialization only
    CWzdTreeView();
    DECLARE_DYNCREATE( CWzdTreeView )

// Attributes
public:
    CWzdDoc* GetDocument();

// Operations
public:

// Overrides
    // ClassWizard generated virtual function overrides
    // {{AFX_VIRTUAL( CWzdTreeView )
public:
    virtual void OnDraw( CDC* pDC );    // overridden to draw this view
    virtual BOOL PreCreateWindow( CREATESTRUCT& cs );
    virtual void OnInitialUpdate();
protected:
    virtual BOOL OnPreparePrinting( CPrintInfo* pInfo );
    virtual void OnBeginPrinting( CDC* pDC, CPrintInfo* pInfo );
    virtual void OnEndPrinting( CDC* pDC, CPrintInfo* pInfo );
    virtual void OnUpdate( CView* pSender, LPARAM lHint, CObject* pHint );
    // }}AFX_VIRTUAL

// Implementation
public:
    virtual ~CWzdTreeView();
#ifdef _DEBUG
    virtual void AssertValid() const;
    virtual void Dump( CDumpContext& dc ) const;
#endif

protected:

// Generated message map functions
protected:
    // {{AFX_MSG( CWzdTreeView )
    afx_msg void OnSelchanged( NMHDR* pNMHDR, LRESULT* pResult );
    // }}AFX_MSG
```

Example 43   Creating an Explorer Interface   **435**

```
    DECLARE_MESSAGE_MAP()
private:
    CImageList m_ImageList;
    void AddBranch( HTREEITEM hTreeItem, CList<CWzdInfo*,CWzdInfo*> *pList );
    HTREEITEM AddLeaf( HTREEITEM hTreeItem, CWzdInfo *pInfo );
    HTREEITEM FindTreeItem( HTREEITEM hTreeItem, long nObjectID );
};

#ifndef _DEBUG    // debug version in WzdTreeView.cpp
inline CWzdDoc* CWzdTreeView::GetDocument()
    { return (CWzdDoc*)m_pDocument; }
#endif


//////////////////////////////////////////////////////////////////////


// {{AFX_INSERT_LOCATION}}
// Microsoft Developer Studio will insert additional declarations immediately
//     before the previous line.

#endif
// WzdTreeView.cpp : implementation of the CWzdTreeView class
//

#include "stdafx.h"
#include "Wzd.h"

#include "WzdDoc.h"
#include "WzdTreeView.h"

#ifdef _DEBUG
#define new DEBUG_NEW
#undef THIS_FILE
static char THIS_FILE[] = __FILE__;
#endif


//////////////////////////////////////////////////////////////////////
// CWzdTreeView

IMPLEMENT_DYNCREATE( CWzdTreeView, CTreeView )
```

```
BEGIN_MESSAGE_MAP( CWzdTreeView, CTreeView )
    // {{AFX_MSG_MAP( CWzdTreeView )
    ON_NOTIFY_REFLECT( TVN_SELCHANGED, OnSelchanged )
    // }}AFX_MSG_MAP
    // Standard printing commands
    ON_COMMAND( ID_FILE_PRINT, CView::OnFilePrint )
    ON_COMMAND( ID_FILE_PRINT_DIRECT, CView::OnFilePrint )
    ON_COMMAND( ID_FILE_PRINT_PREVIEW, CView::OnFilePrintPreview )
END_MESSAGE_MAP()


/////////////////////////////////////////////////////////////////////////////
// CWzdTreeView construction/destruction

CWzdTreeView::CWzdTreeView()
{
    // set the CTreeCtrl attributes
    m_dwDefaultStyle = WS_CHILD | WS_VISIBLE | WS_BORDER | TVS_HASLINES |
        TVS_SHOWSELALWAYS | TVS_LINESATROOT | TVS_HASBUTTONS;
}


CWzdTreeView::~CWzdTreeView()
{
}


BOOL CWzdTreeView::PreCreateWindow( CREATESTRUCT& cs )
{
    // TODO: Modify the Window class or styles here by modifying
    //       the CREATESTRUCT cs

    return CTreeView::PreCreateWindow(cs);
}


/////////////////////////////////////////////////////////////////////////////
// CWzdTreeView drawing

void CWzdTreeView::OnDraw( CDC* pDC )
{
    CWzdDoc* pDoc = GetDocument();
    ASSERT_VALID(pDoc);
```

Example 43   Creating an Explorer Interface   **437**

```
    // TODO: add draw code for native data here
}


///////////////////////////////////////////////////////////////////////////
// CWzdTreeView printing

BOOL CWzdTreeView::OnPreparePrinting( CPrintInfo* pInfo )
{
    // default preparation
    return DoPreparePrinting( pInfo );
}


void CWzdTreeView::OnBeginPrinting( CDC* /*pDC*/, CPrintInfo* /*pInfo*/ )
{
    // TODO: add extra initialization before printing
}


void CWzdTreeView::OnEndPrinting( CDC* /*pDC*/, CPrintInfo* /*pInfo*/
{
    // TODO: add cleanup after printing
}


///////////////////////////////////////////////////////////////////////////
// CWzdTreeView diagnostics

#ifdef _DEBUG
void CWzdTreeView::AssertValid() const
{
    CView::AssertValid();
}


void CWzdTreeView::Dump( CDumpContext& dc ) const
{
    CView::Dump( dc );
}


CWzdDoc* CWzdTreeView::GetDocument()    // non-debug version is inline
{
    ASSERT( m_pDocument -> IsKindOf( RUNTIME_CLASS( CWzdDoc ) ) );
    return ( CWzdDoc* )m_pDocument;
```

II

12

```
}
#endif //_DEBUG


/////////////////////////////////////////////////////////////////////////////
// CWzdTreeView message handlers

void CWzdTreeView::OnInitialUpdate()
{

    CTreeView::OnInitialUpdate();


    // set icons
    m_ImageList.Create( IDB_SMALL_BITMAP, 16, 1, RGB( 0,0,0 ) );
    GetTreeCtrl().SetImageList( &m_ImageList, TVSIL_NORMAL );


}


void CWzdTreeView::OnUpdate( CView* pSender, LPARAM lHint, CObject* pHint )
{

    if ( lHint&LIST_VIEW_ONLY )
    {
        return;
    }



    HTREEITEM hitem;
    CWzdInfo *pInfo;
    switch ( lHint )
    {
        case TREE_VIEW_SELECT:
            pInfo = GetDocument() -> GetSelection();
            if ( ( hitem = FindTreeItem( GetTreeCtrl().GetChildItem(
                GetTreeCtrl().GetRootItem() ), pInfo -> m_nObjectID ) ) != NULL )
            {
                GetTreeCtrl().SelectItem(hitem);
            }
            break;

        case TREE_VIEW_UP_LEVEL:
            hitem = GetTreeCtrl().GetSelectedItem();
            hitem = GetTreeCtrl().GetParentItem( hitem );
```

Example 43   Creating an Explorer Interface   **439**

```
                GetTreeCtrl().SelectItem(hitem);
            break;

        default:
            GetTreeCtrl().DeleteAllItems();
            AddBranch(TVI_ROOT, GetDocument() -> GetWzdList());
            break;
    }
}


// recurse until all CWzdInfo items are in a leaf
void CWzdTreeView::AddBranch( HTREEITEM hTreeItem,
    CList<CWzdInfo*,CWzdInfo*> *pList )
{
    for ( POSITION pos = pList -> GetHeadPosition(); pos; )
    {
        CWzdInfo *pInfo = pList -> GetNext( pos );
        if ( pInfo -> m_nCategory != CWzdInfo::FILE )
        {
            HTREEITEM hTreeItemx = AddLeaf( hTreeItem, pInfo );
            if ( pList -> GetCount() )
            {
                AddBranch( hTreeItemx, &pInfo -> m_list );
            }
        }
    }
}


// add leaf to tree
HTREEITEM CWzdTreeView::AddLeaf( HTREEITEM hTreeItem, CWzdInfo *pInfo )
{
    TV_INSERTSTRUCT tvstruct;
    tvstruct.hParent = hTreeItem;
    tvstruct.hInsertAfter = TVI_LAST;
    tvstruct.item.iImage = tvstruct.item.iSelectedImage = pInfo -> m_nObjectID;
    tvstruct.item.pszText = ( char * )LPCTSTR( pInfo -> m_sName );
    tvstruct.item.mask =
        TVIF_IMAGE |TVIF_SELECTEDIMAGE | TVIF_PARAM | TVIF_TEXT;
    tvstruct.item.lParam = ( DWORD )pInfo;
    return GetTreeCtrl().InsertItem( &tvstruct );
```

II

12

```
}

// recurse until item found or items expended
HTREEITEM CWzdTreeView::FindTreeItem( HTREEITEM hTreeItem, long nObjectID )
{
    while ( hTreeItem != NULL )
    {
        CWzdInfo *pInfo = ( CWzdInfo * )GetTreeCtrl().GetItemData( hTreeItem );
        if ( pInfo -> m_nObjectID == nObjectID )
        {
            return hTreeItem;
        }
        else if ( GetTreeCtrl().ItemHasChildren( hTreeItem ) )
        {
            HTREEITEM hTreeItemx =
                FindTreeItem( GetTreeCtrl().GetChildItem( hTreeItem ),
                nObjectID );
            if ( hTreeItemx )
            {
                return hTreeItemx;
            }
        }
        else
        {
            hTreeItem = GetTreeCtrl().GetNextItem( hTreeItem, TVGN_NEXT );
        }
    }
    return NULL;
}

// new selection, change listview!
void CWzdTreeView::OnSelchanged( NMHDR* pNMHDR, LRESULT* pResult )
{
    NM_TREEVIEW* pNMTreeView = ( NM_TREEVIEW* )pNMHDR;

    GetDocument() -> SaveSelection( ( CWzdInfo * )pNMTreeView ->
        itemNew.lParam,LIST_VIEW_ONLY );

    *pResult = 0;
}
```

Example 43   Creating an Explorer Interface   **441**

List View for Example 43

```
// WzdListView.h : interface of the CWzdListView class
//
/////////////////////////////////////////////////////////////////////////////

#if !defined WZDLISTVIEW_H
#define WZDLISTVIEW_H

#if _MSC_VER >= 1000
#pragma once
#endif     // _MSC_VER >= 1000

#include <afxcview.h>

class CWzdListView : public CListView
{
protected:    // create from serialization only
    CWzdListView();
    DECLARE_DYNCREATE( CWzdListView )

// Attributes
public:
    CWzdDoc* GetDocument();

// Operations
public:

// Overrides
    // ClassWizard generated virtual function overrides
    // {{AFX_VIRTUAL( CWzdListView )
public:
    virtual void OnDraw( CDC* pDC );    // overridden to draw this view
    virtual BOOL PreCreateWindow( CREATESTRUCT& cs );
    virtual void OnInitialUpdate();
protected:
    virtual BOOL OnPreparePrinting( CPrintInfo* pInfo );
    virtual void OnBeginPrinting( CDC* pDC, CPrintInfo* pInfo );
    virtual void OnEndPrinting( CDC* pDC, CPrintInfo* pInfo );
    virtual void OnUpdate( CView* pSender, LPARAM lHint, CObject* pHint );
```

II

12

```
    // }}AFX_VIRTUAL

// Implementation
public:
    virtual ~CWzdListView();
#ifdef _DEBUG
    virtual void AssertValid() const;
    virtual void Dump( CDumpContext& dc ) const;
#endif

protected:

// Generated message map functions
protected:
    // {{AFX_MSG( CWzdListView )
    afx_msg void OnLButtonDblClk( UINT nFlags, CPoint point );
    // }}AFX_MSG
    DECLARE_MESSAGE_MAP()
private:
    CImageList m_ImageList;
    CImageList m_ImageSmallList;
    void AddLine( CWzdInfo *pInfo );
};

#ifndef _DEBUG    // debug version in WzdListView.cpp
inline CWzdDoc* CWzdListView::GetDocument()
    { return (CWzdDoc*)m_pDocument; }
#endif

///////////////////////////////////////////////////////////////////////////

// {{AFX_INSERT_LOCATION}}
// Microsoft Developer Studio will insert additional declarations immediately
//     before the previous line.

#endif
    // !defined( AFX_WzdListView_H__CA9038F0_B0DF_11D1_A18C_DCB3C85EBD34__INCLUDED_ )
// WzdListView.cpp : implementation of the CWzdListView class
//
```

Example 43   Creating an Explorer Interface   **443**

```
#include "stdafx.h"
#include "Wzd.h"

#include "WzdDoc.h"
#include "WzdListView.h"

#ifdef _DEBUG
#define new DEBUG_NEW
#undef THIS_FILE
static char THIS_FILE[] = __FILE__;
#endif

/////////////////////////////////////////////////////////////////////////////
// CWzdListView

IMPLEMENT_DYNCREATE( CWzdListView, CListView )

BEGIN_MESSAGE_MAP( CWzdListView, CListView )
    // {{AFX_MSG_MAP( CWzdListView )
    ON_WM_LBUTTONDBLCLK()
    // }}AFX_MSG_MAP
    // Standard printing commands
    ON_COMMAND( ID_FILE_PRINT, CView::OnFilePrint )
    ON_COMMAND( ID_FILE_PRINT_DIRECT, CView::OnFilePrint )
    ON_COMMAND( ID_FILE_PRINT_PREVIEW, CView::OnFilePrintPreview )
END_MESSAGE_MAP()

/////////////////////////////////////////////////////////////////////////////
// CWzdListView construction/destruction

CWzdListView::CWzdListView()
{
    // TODO: add construction code here

}


CWzdListView::~CWzdListView()
{
}
```

II

12

```
BOOL CWzdListView::PreCreateWindow(CREATESTRUCT& cs)
{
    // TODO: Modify the Window class or styles here by modifying
    //       the CREATESTRUCT cs

    return CListView::PreCreateWindow(cs);
}


/////////////////////////////////////////////////////////////////////////////
// CWzdListView drawing

void CWzdListView::OnDraw( CDC* pDC )
{
    CWzdDoc* pDoc = GetDocument();
    ASSERT_VALID( pDoc );

    // TODO: add draw code for native data here
}


/////////////////////////////////////////////////////////////////////////////
// CWzdListView printing

BOOL CWzdListView::OnPreparePrinting( CPrintInfo* pInfo )
{
    // default preparation
    return DoPreparePrinting( pInfo );
}

void CWzdListView::OnBeginPrinting( CDC* /*pDC*/, CPrintInfo* /*pInfo*/ )
{
    // TODO: add extra initialization before printing
}

void CWzdListView::OnEndPrinting( CDC* /*pDC*/, CPrintInfo* /*pInfo*/ )
{
    // TODO: add cleanup after printing
}


/////////////////////////////////////////////////////////////////////////////
// CWzdListView diagnostics
```

Example 43   Creating an Explorer Interface   **445**

```
#ifdef _DEBUG
void CWzdListView::AssertValid() const
{
    CView::AssertValid();
}


void CWzdListView::Dump( CDumpContext& dc ) const
{
    CView::Dump( dc );
}


CWzdDoc* CWzdListView::GetDocument()     // non-debug version is inline
{
    ASSERT( m_pDocument -> IsKindOf( RUNTIME_CLASS( CWzdDoc ) ) );
    return ( CWzdDoc* )m_pDocument;
}
#endif     //_DEBUG


///////////////////////////////////////////////////////////////////////////
// CWzdListView message handlers

void CWzdListView::OnInitialUpdate()
{
    CListView::OnInitialUpdate();


    // attach image lists to list control
    m_ImageList.Create( IDB_NORMAL_BITMAP, 32, 1, RGB( 0,0,0 ) );
    GetListCtrl().SetImageList( &m_ImageList, LVSIL_NORMAL );
    m_ImageSmallList.Create( IDB_SMALL_BITMAP, 16, 1, RGB( 0,0,0 ) );
    GetListCtrl().SetImageList( &m_ImageSmallList, LVSIL_SMALL );


    // initialize list control
    CClientDC dc( this );
    TEXTMETRIC tm;
    dc.GetTextMetrics( &tm );
    GetListCtrl().ModifyStyle( 0,LVS_REPORT|LVS_SHOWSELALWAYS );
    GetListCtrl().SendMessage( LVM_SETEXTENDEDLISTVIEWSTYLE,0,
        LVS_EX_FULLROWSELECT );
    GetListCtrl().InsertColumn( 0,"Name",LVCFMT_LEFT,20*tm.tmAveCharWidth,0 );
```

II

12

```
        GetListCtrl().InsertColumn( 1,"Comment",LVCFMT_LEFT,
            30*tm.tmAveCharWidth,1 );


}


void CWzdListView::OnUpdate( CView* pSender, LPARAM lHint, CObject* pHint )
{
    if ( lHint&TREE_VIEW_ONLY )
    {
        return;
    }


    GetListCtrl().DeleteAllItems();

    switch (lHint)
    {
        case ICON_LIST_VIEW:
            GetListCtrl().ModifyStyle(LVS_LIST|LVS_SMALLICON|LVS_REPORT,
                LVS_ICON);
            break;

        case LIST_LIST_VIEW:
            GetListCtrl().ModifyStyle(LVS_ICON|LVS_SMALLICON|LVS_REPORT,
                LVS_LIST);
            break;

        case SMALL_ICON_LIST_VIEW:
            GetListCtrl().ModifyStyle(LVS_LIST|LVS_ICON|LVS_REPORT,
                LVS_SMALLICON);
            break;

        case REPORT_LIST_VIEW:
            GetListCtrl().ModifyStyle(LVS_LIST|LVS_SMALLICON|LVS_ICON,
                LVS_REPORT);
            break;
    }

    // list all folders in selection
    POSITION pos;
    CWzdInfo *pSelInfo = GetDocument() -> GetSelection();
    for ( pos = pSelInfo -> m_list.GetHeadPosition(); pos; )
```

Example 43   Creating an Explorer Interface   **447**

```
        {
            CWzdInfo *pInfo = pSelInfo -> m_list.GetNext( pos );
            if ( pInfo -> m_nCategory == CWzdInfo::FOLDER )
            {
                AddLine( pInfo );
            }
        }

    // list all files in selection
    for ( pos = pSelInfo -> m_list.GetHeadPosition(); pos; )
    {
        CWzdInfo *pInfo = pSelInfo -> m_list.GetNext( pos );
        if ( pInfo -> m_nCategory == CWzdInfo::FILE )
        {
            AddLine( pInfo );
        }
    }
}

void CWzdListView::AddLine( CWzdInfo *pInfo )
{
    LV_ITEM item;
    item.mask = LVIF_TEXT | LVIF_IMAGE | LVIF_PARAM;
    item.iItem = GetListCtrl().GetItemCount();
    item.iSubItem = 0;
    item.pszText = ( char * )LPCTSTR( pInfo -> m_sName );
    item.iImage = pInfo -> m_nCategory;
    item.lParam = ( DWORD )pInfo;
    int ndx = GetListCtrl().InsertItem( &item );

    // if report style add comment
    if ( GetListCtrl().GetStyle() & LVS_REPORT )
    {
        GetListCtrl().SetItemText( ndx, 1, pInfo -> m_sComment );
    }
}

void CWzdListView::OnLButtonDblClk( UINT nFlags, CPoint point )
{
    int ndx;
```

II

12

```
    if ( GetListCtrl().GetSelectedCount() == 1 &&
        ( ndx = GetListCtrl().GetNextItem( -1, LVIS_SELECTED ) ) != -1 )
    {

        GetDocument() ->
            SaveSelection( ( CWzdInfo * )GetListCtrl().GetItemData( ndx ),
            TREE_VIEW_SELECT );
    }


    CListView::OnLButtonDblClk( nFlags, point );
}
```

Example 44   Creating a Simple ODBC Database Editor

Objective

You would like to create a simple ODBC database editor that would allow you to scroll through an ODBC-compliant database and edit its records, as seen in Figure 12.4.

Figure 12.4   A Simple ODBC Editor

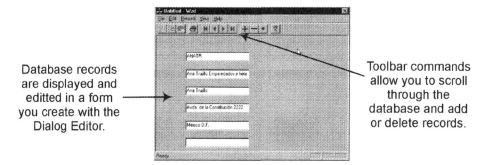

Database records are displayed and edited in a form you create with the Dialog Editor.

Toolbar commands allow you to scroll through the database and add or delete records.

Strategy

Most of the work for this example will be done by the AppWizard, the Dialog Editor, and the ClassWizard. The AppWizard will create an SDI application for us using a CRecordView class. This view class is derived from CFormView and uses a dialog box for its view. We will then use the Dialog

Example 44   Creating a Simple ODBC Database Editor   **449**

Editor to add fields to this view and the ClassWizard to bind these fields to columns in our database. The View Class and other commands added by the AppWizard provide all the horsepower necessary to scroll through and edit these database fields. However, we will add three additional commands ourselves to add, delete, and clear a database record.

Steps

Create an ODBC Database Editor with the AppWizard

1. Use the AppWizard to create an SDI application. In Step Two, add "Database view without file support" to your application. Then, click on the "Data Source" button and pick "ODBC". Use the drop-down box to identify the database you will be editing with this application. If the App-Wizard successfully opens this database, it will then prompt you for the table(s) you would like to edit with this application. Pick the defaults for the rest of this application.

Note: If you pick "Database view with file support", the AppWizard will allow you to access a database and serialized file documents in the same application.

---

2. The AppWizard will create two dialog box templates for your application, one of which is for your view. (The other is that ubiquitous About dialog template.) Use the Dialog Box Editor to add fields to this template that will represent the columns in your database table.

3. Bring up the ClassWizard and select your View Class. Click on the "Member Variables" tab and you will find the control IDs of the fields you just added to this view's dialog box listed here. Select the first control ID and then click the "Add Variable" button to open the "Add Member Variable" dialog box. Locate the "Member variable name" combo box in this dialog and click its drop-down button to reveal a list of all of the columns in this database table. By choosing one of these columns, you are effectively binding this record column to this control. Repeat this process for each control field, assigning a record column to each.

II

12

4. To make your view conform to the size of the dialog box you create, use the ClassWizard to override your view class's `OnInitialUpdate()` function and make it look like the following.

```
void CWzdView::OnInitialUpdate()
{
    m_pSet = &GetDocument() -> m_wzdSet;
    CRecordView::OnInitialUpdate();

    // make frame the size of the original dialog box
    ResizeParentToFit();

    // get rid of those pesky scroll bars
    SetScrollSizes( MM_TEXT, CSize( 20,20 ) );
}
```

Add an "Add" Command to This Editor

1. Use the Menu and Toolbar Editors to add an "Add" command to your menu and toolbar.
2. Use the ClassWizard to handle this command in your View Class, as follows.

```
void CWzdView::OnRecordAdd()
{
    m_pSet -> AddNew();
    UpdateData( FALSE );
}
```

Notice that the `m_pSet` member variable of your View Class is a pointer to the record set that this editor is accessing.

Add a "Clear" Command to This Editor

1. Use the Menu and Toolbar Editors to add an "Clear" command to your menu and toolbar.

Example 44   Creating a Simple ODBC Database Editor   **451**

2. Use the ClassWizard to handle this command in your View Class, as follows.

```
void CWzdView::OnRecordCanceledit()
{
    m_pSet -> CancelUpdate();
    m_pSet -> Move( 0 );
    UpdateData( FALSE );
}
```

Add a "Delete" Command to This Editor

1. Use the Menu and Toolbar Editors to add a "Delete" command to your menu and toolbar.
2. Use the ClassWizard to handle this command in your View Class, as follows.

```
void CWzdView::OnRecordDelete()
{
    try
    {
        m_pSet -> Delete();
    }

    catch( CDBException* e )
    {
        AfxMessageBox( e -> m_strError );
        e -> Delete();
    }

    // Move off the record we just deleted
    m_pSet -> MoveNext();

    // If we moved off the end of file, move back to last record
    if ( m_pSet -> IsEOF() )
        m_pSet -> MoveLast();

    // If the recordset is now empty, clear the record we just deleted
```

II

12

```
    if ( m_pSet -> IsBOF() )
        m_pSet -> SetFieldNull( NULL );
        UpdateData( FALSE );
}
```

Notes

- To edit a field using this editor, you first make your change to the appropriate field and then either scroll forward or backward away from this record. Before the next record is shown, the previous record is updated. A "Clear" command would undo any changes you make to the current record so that the record it represents is not changed.

- The price for all the automation afforded you by using the AppWizard and ClassWizard is in a somewhat inflexible end result. If you don't pick "Database view with file support", you are forced to create an SDI application. You are also forced to use a form view, rather than another type of view. If you select "Header files only" for your database support, you lose all of that automation, but your application becomes much more flexible. See my previous book, *Visual C++ MFC Programming by Example,* for how to access an ODBC database directly using the MFC library.

CD Notes

- When executing the project on the accompanying CD, you will notice that you can scroll through an ODBC database.

Example 45   Creating a Simple DAO Database Editor

Objective

You would like to create a simple DAO database editor that would allow you to scroll through a DAO database and edit its records, as seen in the previous example.

Example 45   Creating a Simple DAO Database Editor   **453**

Strategy

The strategy for this example is identical to the previous example. The only difference comes when we handle the "Add", "Clear", and "Delete" editor commands, which are handled differently for a DAO database.

Steps

Create a DAO Database Editor

1. Follow the steps in the previous example to create your application—except, of course, you should define a DAO database file instead of an ODBC database.

Add an "Add" Command to This Editor

1. Use the Menu and Toolbar Editors to add an "Add" command to your menu and toolbar.
2. Use the ClassWizard to handle this command in your View Class, as follows.

```
void CWzdView::OnRecordAdd()
{
    m_bAddingRecord = TRUE;    // keep track of what mode we're in
    m_pSet -> AddNew();
    UpdateData( FALSE );
}
```

Notice that the m_pSet member variable of your View Class is a pointer to the record set that this editor is accessing.

Before we can create a "Clear" command, we need to add one more override to our View Class to help us determine whether or not we're still in a record-adding mode. When clearing a record in the ODBC database editor, we could blindly use CancelUpdate() to cancel a change. However, if we tried to do the same thing with a DAO database when in the middle of a adding a new record, the MFC DAO class will throw an exception. So, we set a flag in the "Add" command to indicate we are in a record-adding mode, and then we clear that flag in our View Class's OnMove() function.

`OnMove()` is the function that scrolls us away from the current record and saves our changes.

3. Use the ClassWizard to override the your View Class's `OnMove()` function. There, you should clear the `m_bAddingRecord` flag you set in the "Add" command.

```
BOOL CWzdView::OnMove( UINT nIDMoveCommand )
{

    m_bAddingRecord = FALSE;


    return CDaoRecordView::OnMove( nIDMoveCommand );
}
```

Add a "Clear" Command to This Editor

1. Use the Menu and Toolbar Editors to add a "Clear" command to your menu and toolbar.

2. Use the ClassWizard to handle this command in your View Class, as follows.

```
void CWzdView::OnRecordCanceledit()
{

    if ( m_bAddingRecord )
        m_pSet -> CancelUpdate();
    m_pSet -> Move( 0 );
    m_bAddingRecord = FALSE;
    UpdateData( FALSE );


}
```

Add a "Delete" Command to This Editor

1. Use the Menu and Toolbar Editors to add a "Delete" command to your menu and toolbar.

2. Use the ClassWizard to handle this command in your View Class, as follows.

```
void CWzdView::OnRecordDelete()
{
    try
    {
        m_pSet -> Delete();
    }

    catch( CDaoException* e )
    {
        AfxMessageBox( e -> m_pErrorInfo -> m_strDescription );
        e -> Delete();
    }

    // Move off the record we just deleted
    m_pSet -> MoveNext();

    // If we moved off the end of file, move back to last record
    if ( m_pSet -> IsEOF() )
        m_pSet -> MoveLast();

    // If the recordset is now empty, clear the record we just deleted
    if ( m_pSet -> IsBOF() )
        m_pSet -> SetFieldNull( NULL );
    UpdateData( FALSE );
}
```

Notes

- To edit a field using this editor, you first make your change to the appropriate field and then either scroll forward or backward away from this record. Before the next record is shown, the previous record is updated. A "Clear" command would undo any changes you make to the current record, so that the record it represents is not changed.

- If would prefer to access a DAO database directly from a nonform view application, you should specify "Header files only" when creating your application using the AppWizard. Then, refer to my previous book, *Visual C++ MFC Programming by Example*, to access the database.

CD Notes

- When executing the project on the accompanying CD, you will notice that you can scroll through a DAO database.

Example 46 Creating a Simple Wizard

Objective

You would like to create a simple wizard using the conventions found in most of the wizards thoughout Windows, as seen in Figure 12.5.

Figure 12.5 A Simple Wizard

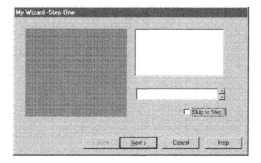

Strategy

We will use CPropertySheet's wizard mode to create our wizard. Moreover, although a wizard can be part of another application, we will be making our wizard a stand-alone application. To do this, we will create a Dialog Application, but then substitute a property sheet for the modal dialog box.

Example 46   Creating a Simple Wizard   **457**

Steps

Create Your Project

1. Use the AppWizard to create a Dialog Application. You can then delete the dialog template and class files it creates.

Create Your Wizard Pages

II

1. Use the Dialog Editor to create the pages in your wizard. Wizard pages are similar to property pages, except that your user scrolls through them sequentially. It doesn't really matter what properties you give each dialog template—the CPropertySheet class will override it. The title for each dialog template will become the page title you see in the wizard. By convention, you should append each of your titles with "—Step One", "—Step Two", etc.

2. Use the ClassWizard to create a property page class for each of these dialog pages, deriving it from CPropertyPage. In each class, you can use the ClassWizard to override up to four member functions of CPropertyPage: OnSetActive(), OnWizardBack(), OnWizardNext(), and OnWizardFinish().

3. Use the ClassWizard to override the OnSetActive() member function of each of your property pages. There, you will set which wizard buttons will be visible or enabled for this page.

```
BOOL CPage1::OnSetActive()
{
    ( ( CPropertySheet* )GetOwner() ) -> SetWizardButtons(
        PSWIZB_NEXT |             // the "Next" button is visible
        PSWIZB_BACK |             // the "Back" button is visible
        PSWIZB_FINISH |           // the "Finish" button is visible
        PSWIZB_DISABLEDFINISH );  // disabled "Finish" button
                                  //    is visible

    return CPropertyPage::OnSetActive();
}
```

12

The "Back", "Next", and "Finish" buttons are handled automatically by the CPropertySheet Class. If you need to take control, use the ClassWizard to override the OnWizardBack(), OnWizardNext(), or OnWizardFinish() function(s).

In each of these functions, you can start by calling `UpdateData(TRUE)` to retrieve any changes to the page. The value you return from `OnWizardBack()` or `OnWizardNext()` determines which page the wizard displays next. Returning a default value of zero (0) causes the next or previous sequential page to appear. If, however, you return the ID of the dialog template of another property page in your wizard, the wizard will display that page next.

You can conditionally terminate your wizard by overriding the `OnWizardFinish()` and then choosing whether or not to call the base class's `CPropertyPage::OnWizardFinish()`. Not calling the base class function will allow your wizard to continue processing.

4. In the following example, if the user clicks the "Page 3 Next" checkbox, the `IDD_PAGE3` page will be displayed next.

```
LRESULT CPage1::OnWizardNext()
{
    UpdateData( TRUE );
    if ( m_bPage3Check )
    {
        return IDD_PAGE3;
    }
    else
    {
        return CPropertyPage::OnWizardNext();
    }
}
```

Note: Both `CPropertyPage::OnWizardNext()` and `CPropertyPage::OnWizardBack()` simply return zeros (0), which causes the wizard to go to the next or previous page, respectively.

Create the Wizard

1. Embed your wizard's property page classes in your Application Class.

```
CPageOne m_Page1;
CPageTwo m_Page2;
    :     :     :
```

Example 46　Creating a Simple Wizard　**459**

2. In the `InitInstance()` function of the Application Class, create an instance of the `CPropertySheet` class on your stack and add your property pages to it in the sequence you would like them to appear to your user. At this point you can also initialize the member variables of your property page classes. Then, set the almighty wizard style and call `CProperty-Sheet::DoModal()`.

```
CWzdSheet wzd( "" );
m_pMainWnd = &wzd;
AddPage( &m_Page1 );
AddPage( &m_Page2 );
AddPage( &m_Page3 );
SetWizardMode();
if ( wzd.DoModal() == ID_WIZFINISH )
{
    // process results
}
```

3. You can optionally process the results of your wizard either in the `OnWiz-ardFinish()` function of your last page or here in your Application Class if the `DoModal()` function returns a value of `ID_WIZFINISH`. By processing your results here, you have easy access to the member variables of all of your pages. However, at this point, the wizard has closed and you can't go back.

Notes

- The wizard mode of a property sheet currently doesn't display anything other than the title in its caption bar—in other words, no close button or minimize button. If you'd really like to add these buttons to your caption bar, please refer to Example 27 to embed your wizard in the template of a Dialog Application. To get rid of your wizard's title bar, you will probably also need to derive a class from `CPropertySheet` and override the `WM_NCPAINT` message to prevent it from getting to the property sheet.

- Another approach to adding a close button to your wizard's title bar would be to draw the close button yourself by referring to Example 29. Rather than create your own button bitmaps as outlined in that example, use `CDC::DrawFrameControl()` with the appropriate type and state.

- Why doesn't the wizard mode of the property sheet include any controls in its caption bar? After all, the AppWizard has a close button *and* a

II

12

Context Help button in its caption bar. The reason is a little shocking—the AppWizard doesn't use the wizard mode of a property sheet. It doesn't use a property sheet at all. The AppWizard is, in fact, a Dialog Application with a single main dialog template and, therefore, it can have all the caption buttons that any Dialog Application can automatically have. How, then, does it create the illusion of paging from step to step? Please refer to Example 26.

- The Developer Studio has a wizard that lets you create your own customized AppWizard. A fairly detailed example is included in your MFC distribution.

CD Notes

- When executing the project on the accompanying CD, you will notice that it behaves like a standard wizard.

Property Page Class for Example 46

```
#if !defined( AFX_PAGE1_H__A846F4D3_ECA0_11D1_A18D_DCB3C85EBD34__INCLUDED_ )
#define AFX_PAGE1_H__A846F4D3_ECA0_11D1_A18D_DCB3C85EBD34__INCLUDED_

#if _MSC_VER >= 1000
#pragma once
#endif    // _MSC_VER >= 1000
// Page1.h : header file
//


/////////////////////////////////////////////////////////////////////////////
// CPage1 dialog

class CPage1 : public CPropertyPage
{
    DECLARE_DYNCREATE( CPage1 )

// Construction
public:
    CPage1();
    ~CPage1();
```

Example 46   Creating a Simple Wizard   **461**

```
// Dialog Data
    // {{AFX_DATA( CPage1 )
    enum { IDD = IDD_PAGE1 };
    BOOL m_bPage3Check;
    // }}AFX_DATA


// Overrides
    // ClassWizard generate virtual function overrides
    // {{AFX_VIRTUAL( CPage1 )
public:
    virtual LRESULT OnWizardBack();
    virtual LRESULT OnWizardNext();
    virtual BOOL OnWizardFinish();
    virtual BOOL OnSetActive();
protected:
    virtual void DoDataExchange( CDataExchange* pDX );    // DDX/DDV support
    // }}AFX_VIRTUAL


// Implementation
protected:
    // Generated message map functions
    // {{AFX_MSG( CPage1 )
    virtual BOOL OnInitDialog();
    // }}AFX_MSG
    DECLARE_MESSAGE_MAP()
private:
    CWzdSheet *m_pSheet;
};


// {{AFX_INSERT_LOCATION}}
// Microsoft Developer Studio will insert additional declarations immediately
//      before the previous line.

#endif
    // !defined( AFX_PAGE1_H__A846F4D3_ECA0_11D1_A18D_DCB3C85EBD34__INCLUDED_ )
// Page1.cpp : implementation file
//
```

II

12

```
#include "stdafx.h"
#include "wzd.h"
#include "WzdSheet.h"
#include "Page1.h"

#ifdef _DEBUG
#define new DEBUG_NEW
#undef THIS_FILE
static char THIS_FILE[] = __FILE__;
#endif


/////////////////////////////////////////////////////////////////////////////
// CPage1 property page

IMPLEMENT_DYNCREATE( CPage1, CPropertyPage )

CPage1::CPage1() : CPropertyPage( CPage1::IDD )
{
    // {{AFX_DATA_INIT( CPage1 )
    m_bPage3Check = FALSE;
    // }}AFX_DATA_INIT
}


CPage1::~CPage1()
{
}


void CPage1::DoDataExchange( CDataExchange* pDX )
{
CPropertyPage::DoDataExchange( pDX );
    // {{AFX_DATA_MAP( CPage1 )
    DDX_Check( pDX, IDC_PAGE3_CHECK, m_bPage3Check );
    // }}AFX_DATA_MAP
}

BEGIN_MESSAGE_MAP( CPage1, CPropertyPage )
    // {{AFX_MSG_MAP( CPage1 )
    // }}AFX_MSG_MAP
END_MESSAGE_MAP()
```

Example 46   Creating a Simple Wizard   **463**

```
/////////////////////////////////////////////////////////////////////////
// CPage1 message handlers

BOOL CPage1::OnInitDialog()
{
    CPropertyPage::OnInitDialog();

    m_pSheet = ( CWzdSheet* )GetOwner();

    return TRUE;    // return TRUE unless you set the focus to a control
                    //      EXCEPTION: OCX Property Pages should return FALSE
}

BOOL CPage1::OnSetActive()
{
    m_pSheet -> SetWizardButtons( PSWIZB_NEXT );
        // PSWIZB_BACK,PSWIZB_FINISH,PSWIZB_DISABLEDFINISH

    return CPropertyPage::OnSetActive();
}

LRESULT CPage1::OnWizardBack()
{

    return CPropertyPage::OnWizardBack();
}

LRESULT CPage1::OnWizardNext()
{
    UpdateData( TRUE );
    if ( m_bPage3Check )
    {
        return IDD_PAGE3;
    }
    else
    {
        return CPropertyPage::OnWizardNext();
    }
}
```

II

12

```
BOOL CPage1::OnWizardFinish()
{


    return CPropertyPage::OnWizardFinish();
}
```

Internal Processing Examples

Your application is not all user interface. A great deal goes on under the hood, from reading and writing files to timing events and multitasking. Even though MFC is mostly known as an interface development system, there are several classes in the MFC library that provide support for the noninterface part of your application, as well.

The examples in this section relate to a broad range of processing inside of your application. This includes multitasking within your program and sending messages within and outside of your application. Also included are playing sounds and playing with timers and the time.

Messaging and Communication

The examples in Chapter 13 all relate to sending data and messages within and outside of your application. There are also two example ways of manually processing the messages in your message queue. Chapter 13 also presents three ways, other than using a window message, available to your application to communicate wih the outside world. This includes Window Sockets and lowly serial I/O.

Multitasking

The examples in Chapter 14 relate to the processing choices you have for your application, including multitasking, background processing, and executing other applications.

Potpourri

Chapter 15 is the repository of "everything else internal". In this case, that turns out to be timers, binary strings, and VC++ macros, among others.

Chapter 13

Messaging and Communication

Beyond the window messages that your application can send to another application, there are several other avenues available for communicating with the environment. These include everything from serial I/O to using window sockets to talk over a network. Most of these topics are covered in depth in Chapters 1 and 3. The examples in this chapter present three practical applications of those topics.

Example 47   Waiting for a Message   We will look at how to pause our application within a message handler until another message comes in.

Example 48   Clearing Messages   We will look at how to clear out our application's message queue, so that the next message is fresh.

Example 49   Sending a Message to Another Application   We   will create a message so unique that it can be sent to another application in the system.

Example 50   Sharing Data with Another Application   We will look at how to send large amounts of data to another application.

Example 51   Communicating with Any Application Using Sockets
We will use Window Sockets to communicate with another Windows application or any other application that supports sockets, such as UNIX applications.

Example 52   Using Serial or Parallel I/O   We will communicate with a serial device from your MFC application.

Example 47   Waiting for a Message

Objective

You would like to pause your application within a function until someone presses either a mouse button or keyboard key.

Strategy

When your application is idle, it is, in fact, executing the function `CWinApp::Run()`. Not only does this function have logic that looks for new posted messages, but it also does a lot of application maintenance. `CWinApp::Run()` updates the status of your user interface and cleans up temporary memory objects. Therefore, if we were to stop execution somewhere else in our application to wait for a message, we would also need to perform this application maintenance. Since Microsoft has been thoughtful enough to provide the source code for `CWinApp::Run()`, we will simply create a version of it that we can use anywhere else in our application.

Why would you ever want to pause your application somewhere other then `CWinApp`? Please refer to the section "Porting Issues" on page 471 for one possible use.

Example 47   Waiting for a Message   **469**

Steps

Wait for a Window Message

1. Locate the point at which you would like to pause your application to wait for a particular message and add the following code there.

```
// wait till user clicks on status bar before proceeding
MSG msg;
BOOL bIdle = TRUE;
LONG lIdleCount = 0;
CWinApp* pApp = AfxGetApp();

AfxMessageBox( "Into wait loop." );
m_bWait = TRUE;
while ( m_bWait )
{
    // idle loop waiting for messages
    while ( bIdle && !::PeekMessage( &msg, NULL, NULL, NULL,
        PM_NOREMOVE) )
    {
        if ( !pApp -> OnIdle( lIdleCount++ ) ) bIdle = FALSE;
    }

    // process new messages
    do {
        // pump messages
        pApp -> PumpMessage();

        // if we're done, let's go...
        if ( !m_bWait )
            break;

        // otherwise keep looping
        if ( pApp -> IsIdleMessage( &msg ) )
        {
            bIdle = TRUE;
```

III 13

```
            lIdleCount = 0;
        }
    } while ( ::PeekMessage( &msg, NULL, NULL, NULL,
        PM_NOREMOVE ) );
}
```

This routine will wait here, processing messages and application mainte-nance, until m_bWait is TRUE.

2. Use the ClassWizard to add a message handler for the message for which you are waiting such as WM_LBUTTONDOWN message. There, you will set m_bWait to be TRUE.

Notes

- Notice that not only does this waiting routine pass messages onto their intended windows, but it also does background processing. If you were to omit the background processing, you would find that you could still change the size of your application's windows and move them around, but that they wouldn't completely redraw themselves and would leave screen garbage. Toolbars and the status bar would also stop updating their status.

- Rather than wait for m_bWait to be TRUE, you could also examine mes-sages as they go through the message pump in this waiting routine. Be aware that you will only intercept posted messages here. Messages that are sent with SendMessage() go directly to the intended window. Posted messages generally include only mouse and keyboard activity—every-thing else is usually sent. (If you would like to intercept SendMessage()'s messages, try using SetWindowsHookEx() using WH_CALLWNDPROC.)

- For more on messaging in general and the message pump in particular, please see Chapter 1.

CD Notes

- When executing the project on the accompanying CD, click on the "Test" then "Wzd1" menu commands and notice that the application will not leave the OnWzd1Test() function until you then click on the "Wzd2" menu command.

Example 48   Clearing Messages   **471**

Porting Issues

This example lends itself ideally to porting a DOS application to Windows. Window applications are event driven. Functions are usually called only when a mouse has been clicked or a key has been pressed and, when finished, the function relinquishes control until the next time. DOS applications, on the other hand, typically hog the system resources like the keyboard and the mouse until their function has been accomplished.

A typical function in a DOS application, when initiated by the user, might ask a series of questions, pausing each time to wait for a reply and holding the system hostage until the user does. To port this type of function to Windows, you can use the code in this example to allow the system to continue to process messages until the user replies. You can then process the user's reply in another message handler, which then sets m_bWait to TRUE so that the next question can be asked or the function can terminate.

Example 48   Clearing Messages

Objective

You would like to clear all of the queued up mouse clicks and keystrokes for a particular window before continuing.

Strategy

We will write a mini-message pump based on the last example that will process all of the messages currently in your application's message queue—except for the window that you choose. Messages for that window will, instead, be discarded.

III 13

Steps

Clear Messages

1. Add the following code in the point at which you would like to clear the message queue for a particular window.

```
MSG msg;
CWinApp* pApp = AfxGetApp();
while ( ::PeekMessage( &msg, NULL, 0, 0, PM_REMOVE ) )
{
    // kill any mouse messages for this window
    if ( ( msg.hwnd != m_hWnd || ( msg.message < WM_MOUSEFIRST ||
        msg.message > WM_MOUSELAST ) ) && !pApp ->
        PreTranslateMessage( &msg ) )
    {
        ::TranslateMessage( &msg );
        ::DispatchMessage( &msg );
    }
}
```

Here, we clear all mouse messages. To also clear all keyboard messages, include a range check for WM_KEYFIRST to WM_KEYLAST. To clear any posted message for a window, don't include any range check at all—just check to see if this message was intended for this window. To clear all posted messages to all windows, simply remove everything inside the while statement.

Notes

- This example can be especially useful when doing anything unusual with popup menus. Since popup menus are immediately closed when you click outside of them, any left-over mouse clicks might close your popup menu before it even appears. To eliminate any errant clicks, put this code just before you open the popup menu.
- Please see Chapter 1 for more on MFC messaging.

Example 49   Sending a Message to Another Application   **473**

CD Notes

- When executing the project on the accompanying CD, notice that any mouse messages for the view are removed from the message queue in OnTestWzd() before it pops up a menu.

Example 49   Sending a Message to Another Application

Objective

You would like to send a message to another application.

Strategy

We will be using the Windows API ::SendMessage() function to send a message to the handle of one of the windows belonging to our destination application. The trick will be in getting that window handle.

We will also use another Windows API function called ::RegisterWindow-Message() to create a unique message that will be universally understood between our two applications.

We will also look at yet another Windows API function, ::BroadcastSystemMessage(), which will allow us to send messages to the main window of every application in the system, thus eliminating the problem of getting the handle to another application's window.

Steps

We start by registering our own window message. The advantage of registering a window message, rather than using the WM_USER+1 technique shown in previous examples, is that you don't have the headache of trying to keep track of what WM_USER+x means across project boundaries. Instead, messages are registered with a text string you use in both projects. Since this text string also has to be unique throughout the system, we will be using a feature of COM technology called the GUID to name our message. The GUID name generator can be found in your MFC's \BIN directory under the name GUIDGEN.EXE and will generate a text string that is supposedly unique throughout the known universe. So, it's certainly good enough for this application.

III   13

Register a Unique Window Message

1. Generate a GUID using GUIDGEN.EXE.
2. Define this GUID as a window message text string in your application.

```
#define HELLO_MSG "{6047CCB1-E4E7-11d1-9B7E-00AA003D8695}"
```

3. Register this windows text string using ::RegisterWindowMessage().

```
idHelloMsg = ::RegisterWindowMessage( HELLO_MSG );
```

Save the message ID, idHelloMsg, for later use.

Send a Message to Another Application

1. To send a message using the message ID returned by ::RegisterWindow-Message(), you can use the following.

```
::SendMessage(
    hWnd,           // handle of a window belonging to destination app
    idHelloMsg,     // registered message id
    wParam,         // as usual
    lParam          // as usual
    );
```

This code assumes that you have somehow been able to acquire the window handle of one of the windows of your target application. A pointer to a CWnd class won't work across program boundaries, although you could wrap the acquired window handle in a CWnd class and send it.

```
CWnd wnd;
wnd.Attach( hWnd );
wnd.SendMessage( idHelloMsg,wParam,lParam );
```

Example 49   Sending a Message to Another Application   **475**

Receive a Registered Window Message

1. To receive a registered message, you will need to manually add an `ON_REGISTERED_MESSAGE` macro to the message map of the receiving window class, typically `CMainFrame`.

```
BEGIN_MESSAGE_MAP( CMainFrame, CMDIFrameWnd )
    // {{AFX_MSG_MAP( CMainFrame )
    // }}AFX_MSG_MAP
    ON_REGISTERED_MESSAGE( idHelloMsg,OnHelloMsg )
END_MESSAGE_MAP()
```

2. The syntax for the message handler of a registered message is:

```
LRESULT CMainFrame::OnHelloMsg( WPARAM wParam,LPARAM lParam )
{
    // process message
    return 0;
}
```

This example so far assumes that you have somehow been able to acquire the window handle of the target application. This, however, can be a difficult task. An easier approach might be to simply broadcast a message to every application in the system and hope your intended application is listening. Because you registered a unique message with the system, only your target application should respond to your message. The message your application can broadcast can be its own window handle so that the receiving application can then use `::SendMessage()` to send a reply, possibly its own window handle to complete the loop.

Broadcast a Window Message

1. To broadcast a window message to all of the applications on a system, you can use

```
WPARAM wParam = xxx;    // your definition
LPARAM lParam = xxx;    // your definition
DWORD dwRecipients = BSM_APPLICATIONS;
::BroadcastSystemMessage( BSF_IGNORECURRENTTASK,&dwRecipients,
    idHelloMsg,            // registered window message
    wParam,lParam );      // user defined parameters
```

Notes

- The `::BroadcastSystemMessage()` function offers an additional flag called `BSF_LPARAMPOINTER` that supposedly converts the pointer you put into `lParam` into a pointer that can be used by the target application to access your program space, However, this flag is suspiciously under-documented.

- The biggest use of interprocess messages is among the child processes of a shell application. Typically, the shell application creates a plain, but invisible, uniquely named window through which the child processes communicate with the shell. The child processes can use `CWnd::FindWindow()` to get the handle of this communication window using that unique window name. Then, the child windows can send an "I'm here" message to the shell application with its own window handle, thus completing the connection. To create your own plain window, please see Example 38.

CD Notes

- Execute two instances of the project, one in debug mode and one in release mode. Then, click on the "Test" and "Wzd" menu command on one application and notice that it exchanges messages with the other application.

Example 50   Sharing Data with Another Application

Objective

You would like to share globs of data with another application.

Strategy

We will be using the `::CreateFileMapping()` Windows API to carve off a segment of the swap file that we can share with other applications.

Example 50   Sharing Data with Another Application   **477**

Steps

Create Shared Memory

1. To open a segment of the swap file for shared memory you can use the following.

```
m_hMap = ::CreateFileMapping(
    ( HANDLE )0xffffffff,    // or can be an open file handle
    0,                       // security
    PAGE_READWRITE,          // or PAGE_READONLY or PAGE_WRITECOPY
    0,                       // size -- high order
                             //     (required if no file handle)
    0x1000,                  // size -- low order
                             //     (required if no file handle)
    MAP_ID                   // unique id--required if no file handle
    );
```

If you would prefer to share data through your own file, rather than the swap file, you can provide a file handle that has already been opened using the same file attributes you use here (e.g., read, write, etc.). To create a unique MAP_ID, you can use GUIDGEN.EXE in your VC++'s \BIN directory.

Note: The first application to call ::CreateFileMapping() with this MAP_ID will create the mapped file. All subsequent calls using this same MAP_ID will simply return a handle to the existing file.

---

2. To get a memory pointer to this mapped file handle, you can use the following.

```
m_pSharedData = ::MapViewOfFile(
    m_hMap,
    FILE_MAP_WRITE,    // or FILE_MAP_READ, FILE_MAP_COPY
                       //     (FILE_MAP_WRITE is read/write)
    0,                 // offset -- high order
    0,                 // offset -- low order
    0                  // number of bytes (zero maps entire file)
    );
```

III **13**

Use Shared Memory

1. You can now use the pointer returned from `::MapViewOfFile()` to copy data to or from the shared memory file.

```
// writing to shared memory
memcpy( ( LPBYTE )m_pSharedData,pWrite,10 );

// reading from shared memory
memcpy( pRead, ( LPBYTE )m_pSharedData,10 );
```

2. If you would like to access this data using file functions, you can also wrap this memory pointer with a `CMemFile` class.

```
CMemFile file;
file.Attach( ( LPBYTE )m_pSharedData,size );
file.Write( pBuffer,100 );    // write 100 bytes to shared memory
```

Closing Shared Memory

To close shared memory you must first unmap your view to it, then close the handle.

```
::UnmapViewOfFile( m_pSharedData );
::CloseHandle( m_hMap );
```

Notes

- The system synchronizes all access to shared memory created with `::CreateFileMapping()`. That means you don't have to worry that applications A and B are writing to the same shared memory at the same time. There can be a problem, though, when application A and B are sharing data over a network because they first write to a local buffer before it's transferred to shared memory. In this case, you will have to provide your own file synchronization.

- When using a file offset other than zero (0) in `::MapViewOfFile()`, it must be a multiple of the system's memory allocation granularity. To find out what this is, you can use `::GetSystemInfo()` to get the `SYSTEM_INFO` structure. The granularity is in this structure's `dwAllocationGranularity` member. This value has been hard coded to 64Kb in the past, but may change in the future.

Example 51   Communicating with Any Application Using Sockets   **479**

Example 51   Communicating with Any Application Using Sockets

Objective

You would like to use Window Sockets to communicate with another Windows application or any other application that supports sockets, such as UNIX applications.

Strategy

We will be using MFC's CSocket class to set up communications between two or more applications. With Window Sockets, a server application creates one special socket that simply listens for client applications to request a connection to their own socket. The server then creates a new socket to complete the connection. Reads are then made to these connections from both the server and client until a message comes in and requires processing. Please refer to Chapter 3 for more on Window Sockets.

We will encapsulate our functionality so that all our application has to do is create a socket connection and then process incoming messages. This will include creating a new server socket class, a new client socket class, and a new message queue class. The message queue class will allow us to add messages to a queue (actually a CObList) using synchronization so that two messages don't clobber each other in a multitasking environment. We need this protection because each socket will be doing a synchronous read in its very own thread.

Steps

III 13

Create a New Server Socket Class

1. Use the ClassWizard to create a new class derived from CSocket. You should use the word "Server" somewhere in the name of this new class.

2. Start this class by creating a new function called Open(), which will simply use CSocket's Create() function to create the socket. We do this, rather

than use `CSocket::Create()` directly, to make opening sockets similar to opening a port or file.

```
BOOL CWzdServer::Open( UINT nPort )
{
    return Create( nPort );
}
```

Our objective now is to write a set of functions that will listen for client applications wanting to connect and make that connection by creating a new socket. Since several clients can make a connection, we will keep track of these new socket objects in a map of objects. Once a socket is created, we will immediately start reading from it.

In this example, this functionality requires three functions and a structure. The first function, `ListenEx()`, tells our socket to start listening for client applications. The second function, `OnAccept()`, is called by our socket when it receives a request to connect. There, we create a new socket and immediately set it up to start reading messages from the client application by calling the third function, `RecvThread()`, which is in its own thread.

So, let's start by creating the `ListenEx()` function.

3. Add a `ListenEx()` function that will listen for connection requests from client applications by calling `CSocket`'s `Listen()` function. `ListenEx()` also sets up its calling arguments in a structure that will eventually be passed to the `RecvThread()` function for reading.

```
void CWzdServer::ListenEx( int hdrSz, int bodyPos, CWzdQueue *pQueue,
    CWnd *pWnd, UINT id )
{
    // initialize receive data
    m_RecvData.hdrSz = hdrSz;
    m_RecvData.bodyPos = bodyPos;
    m_RecvData.pQueue = pQueue;
    m_RecvData.pWnd = pWnd;
    m_id = id;                              // starting id

    // start listening
    Listen();
}
```

4. Next, use the Text Editor to override `CSocket`'s `OnAccept()` function. There, you will create a new socket to connect to the client and save that

Example 51   Communicating with Any Application Using Sockets   **481**

new socket in an object map using a user defined ID as a key. Then, you will put the socket into synchronous mode and create a thread to start reading from this socket.

```
void CWzdServer::OnAccept ( int nErrorCode )
{
    if ( nErrorCode == 0 )
    {
        // create a new socket and add to map
        CSocket *pSocket = new CSocket;
        m_mapSockets[m_id] = pSocket;

        // use this new socket to connect to client
        Accept( ( CAsyncSocket& )*pSocket );

        // put socket into synchronous mode
        DWORD arg = 0;
        pSocket -> AsyncSelect( 0 );
        pSocket -> IOCtl( FIONBIO, &arg );

        // setup this socket to listen for client messages
        m_RecvData.pSocket = pSocket;
        m_RecvData.id = m_id++;

        // start the thread
        AfxBeginThread( RecvThread,&m_RecvData );
    }
}
```

III 13

5. Finally, add the thread routine, RecvThread(), which will sit and wait until a new message comes over the socket using CSocket's Receive() function. This thread assumes that each message will consist of a fixed byte length header and a variable length body. Once a message comes in, it's put in a message queue, which we will design later. For every new socket message, RecvThread() will also send a WM_NEW_MESSAGE to your application telling it a new message is ready for processing. If the socket

closes, the thread terminates, but not before it also sends a WM_DONE_MESSAGE to your application.

```
UINT RecvThread( LPVOID pParam )
{
    // get data from thread creator
    RECVDATA *pRecv = ( RECVDATA * )pParam;

    int len = 1;
    int error = 0;
    char *pBody = NULL;
    char *pHdr = NULL;
    // while both sockets are open
    while (TRUE)
    {
        // read the header
        int res;
        pBody = NULL;
        pHdr = new char[pRecv -> hdrSz];
        if ( ( res = pRecv -> pSocket ->
            CAsyncSocket::Receive( pHdr, pRecv -> hdrSz ) )
            == SOCKET_ERROR )
            error = ::GetLastError();
        else
            len = res;

        // if closing down, exit thread
        if ( len == 0 || error == WSAECONNRESET ||
            error == WSAECONNABORTED ) break;

        // read the body???
        if ( !error && len && pRecv -> bodyPos != -1 )
        {
            int bodyLen = *( ( short * )pHdr+pRecv -> bodyPos );
            pBody = new char[bodyLen];
            if ( ( res = pRecv -> pSocket ->
                CAsyncSocket::Receive( pBody, bodyLen ) ) ==
                SOCKET_ERROR )
                error = ::GetLastError();
```

Example 51   Communicating with Any Application Using Sockets   **483**

```
        else
            len+ = res;

        // if closing down, exit thread
        if ( len == 0 || error == WSAECONNRESET ||
            error == WSAECONNABORTED ) break;
    }


    // put message in queue
    pRecv -> pQueue ->
        Add( new CWzdMsg( pRecv -> id,pHdr,pBody,len,error ) );

    // post message to window to process this new message
    pRecv -> pWnd -> PostMessage( WM_NEW_MESSAGE );


    }

    // cleanup anything we started
    delete []pHdr;
    delete []pBody;

    // tell somebody we stopped
    pRecv -> pWnd -> SendMessage( WM_DONE_MESSAGE,
        ( WPARAM )pRecv -> id,( LPARAM )error );

    return 0;
}
```

6. Next, add a function called SendEx() to send messages back to the client application. This function will pull a socket object out of the object map based on the user defined ID and then call a thread function to send the message to this socket.

```
void CWzdServer::SendEx( int id, LPSTR lpBuf, int len )
{
    // locate the socket for this id
    CSocket *pSocket = m_mapSockets[id];
    if ( pSocket )
    {
```

III 13

```
        m_SendData.pSocket = pSocket;
        m_SendData.lpBuf = lpBuf;
        m_SendData.len = len;

        // start the thread
        AfxBeginThread( SendThread,&m_SendData );
    }
}
```

7. The SendThread() function then sends this data out using CSocket's Send() function.

```
UINT SendThread( LPVOID pParam )
{
    // get data from thread creator
    SENDDATA *pSend = ( SENDDATA * )pParam;

    // do the write
    pSend -> pSocket -> Send( pSend -> lpBuf, pSend -> len );

    return 0;
}
```

8. Our final stop in this class is to create a close function that will not only close our listening socket, but also each of the client sockets we created along the way.

```
void CWzdServer::CloseEx()
{
    int id;
    CSocket *pSocket;
    for ( POSITION pos = m_mapSockets.GetStartPosition(); pos; )
    {
        m_mapSockets.GetNextAssoc( pos,id,pSocket );
        pSocket -> Close();
    }
    Close();

}
```

Example 51   Communicating with Any Application Using Sockets   **485**

Create a New Client Socket Class

We will actually add this class to the same .cpp and .h file in which we created our server socket class. Why? Because the two thread functions we created previously (SendThread() and RecvThread()) can be reused in our client socket class.

1. Start by adding an Open() function that will use CSocket's Open() and Connect() functions to both create a socket and attempt to connect it to a server application.

```
BOOL CWzdClient::Open( LPCTSTR lpszHostAddress, UINT nHostPort )
{
    if ( Create() && Connect( lpszHostAddress, nHostPort ) )
    {
        // put socket into synchronous mode
        DWORD arg = 0;
        AsyncSelect( 0 );
        IOCtl( FIONBIO, &arg );
        return TRUE;
    }
    return FALSE;
}
```

2. Next, add a send function. Since there is only one server, this send function will not require an ID. Add the appropriate data to the send structure and call SendThread() to do the actual send.

```
void CWzdClient::SendEx( LPSTR lpBuf, int len )
{
    // initialize the structure we will pass to thread
    m_SendData.pSocket = this;
    m_SendData.lpBuf = lpBuf;
    m_SendData.len = len;

    // start the thread
    AfxBeginThread( SendThread,&m_SendData );
}
```

III 13

Note: The SendThread() function is the same one used previously in the server socket class.

---

3. Next, add a ListenEx() function to this class, which will cause your client socket to start listening for **messages**, unlike the server socket that was listening for **connections**. ListenEx() will call RecvThread() to do the actual listening.

```
void CWzdClient::ListenEx( int hdrSz, int bodyPos, CWzdQueue *pQueue,
    CWnd *pWnd, UINT id )
{
    // initialize receive data
    m_RecvData.pSocket = this;
    m_RecvData.hdrSz = hdrSz;
    m_RecvData.bodyPos = bodyPos;
    m_RecvData.pQueue = pQueue;
    m_RecvData.pWnd = pWnd;
    m_RecvData.id = id;


    // start the thread
    AfxBeginThread( RecvThread,&m_RecvData );
}
```

Note: The RecvThread() function is the same one used previously in the server socket class.

Create a Message Queue Class

The message queue class is responsible for holding onto messages from one or more sockets until they can be processed. Not only does it contain a general purpose message class for holding message information, but it also uses synchronization to prevent one thread from adding a message to the queue while another one is pulling a message out. Both the server and client classes we created use this class.

1. Use the ClassWizard to create a new class derived from CObList.

Example 51   Communicating with Any Application Using Sockets   **487**

2. Embed in this class a `CMutex` class variable.

```
CMutex     m_mutex;
```

3. Add an `Add()` function to this class that will use `CObList`'s `AddTail()` function to add a message object to the queue. Synchronize access to this function using `m_mutex` and the `CSingleLock` class.

```
void CWzdQueue::Add( CWzdMsg *pMsg )
{
    CSingleLock slock( &m_mutex );

    if ( slock.Lock( 1000 ) )    // timeout in milliseconds,
                                 //     default = INFINITE

    {
        AddTail( pMsg );         // fifo
    }
}
```

4. Add a `Remove()` function to this class, which will use `CObList`'s `RemoveHead()` function to remove message objects from this queue. Again, protect this access with `m_mutex` and `CSingleLock`.

```
CWzdMsg *CWzdQueue::Remove()
{
    CSingleLock slock( &m_mutex ) ;

    if ( slock.Lock( 1000 ) )    // timeout in milliseconds,
                                 //     default = INFINITE

    {
        if ( !IsEmpty() )
            return ( CWzdMsg* )RemoveHead();
    }
    return NULL;
}
```

III **13**

5. To see a complete listing of this class, please refer to the Message Queue Class for Example 51 (page 493).

Use the New Server Socket Class

1. First and foremost, the socket DLL must be initialized by your application by calling `AfxSocketInit()` in your application class's `InitInstance()` function.

2. Next, embed a new socket and message queue class objects in one of your application's window classes, such as `CMainFrame`.

```
CWzdServer    m_server;
CWzdQueue     m_queue;
```

3. Then, open the socket and start listening.

```
if ( m_server.Open(
    1032        // between 1025 and 0xffffffff set by you
                //     to identify this server to your other apps
    ) )
{
    m_server.ListenEx(
    10,         // size of message header
    -1,         // position of size of message body in header
                //     -1 means all message lengths are fixed
    &m_queue,   // CWzdQueue to store new messages
    this,       // pWnd of window to send messages
    32          // starting id--as new connections are made
                //     this number increases
    );
}
```

4. When the socket receives a new message, you will be notified with a `WM_NEW_MESSAGE` message, which you can process as seen here.

```
LRESULT CMainFrame::OnNewMessage( WPARAM,LPARAM )
{
    CWzdMsg *pMsg = NULL;
    while ( pMsg = m_queue.Remove() )
    {
        // pMsg contains:
        // m_nID   -- the user defined id of which port
        //             sent the message
```

Example 51   Communicating with Any Application Using Sockets   **489**

```
        // m_pHdr  -- the message header
        // m_pBody -- the message body
        // m_len   -- the total message length
        // m_error -- any errors

        // make sure to delete the message after processing!
        delete pMsg;
    }
    return 0L;
}
```

5. You will also receive a WM_DONE_MESSAGE if a connection terminates.

```
LRESULT CMainFrame::OnDoneMessage( WPARAM id,LPARAM error )
{
    // gets here if a socket receive thread returns
    // id = id of client port that terminated
    // error = any error that caused the socket to close
    return 0L;
}
```

Note that CWzdServer keeps track of new client connects as they are made. All you have to know is that when a new connection is made, it gets an ID number that will be returned in any messages from that connection. That ID number can start at any value you specify and is continually incremented for each new connection. You can keep your own correlation between ID number and an actual client in a list in your own application. When a connection closes, you can delete that ID from list.

6. To send a message to a client, use

III 13

```
m_server.SendEx(
    32,        // id of client port
    hello,     // buffer to send
    10         // length of buffer to send
    );
```

7. You must also call the server socket's `CloseEx()` function if your application is terminating. Just add a `WM_CLOSE` message handler to your window class and call it there.

```
void CMainFrame::OnClose()
{
    m_server.CloseEx();

    CMDIFrameWnd::OnClose();
}
```

Use the New Client Socket Class

1. You must again initialize the socket DLL by calling `AfxSocketInit()` in your application class's `InitInstance()` function.

2. Next, embed a new socket and message queue class objects in one of your application's window classes, such as `CMainFrame`.

```
CWzdClient    m_client;
CWzdQueue     m_queue;
```

3. Then, open the socket and start listening to the server.

```
if ( m_client.Open(
    "localhost",    // system address of server specified as:
                    //      "ftp.myhost.com" or "128.23.1.22" or
                    //      "localhost" for the same machine
    1032            // the server's port number
    ) )
{
    m_client.ListenEx(
        10,         // size of message header
        -1,         // position of size of message body in header
                    //      -1 means all message lengths are fixed
        &m_queue,   // CWzdQue to store new messages
        this,       // pWnd of window to send messages
        0           // the user defined id of which client socket
                    //      sent the message
        );
}
```

Example 51   Communicating with Any Application Using Sockets   **491**

4. Although the last step allows our application to listen for messages from the server, these messages will only come in response to a message from us. To send a message to the server, use the following.

```
m_client.SendEx(
    hello,    // buffer to send
    10        // length of buffer to send
    );
```

5. Just as with the server application, your client application will receive a WM_NEW_MESSAGE message when a new message is in the queue, and a WM_DONE_MESSAGE when your connection is terminated. You can process these messages as seen previously for the server application.

6. Use the ClassWizard to add a WM_CLOSE message handler to this window class where you will close the socket.

```
void CMainFrame::OnClose()
{
    m_client.Close();

    CMDIFrameWnd::OnClose();
}
```

Notes

- Due to the size of this example, some of the detail has been left out of the steps, such as the exact make up of the message class and the format of the data structures that are passed to the thread functions. You can review these for yourself in the listings at the end of this example (page 493).

- The message header and body are treated as two separate entities so that your messages can be variable length. In this scenario, the length of the body is stored in the message header at some fixed offset that you specify when calling the ListenEx() function. This value is assumed to be a 16-bit variable. For a different size, just change the (short*) type override in the RecvThread() function. If your messages will all be fixed length, you can either chop out the body portion of this example or use -1 when specifying the position of the message length variable.

- This example uses the synchronous mode of sockets, meaning a read or write operation won't return until the entire message has been read or

III 13

written. This is obviously a problem if your application needs to do other things, such as respond to the user. Therefore, put each socket in it's own thread. Sockets can also do asynchronous reads and writes, but their implementation is more complex and doesn't accomplish any more than this solution.

- Sockets also allow you to communicate using Serialization between applications, so that you can easily communicate variable length class objects instead of structures with message headers and bodies. Serialization also takes care of big endian and little endian between Intel and Motorola machines (e.g., between a Windows machine and an iMac), but both applications must be created using MFC. To implement sockets using serialization, you will need to rewrite the SendThread() and RecvThread() functions along these lines.

```
// create a socket file class object
CSocketFile file(
    &sock                  // either the client or server socket class
    );
// construct an archive
CArchive ar(
    &file,                 // the file from above
    CArchive::load         // for RecvThread()
    // CArchive:: store     // for SendThread()
    );
// for RecvThread() to read a message
ar >> object;
// for SendThread() to write a message
ar << object;
```

By using polymorphism, you can even send variable length messages. For more on Serialization, please refer to my previous book, *Visual C++ MFC Programming by Example*. For more on sockets, please refer to Chapter 3.

CD Notes

- Execute two instances of the project, one in debug mode and one in release mode. Then, click on the "Test" menu for each. Notice that you can make either application a server or a client application. There are also menu commands for sending messages between these applications that you can break on to watch in CMainFrame.

Example 51   Communicating with Any Application Using Sockets   **493**

Message Queue Class for Example 51

```
// WzdQue.h: interface for the CWzdQueue class.
//
//////////////////////////////////////////////////////////////////////

#if !defined( AFX_QUEUE_H__81CE0F22_8C16_11D2_A18D_99620BDF6820__INCLUDED_ )
#define AFX_QUEUE_H__81CE0F22_8C16_11D2_A18D_99620BDF6820__INCLUDED_

#if _MSC_VER > 1000
#pragma once
#endif    // _MSC_VER > 1000

#include <afxmt.h>

class CWzdMsg : public CObject
{
public:
    CWzdMsg( int id,LPSTR pHdr,LPSTR pBody,int len,int error );
    virtual ~CWzdMsg();

    int m_nID;
    LPSTR m_pHdr;
    LPSTR m_pBody;
    int m_len;
    int m_error;
};

class CWzdQueue : public CObList
{
public:
    void Add( CWzdMsg *pMsg );
    CWzdMsg *Remove();

private:
    CMutex m_mutex;
};

#endif
    // !defined( AFX_QUEUE_H__81CE0F22_8C16_11D2_A18D_99620BDF6820__INCLUDED_ )
```

III   13

```
// WzdQue.cpp: implementation of the CWzdQueue class.
//
//////////////////////////////////////////////////////////////////////


#include "stdafx.h"
#include "WzdQue.h"


//////////////////////////////////////////////////////////////////////
// Construct Message
//////////////////////////////////////////////////////////////////////

CWzdMsg::CWzdMsg( int id,LPSTR pHdr,LPSTR pBody,int len,int error )
{
    m_nID = id;
    m_pHdr = pHdr;
    m_pBody = pBody;
    m_len = len;
    m_error = error;
}



CWzdMsg::~CWzdMsg()
{
    delete []m_pHdr;
    delete []m_pBody;
}



//////////////////////////////////////////////////////////////////////
// Add Message
//////////////////////////////////////////////////////////////////////

void CWzdQueue::Add( CWzdMsg *pMsg )
{
    CSingleLock slock( &m_mutex );

    if ( slock.Lock( 1000 ) )      // timeout in milliseconds, default = INFINITE
    {
        AddTail(pMsg);              // fifo
    }
```

Example 51   Communicating with Any Application Using Sockets   **495**

```
}

/////////////////////////////////////////////////////////////////////
// Remove Message
/////////////////////////////////////////////////////////////////////

CWzdMsg *CWzdQueue::Remove()
{
    CSingleLock slock( &m_mutex );

    if ( slock.Lock( 1000 ) )    // timeout in milliseconds, default = INFINITE
    {
        if ( !IsEmpty() )
            return ( CWzdMsg* )RemoveHead();
    }
    return NULL;
}
```

Socket Class for Example 51

```
// WzdSock.h: interface for the CWzdServer and CWzdClient class.
//
/////////////////////////////////////////////////////////////////////

#if !defined( AFX_WZDSOCK_H__81CE0F20_8C16_11D2_A18D_99620BDF6820__INCLUDED_ )
#define AFX_WZDSOCK_H__81CE0F20_8C16_11D2_A18D_99620BDF6820__INCLUDED_

#if _MSC_VER > 1000
#pragma once
#endif    // _MSC_VER > 1000

#include <afxsock.h>
#include <afxtempl.h>
#include "WzdQue.h"

#define WM_NEW_MESSAGE WM_USER+1
#define WM_DONE_MESSAGE WM_USER+2


/////////////////////////////////////////////////////////////////////
// Thread data
```

III **13**

```
///////////////////////////////////////////////////////////////////////////
typedef struct t_SENDDATA
{
    CSocket *pSocket;
    LPSTR lpBuf;
    int len;
} SENDDATA;


typedef CMap<int,int,CSocket *,CSocket *> SOCKMAP;
typedef struct t_RECVDATA
{
    CSocket *pSocket;
    int hdrSz;
    int bodyPos;
    CWzdQueue *pQueue;
    CWnd *pWnd;
    UINT id;
} RECVDATA;


///////////////////////////////////////////////////////////////////////////
// CWzdServer
///////////////////////////////////////////////////////////////////////////


class CWzdServer : public CSocket
{
public:
    CWzdServer(){};
    virtual ~CWzdServer();

    BOOL Open( UINT nPort );
    void CloseEx();
    void SendEx( int id, LPSTR lpBuf, int len );
    void ListenEx( int hdrSz, int bodyPos, CWzdQueue *pQueue, CWnd *pWnd,
        UINT id );

// Overrides
    virtual void OnAccept ( int nErrorCode );

private:
```

Example 51   Communicating with Any Application Using Sockets   **497**

```
    int m_id;
    SENDDATA m_SendData;
    RECVDATA m_RecvData;
    SOCKMAP m_mapSockets;
};


//////////////////////////////////////////////////////////////////////
// CWzdClient
//////////////////////////////////////////////////////////////////////


class CWzdClient : public CSocket
{
public:
    CWzdClient() {};
    virtual ~CWzdClient() {};

    BOOL Open( LPCTSTR lpszHostAddress, UINT nHostPort );
    void SendEx( LPSTR lpBuf, int len );
    void ListenEx( int hdrSz, int bodyPos, CWzdQueue *pQueue, CWnd *pWnd,
        UINT id );
private:
    SENDDATA m_SendData;
    RECVDATA m_RecvData;
};


//////////////////////////////////////////////////////////////////////
// Threads
//////////////////////////////////////////////////////////////////////
UINT SendThread( LPVOID pParam );
UINT RecvThread( LPVOID pParam );

#endif
    // !defined( AFX_WZDSOCK_H__81CE0F20_8C16_11D2_A18D_99620BDF6820__INCLUDED_ )
// WzdSock.cpp: implementation of the CWzdServer and CWzdClient classes.
//
//////////////////////////////////////////////////////////////////////


#include "stdafx.h"
#include "WzdSock.h"
```

III 13

```
/////////////////////////////////////////////////////////////////////////
// CWzdServer
/////////////////////////////////////////////////////////////////////////


///////////////////////////
// Cleanup
///////////////////////////

CWzdServer::~CWzdServer()
{
    // cleanup all created sockets
    int id;
    CSocket *pSocket;
    for ( POSITION pos = m_mapSockets.GetStartPosition(); pos; )
    {
        m_mapSockets.GetNextAssoc( pos,id,pSocket );
        delete pSocket;
    }
}


///////////////////////////
// Open Socket
///////////////////////////

BOOL CWzdServer::Open( UINT nPort )
{
    return Create( nPort );
}


///////////////////////////
// Send to Socket
///////////////////////////

void CWzdServer::SendEx( int id, LPSTR lpBuf, int len )
{
    // locate the socket for this id
    CSocket *pSocket = m_mapSockets[id];
    if ( pSocket )
    {
        m_SendData.pSocket = pSocket;
```

Example 51   Communicating with Any Application Using Sockets   **499**

```
        m_SendData.lpBuf = lpBuf;
        m_SendData.len = len;

        // start the thread
        AfxBeginThread(SendThread,&m_SendData);
    }
}


/////////////////////////////
// Listen to Socket
/////////////////////////////

void CWzdServer::ListenEx( int hdrSz, int bodyPos, CWzdQueue *pQueue,
    CWnd *pWnd, UINT id )
{
    // initialize receive data
    m_RecvData.hdrSz = hdrSz;
    m_RecvData.bodyPos = bodyPos;
    m_RecvData.pQueue = pQueue;
    m_RecvData.pWnd = pWnd;
    m_id = id;                              // starting id

    // start listening
    Listen();
}

// Listen() calls OnAccept() when a new client is attempting to connect
void CWzdServer::OnAccept ( int nErrorCode )
{
    if ( nErrorCode == 0 )
    {
        // create a new socket and add to map
        CSocket *pSocket = new CSocket;
        m_mapSockets[m_id] = pSocket;

        // use this new socket to connect to client
        Accept( ( CAsyncSocket& )*pSocket );

        // put socket into synchronous mode
        DWORD arg = 0;
```

III 13

```
                pSocket -> AsyncSelect( 0 );
                pSocket -> IOCtl( FIONBIO, &arg );

                // setup this socket to listen for client messages
                m_RecvData.pSocket = pSocket;
                m_RecvData.id = m_id++;

                // start the thread
                AfxBeginThread( RecvThread,&m_RecvData );
        }
}


/////////////////////////////
// Close Sockets
/////////////////////////////

void CWzdServer::CloseEx()
{
        int id;
        CSocket *pSocket;
        for ( POSITION pos = m_mapSockets.GetStartPosition(); pos; )
        {
                m_mapSockets.GetNextAssoc( pos,id,pSocket );
                pSocket -> Close();
        }
        Close();

}


/////////////////////////////////////////////////////////////////////////
// CWzdClient
/////////////////////////////////////////////////////////////////////////


/////////////////////////////
// Open Socket
/////////////////////////////

BOOL CWzdClient::Open( LPCTSTR lpszHostAddress, UINT nHostPort )
{
        if ( Create() && Connect( lpszHostAddress, nHostPort ) )
```

Example 51   Communicating with Any Application Using Sockets   **501**

```
    {
        // put socket into synchronous mode
        DWORD arg = 0;
        AsyncSelect( 0 );
        IOCtl( FIONBIO, &arg );
        return TRUE;
    }
    return FALSE;
}


///////////////////////////
// Send to Socket
///////////////////////////

void CWzdClient::SendEx( LPSTR lpBuf, int len )
{
    // initialize the structure we will pass to thread
    m_SendData.pSocket = this;
    m_SendData.lpBuf = lpBuf;
    m_SendData.len = len;

    // start the thread
    AfxBeginThread( SendThread,&m_SendData );
}



///////////////////////////
// Listen to Socket
///////////////////////////

void CWzdClient::ListenEx( int hdrSz, int bodyPos, CWzdQueue *pQueue,
    CWnd *pWnd, UINT id )
{
    // initialize receive data
    m_RecvData.pSocket = this;
    m_RecvData.hdrSz = hdrSz;
    m_RecvData.bodyPos = bodyPos;
    m_RecvData.pQueue = pQueue;
    m_RecvData.pWnd = pWnd;
    m_RecvData.id = id;
```

III 13

```
    // start the thread
    AfxBeginThread( RecvThread,&m_RecvData );
}



//////////////////////////////////////////////////////////////////////////
// Threads
//////////////////////////////////////////////////////////////////////////

UINT SendThread( LPVOID pParam )
{
    // get data from thread creator
    SENDDATA *pSend = ( SENDDATA * )pParam;

    // do the write
    pSend -> pSocket -> Send( pSend -> lpBuf, pSend -> len );

    return 0;
}


UINT RecvThread( LPVOID pParam )
{
    // get data from thread creator
    RECVDATA *pRecv = ( RECVDATA * )pParam;

    int len = 1;
    int error = 0;
    char *pBody = NULL;
    char *pHdr = NULL;
    // while both sockets are open
    while ( TRUE )
    {
        // read the header
        int res;
        pBody = NULL;
        pHdr = new char[pRecv -> hdrSz];
        if ( ( res = pRecv -> pSocket ->
            CAsyncSocket::Receive( pHdr, pRecv -> hdrSz)) == SOCKET_ERROR )
```

Example 51   Communicating with Any Application Using Sockets   **503**

```
                error = ::GetLastError();
        else
            len = res;


        // if closing down, exit thread
        if ( len == 0 || error == WSAECONNRESET || error == WSAECONNABORTED )
            break;


        // read the body???
        if ( !error && len && pRecv -> bodyPos != -1 )
        {
            int bodyLen = *( ( short * )pHdr+pRecv -> bodyPos );
            pBody = new char[bodyLen];
            if ( ( res = pRecv -> pSocket ->
                CAsyncSocket::Receive( pBody, bodyLen ) ) == SOCKET_ERROR )
                error = ::GetLastError();
            else
                len += res;

            // if closing down, exit thread
            if (len == 0 || error == WSAECONNRESET || error == WSAECONNABORTED)
                break;
        }


        // put message in queue
        pRecv -> pQueue ->
            Add(new CWzdMsg( pRecv -> id,pHdr,pBody,len,error ) );


        // post message to window to process this new message
        pRecv -> pWnd -> PostMessage( WM_NEW_MESSAGE );


    }


    // cleanup anything we started
    delete []pHdr;
    delete []pBody;


    // tell somebody we stopped
    pRecv -> pWnd ->
        SendMessage( WM_DONE_MESSAGE,( WPARAM )pRecv -> id,( LPARAM )error );
```

III 13

```
      return 0;
}
// WzdQue.h: interface for the CWzdQueue class.
//
//////////////////////////////////////////////////////////////////////

#if !defined( AFX_QUEUE_H__81CE0F22_8C16_11D2_A18D_99620BDF6820__INCLUDED_ )
#define AFX_QUEUE_H__81CE0F22_8C16_11D2_A18D_99620BDF6820__INCLUDED_

#if _MSC_VER > 1000
#pragma once
#endif    // _MSC_VER > 1000

#include <afxmt.h>

class CWzdMsg : public CObject
{
public:
    CWzdMsg( int id,LPSTR pHdr,LPSTR pBody,int len,int error );
    virtual ~CWzdMsg();

    int m_nID;
    LPSTR m_pHdr;
    LPSTR m_pBody;
    int m_len;
    int m_error;
};

class CWzdQueue : public CObList
{
public:
    void Add( CWzdMsg *pMsg );
    CWzdMsg *Remove();

private:
    CMutex m_mutex;
};

#endif
    // !defined( AFX_QUEUE_H__81CE0F22_8C16_11D2_A18D_99620BDF6820__INCLUDED_ )
```

Example 51   Communicating with Any Application Using Sockets   **505**

```cpp
// WzdQue.cpp: implementation of the CWzdQueue class.
//
//////////////////////////////////////////////////////////////////////

#include "stdafx.h"
#include "WzdQue.h"


//////////////////////////////////////////////////////////////////////
// Construct Message
//////////////////////////////////////////////////////////////////////

CWzdMsg::CWzdMsg( int id,LPSTR pHdr,LPSTR pBody,int len,int error )
{
    m_nID = id;
    m_pHdr = pHdr;
    m_pBody = pBody;
    m_len = len;
    m_error = error;
}


CWzdMsg::~CWzdMsg()
{
    delete []m_pHdr;
    delete []m_pBody;
}



//////////////////////////////////////////////////////////////////////
// Add Message
//////////////////////////////////////////////////////////////////////

void CWzdQueue::Add( CWzdMsg *pMsg )
{
    CSingleLock slock( &m_mutex );

    if ( slock.Lock( 1000 ) )      // timeout in milliseconds, default = INFINITE
    {
        AddTail(pMsg);              // fifo
    }
```

```
}

////////////////////////////////////////////////////////////////////////
// Remove Message
////////////////////////////////////////////////////////////////////////

CWzdMsg *CWzdQueue::Remove()
{
    CSingleLock slock( &m_mutex );

    if ( slock.Lock( 1000 ) )     // timeout in milliseconds, default = INFINITE
    {
        if ( !IsEmpty() )
            return (CWzdMsg*)RemoveHead();
    }
    return NULL;
}
```

Example 52   Using Serial or Parallel I/O

Objective

You would like to support a serial or parallel device from your MFC application.

Strategy

Opening and talking to a serial or parallel device is not that much different than interacting with a disk file. In fact, we will be using MFC's CFile class to communicate with these devices using their COM1: or LPT2: designations. The actual work of turning our read and write operations into serial or parallel operations will be performed by our hardware drivers. This example, however, takes this solution one step further and allows your application to communicate with several serial or parallel devices at one time, using multitasking to prevent any one device from freezing things up. In fact, we borrow the message queue class we created in the previous socket example, as well as some of the terminology, to channel incoming messages to one central message processor. Unlike the previous example, serial and parallel communication is so similar that we are able to package

Example 52   Using Serial or Parallel I/O   **507**

all of this functionality under one class, which we will be deriving from CFile.

Steps

Create a New Port I/O Class

1. Use the ClassWizard to create a new class derived from CFile.

2. Use the Text Editor to add an OpenLPT() function to this class to open parallel devices. Here, you simply call CFile's Open() using the CFile::modeReadWrite mode.

```
BOOL CWzdPortIO::OpenLPT( int n,CFileException *e )
{
    CString portName;
    portName.Format( "LPT%d:",n );
    return Open( portName, CFile::modeReadWrite, e );
}
```

3. Next, create an OpenCOM() function for opening a serial device. A serial device also has to contend with baud rates and stop bits, which can be configured with ::SetCommState(), as seen here.

```
BOOL CWzdPortIO::OpenCOM( int n, CFileException *e, int baud,
    int parity, int databits, int stopbits )
{
    CString portName;
    portName.Format( "COM%d:",n );
    if ( Open( portName, CFile::modeReadWrite, e ) )
    {
        DCB dcb;
        ::GetCommState( ( HANDLE )m_hFile, &dcb );
        if ( baud != -1 )     dcb.BaudRate = baud;
        if ( databits != -1 ) dcb.ByteSize = databits;
        if ( stopbits != -1 ) dcb.StopBits = stopbits;
        if ( parity != -1 )   dcb.Parity = parity;
        ::SetCommState( ( HANDLE )m_hFile, &dcb );
```

III 13

```
        return( TRUE );
    }
    return( FALSE );
}
```

4. To send data to this port device, simply use CFile's Write() function, but do it from a thread so that your application isn't tied up until it completes.

```
void CWzdPortIO::Send( LPSTR lpBuf, int len )
{
    // initialize the structure we will pass to thread
    m_SendData.pFile = this;
    m_SendData.lpBuf = lpBuf;
    m_SendData.len = len;

    // start the thread
    AfxBeginThread( SendThread,&m_SendData );
}
```

5. As mentioned, the send thread simply uses CFile's Write() function.

```
UINT SendThread( LPVOID pParam )
{
    // get data from thread creator
    SENDDATA *pSend = ( SENDDATA * )pParam;

    // do the write
    pSend -> pFile -> Write( pSend -> lpBuf, pSend -> len );

    return 0;
}
```

To read from the port, we adopt the window socket concept of listening. We create a function called Listen(), which uses CFile's Read() function from a thread that keeps reading until a certain number of bytes have been received. When the appropriate number of bytes come in, a message object is created and put in a message queue and a WM_NEW_MESSAGE message is sent to the main application in which the message can be processed. In fact, the message queue class we use here comes from the previous socket example.

Example 52   Using Serial or Parallel I/O   **509**

6. Use the Text Editor to add a `Listen()` function in which you will first prevent this port from timing out during a read operation by using `::SetCommTimeout()`. Then, you will call the `RecvThread()` function to make the actual read.

```
void CWzdPortIO::Listen( int hdrSz, int bodyPos, CWzdQueue *pQueue,
    CWnd *pWnd, UINT msg, UINT id )
{

    // cancel timeouts! we want to wait forever until
    //     next message comes in
    COMMTIMEOUTS cto;
    ::GetCommTimeouts( ( HANDLE )m_hFile, &cto );
    cto.ReadIntervalTimeout = 0;
    cto.WriteTotalTimeoutMultiplier = 0;
    cto.WriteTotalTimeoutConstant = 0;
    ::SetCommTimeouts( ( HANDLE )m_hFile, &cto );


    // initialize the structure we will pass to thread
    m_RecvData.pFile = this;
    m_RecvData.hdrSz = hdrSz;
    m_RecvData.bodyPos = bodyPos;
    m_RecvData.pQueue = pQueue;
    m_RecvData.pWnd = pWnd;
    m_RecvData.msg = msg;
    m_RecvData.id = id;


    // start the thread
    AfxBeginThread( RecvThread,&m_RecvData );

}
```

III **13**

To allow messages to be variable length, the `RecvThread()` function uses the concept of a message header and a message body. The message header is always some fixed length and contains the length of the following message body, if any. This way, the `Read()` function knows exactly how many bytes to read before returning.

7. Use the Text Editor to add a `RecvThread()` function to this class, in which you loop forever or until there's an error. If there's an error, you stop reading and send a `WM_DONE_MESSAGE` message to your application. Use `CFile`'s `Read()` function to read message headers and bodies. Place each in

a message queue and then inform your application with a `WM_NEW_MESSAGE` message.

```
UINT RecvThread( LPVOID pParam )
{
    // get data from thread creator
    RECVDATA *pRecv = ( RECVDATA * )pParam;


    while ( TRUE )    // forever
    {
        // read the header
        int len;
        int error = 0;
        char *pHdr = new char[pRecv -> hdrSz];
        try
        {
            len = pRecv -> pFile -> Read( pHdr, pRecv -> hdrSz );
        }
        catch ( CFileException *e )
        {
            error = e -> m_cause;
            e -> Delete();
        }


        // read the body???
        char *pBody = NULL;
        if ( !error && pRecv -> bodyPos != -1 )
        {
            int bodyLen = *( ( short * )pHdr+pRecv -> bodyPos );
            pBody = new char[bodyLen];
            try
            {
                len += pRecv -> pFile -> Read( pBody, bodyLen );
            }
            catch ( CFileException *e )
            {
                error = e -> m_cause;
                e -> Delete();
```

Example 52   Using Serial or Parallel I/O   **511**

```
            }
        }

        // put message in queue
        pRecv -> pQueue ->
            Add( new CWzdMsg( pRecv -> id,pHdr,pBody,len,error ) );

        // post message to window to process this new message
        pRecv -> pWnd -> PostMessage( pRecv -> msg );
    }


    return 0;
}
```

Use This New Port I/O Class for Parallel Communication

1. Embed CWzdPortIO and CWzdQueue in a window class, such as CMainFrame.

```
CWzdPortIO m_parallel;
CWzdQueue m_queue;
```

Note: The CWzdQueue class was created in the previous example to synchronously queue messages. You can find the steps to create it, as well as a complete listing, in that example.

---

2. Open the connection and start listening with

```
CFileException e;
if ( m_parallel.OpenLPT(
        1,                      // LPT number (1,2,etc.)
        &e                      // exception errors (defaults to NULL)
    ) )
{
    m_parallel.Listen(
        10,                     // size of message header
        -1,                     // position of size of message body
                                //     in header
```

III 13

```
                              //     -1 means all message lengths
                              //     are fixed
          &m_queue,           // CWzdQue to store new messages
          this,               // pWnd of window to send
                              //    "new message" message
          WM_NEW_MESSAGE,     // "new message" to send
          0                   // the user defined id of which port
                              //    sent the message

          );
}
```

3. To process new messages as they come in, manually add a WM_NEW_MESSAGE handler to your window class and process it like so.

```
ON_MESSAGE( WM_NEW_MESSAGE,OnNewMessage )
   :    :    :

LRESULT CMainFrame::OnNewMessage( WPARAM,LPARAM )
{
    CWzdMsg *pMsg = NULL;
    while ( pMsg = m_queue.Remove() )
    {
        // pMsg contains:
        // m_nID    -- the user defined id of which port sent
        //             the message
        // m_pHdr   -- the message header
        // m_pBody  -- the message body
        // m_len    -- the total message length
        // m_error  -- any errors

        // make sure to delete the message after processing!
        delete pMsg;
    }
    return 0L;
}
```

Example 52   Using Serial or Parallel I/O   **513**

4. To send a message back to the port, use

```
m_parallel.Send(
    hello,      // buffer to send
    7           // length of buffer to send
    );
```

Use This New Port I/O Class for Serial Communication

1. Embed CWzdPortIO and CWzdQueue in a window class, such as CMainFrame.

```
CWzdPortIO m_serial;
CWzdQueue m_queue;
```

2. Open the connection and start listening with

```
CFileException e;
if ( m_serial.OpenCOM(
    1,                  // COM number (1,2,etc.)
    &e,                 // exception errors (defaults to NULL)
    CBR_19200,          // baud rate, also CBR_1200, CBR_2400, etc.
    NOPARITY,           // parity, also EVENPARITY, ODDPARITY,
                        //     MARKPARITY, SPACEPARITY
    8,                  // number of bits in a byte
    ONESTOPBIT          // stopbits, also ONE5STOPBITS, TWOSTOPBITS
    ) )

{
    m_serial.Listen(
    10,                 // size of message header
    -1,                 // position of size of message body in header
                        //     -1 means all message lengths are fixed
    &m_queue,           // CWzdQue to store new messages
    this,               // pWnd of window to send "new message" message
    WM_NEW_MESSAGE,     // message to send
    1                   // the user defined id of which port sent
                        //     the message
    );
}
```

III 13

3. Processing new messages and sending messages is identical to parallel ports, seen in the last example.

Notes

- This solution allows your application to monitor several devices at once using your own IDs to distinguish which message came from which device.

CD Notes

Execute two instances of the project, one in debug mode and one in release mode. Then, click on the "Test" menu for each. You can now test serial or parallel communication, but you need a cooperating device.

Port I/O Class for Example 52

```cpp
// PortIO.h: interface for the CPortIO class.
//
//////////////////////////////////////////////////////////////////////

#if !defined( AFX_PORTIO_H__81CE0F20_8C16_11D2_A18D_99620BDF6820__INCLUDED_ )
#define AFX_PORTIO_H__81CE0F20_8C16_11D2_A18D_99620BDF6820__INCLUDED_

#if _MSC_VER > 1000
#pragma once
#endif    // _MSC_VER > 1000

#include "WzdQue.h"

// define the send and receive threads
typedef struct t_SENDDATA
{
    CFile *pFile;
    LPSTR lpBuf;
    int len;
} SENDDATA;

typedef struct t_RECVDATA
{
```

Example 52   Using Serial or Parallel I/O   **515**

```
    CFile *pFile;
    int hdrSz;
    int bodyPos;
    CWzdQueue *pQueue;
    CWnd *pWnd;
    UINT msg;
    UINT id;
} RECVDATA;

class CWzdPortIO : public CFile
{
public:
    CWzdPortIO() {};
    virtual ~CWzdPortIO() {};

    BOOL OpenLPT( int n,CFileException *e = NULL );
    BOOL OpenCOM( int n,CFileException *e = NULL, int baud = -1,
        int parity = -1, int databits = -1, int stopbits = -1);

    void Send( LPSTR lpBuf, int len );
    void Listen( int hdrSz, int bodyPos, CWzdQueue *pQueue, CWnd *pWnd,
        UINT msg, UINT id);

private:
    SENDDATA m_SendData;
    RECVDATA m_RecvData;



};
UINT SendThread( LPVOID pParam );
UINT RecvThread( LPVOID pParam );

#endif
    // !defined( AFX_PORTIO_H__81CE0F20_8C16_11D2_A18D_99620BDF6820__INCLUDED_ )
// PortIO.cpp: implementation of the CPortIO class.
//
///////////////////////////////////////////////////////////////////

#include "stdafx.h"
#include "WzdPrtIO.h"
```

III 13

```
/////////////////////////////////////////////////////////////////////
// Open Printer Port
/////////////////////////////////////////////////////////////////////


BOOL CWzdPortIO::OpenLPT( int n,CFileException *e )
{
    CString portName;
    portName.Format( "LPT%d:",n );
    return Open( portName, CFile::modeReadWrite, e );
}


/////////////////////////////////////////////////////////////////////
// Open Serial Port
/////////////////////////////////////////////////////////////////////

BOOL CWzdPortIO::OpenCOM( int n, CFileException *e, int baud, int parity,
    int databits, int stopbits )
{
    CString portName;
    portName.Format( "COM%d:",n );
    if ( Open( portName, CFile::modeReadWrite, e) )
    {
        DCB dcb;
        ::GetCommState( ( HANDLE )m_hFile, &dcb );
        if ( baud != -1 )     dcb.BaudRate = baud;
        if ( databits != -1 ) dcb.ByteSize = databits;
        if ( stopbits != -1 ) dcb.StopBits = stopbits;
        if ( parity != -1 )   dcb.Parity = parity;
        ::SetCommState( ( HANDLE )m_hFile, &dcb );
        return( TRUE );
    }
    return( FALSE );
}


/////////////////////////////////////////////////////////////////////
// Send to Port
/////////////////////////////////////////////////////////////////////

void CWzdPortIO::Send( LPSTR lpBuf, int len )
```

Example 52   Using Serial or Parallel I/O   **517**

```
{
    // initialize the structure we will pass to thread
    m_SendData.pFile = this;
    m_SendData.lpBuf = lpBuf;
    m_SendData.len = len;

    // start the thread
    AfxBeginThread( SendThread,&m_SendData );
}


UINT SendThread( LPVOID pParam )
{
    // get data from thread creator
    SENDDATA *pSend = ( SENDDATA * )pParam;

    // do the write
    pSend -> pFile -> Write( pSend -> lpBuf, pSend -> len );

    return 0;
}




////////////////////////////////////////////////////////////////
// Listen to Port
////////////////////////////////////////////////////////////////

void CWzdPortIO::Listen( int hdrSz, int bodyPos, CWzdQueue *pQueue,
    CWnd *pWnd, UINT msg, UINT id )
{
    // cancel timeouts! we want to wait forever until next message comes in
    COMMTIMEOUTS cto;
    ::GetCommTimeouts( ( HANDLE )m_hFile, &cto );
    cto.ReadIntervalTimeout = 0;
    cto.WriteTotalTimeoutMultiplier = 0;
    cto.WriteTotalTimeoutConstant = 0;
    ::SetCommTimeouts( ( HANDLE )m_hFile, &cto );

    // initialize the structure we will pass to thread
    m_RecvData.pFile = this;
    m_RecvData.hdrSz = hdrSz;
```

III 13

```
    m_RecvData.bodyPos = bodyPos;
    m_RecvData.pQueue = pQueue;
    m_RecvData.pWnd = pWnd;
    m_RecvData.msg = msg;
    m_RecvData.id = id;


    // start the thread
    AfxBeginThread( RecvThread,&m_RecvData );
}


UINT RecvThread( LPVOID pParam )
{
    // get data from thread creator
    RECVDATA *pRecv = ( RECVDATA * )pParam;


    while ( TRUE ) //forever
    {
        // read the header
        int len;
        int error = 0;
        char *pHdr = new char[pRecv -> hdrSz];
        try
        {
            len = pRecv -> pFile -> Read( pHdr, pRecv -> hdrSz );
        }
        catch ( CFileException *e )
        {
            error = e -> m_cause;
            e -> Delete();
        }


        // read the body???
        char *pBody = NULL;
        if ( !error && pRecv -> bodyPos != -1 )
        {
            int bodyLen = *( ( short * )pHdr+pRecv -> bodyPos );
            pBody = new char[bodyLen];
            try
            {
                len += pRecv -> pFile -> Read( pBody, bodyLen );
```

Example 52   Using Serial or Parallel I/O   **519**

```
            }
        catch ( CFileException *e )
        {
            error = e -> m_cause;
            e -> Delete();
        }
    }


    // put message in queue
    pRecv -> pQueue ->
        Add( new CWzdMsg( pRecv -> id,pHdr,pBody,len,error ) );


    // post message to window to process this new message
    pRecv -> pWnd -> PostMessage( pRecv -> msg );
    }

    return 0;
}
```

Chapter 14

Multitasking

Now that Windows is a multitasking operating system, your application has more ways to execute. You can multitask within your own application by creating processing threads that run concurrently with your application. You can still execute other applications from your own, but now they will actually run concurrently with your own application. MFC also provides a convenient member function you can override to perform background processing.

Example 53   Using Background Processing for Cleanup   We   will look at a way to continually clean up after our application in the background, when there are no commands to process.

Example 54   Executing Another Application   We will execute another application from our own.

Example 55   Changing Your Priority   We will change our application or thread's priority to execute faster or slower when no other application is active.

Example 56   Multitasking Within Your Application—Worker Threads   We will create a worker thread to perform some mathematical function while our main application is looking after the wants and needs of its user. Worker threads, unlike their User Interface threads used in the next example, are intended to live for one task and then die.

Example 57   Multitasking Within Your Application—User Interface Threads   We will create a user interface thread with all of the functionality of having executed another application, but with the advantage of being in the same address space allowing easy sharing of data between thread and application.

Example 58   Sending a Message to a User Interface Thread   We will communicate with our user interface threads.

Example 59   Sharing Data with Your Threads   We will review the safe way to share the same data address space amongst several threads without causing conflicts.

Example 53   Using Background Processing for Cleanup

Objective

You would like to perform cleanup chores in your application when it's idle. In this example, we clean up temporarily allocated memory.

Strategy

We will use the ClassWizard to override our Application Class's `OnIdle()` function, which is called by our application repeatedly whenever there are no user commands to process.

Example 53   Using Background Processing for Cleanup   **523**

Steps

Override the `OnIdle()` Function

1. Use the ClassWizard to override the `OnIdle()` member function of your Application Class.
2. In this example, we use our override of `OnIdle()` to delete a list of temporary objects.

```
BOOL CWzdApp::OnIdle( LONG lCount )
{
    // clean up temporary objects
    while ( !m_TempList.IsEmpty() )
    {
        delete m_TempList.RemoveHead();
    }
    return CWinApp::OnIdle( lCount );
}
```

Notes

- Make sure your override of `OnIdle()` continues to call `CWinApp::OnIdle()` or your application will cease to work.
- An application that runs several other applications within its "shell" can use this example to make sure each of its spawned applications is still running and, if not, it can re-execute them.
- `OnIdle()` is not a good place to do any lengthy processing, such as math processing. If `OnIdle()` doesn't return promptly, your application will appear to be sluggish since mouse and keyboard message processing will be delayed. A much better place to do intensive calculations would be in another application or in a thread. Please see the following examples.

III

14

CD Notes

- When executing the project on the accompanying CD, set a breakpoint on the `OnIdle()` function in `Wzd.cpp` (inside the conditional so that it doesn't continually break). Then, click on the "Test" then "Wzd" menu commands to fill up a global list collection. Then, notice that the `OnIdle()` routine cleans up this collection.

Example 54   Executing Another Application

Objective

You would like to execute another application from your application.

Strategy

Since there currently isn't an MFC class that can execute another application, we will use one of two Windows API calls. The first, ::WinExec(), is a simple, straightforward API that Microsoft prefers you don't use. The ::WinExec() function calls the second API, ::CreateProcess(), which gives you greater control over spawning a new application.

Steps

Execute an Application with ::WinExec()

To execute another application using ::WinExec(), please refer to the following code. In this example, we are executing a batch file called Wzd.bat.

```
CString str;
str = "Wzd.bat";

// execute a batch file
if ( ::WinExec(
    str,              // command line
    SW_NORMAL )       // see ShowWindow for other options
    >31 )             // numbers lower then 31 are failures
{
    AfxMessageBox( "Successfully created." );
}
else
{
    AfxMessageBox( "Failed to create process." );
}
```

Example 54   Executing Another Application   **525**

Execute an Application with ::CreateProcess()

For another way to execute the same batch file, but with more options, use the following.

```
CString str;
STARTUPINFO si;
PROCESS_INFORMATION pi;


// specify command
str = "Wzd.bat";


// zero out and initialize STARTUPINFO
memset( &si, 0, sizeof( si ) );
si.cb = sizeof( si );
si.dwFlags = STARTF_USESHOWWINDOW;
si.wShowWindow = SW_SHOW;



if ( CreateProcess(
    NULL,                       // can be name of process unless
                                //    batch file, else must be
                                //    in command line:

    ( char * )LPCSTR( str ),    // command line
    NULL,NULL,                  // security options
    FALSE,                      // if true will inherit all
                                //    inheritable handles
                                //    from this process

    NORMAL_PRIORITY_CLASS,      // can also be HIGH_PRIORITY_CLASS
                                //    or IDLE_PRIORITY_CLASS

    NULL,                       // inherit this process's
                                // environment block

    NULL,                       // specifies working directory
                                //    of created process

    &si,                        // STARTUPINFO specified above
    &pi                         // PROCESS_INFORMATION returned
    ) )
{
    AfxMessageBox( "Successfully created." );
```

III

14

```
        HANDLE pH = pi.hProcess;

        // wait until application is ready for input
        if ( !WaitForInputIdle( pH,1000 ) )
        {
            // send messages, etc.
        }

        // kill process with 0 exit code
        TerminateProcess( pH, 0 );


    }
    else
    {
        AfxMessageBox( "Failed to create process." );
    }
}
```

Get Your Application's Directory

Another Windows API call, `::GetModuleFileName()`, will tell a running application which subdirectory its execution file is located in on disk. This can be especially useful if your application is installed in the same directory with other files that it needs to access at run time.

1. To find the directory where your application is located, use the following.

```
// change directory to this application's .exe file
char szBuffer[128];
::GetModuleFileName( AfxGetInstanceHandle(), szBuffer,
    sizeof( szBuffer ) );
char *p = strrchr( szBuffer, '\\' );
*p = 0;
```

The `szBuffer` variable now contains the path name of your application's home directory.

2. To change your application's current working directory to this path, use

```
_chdir( szBuffer );
```

Example 54   Executing Another Application   **527**

Once this becomes your current working directory, your application can open any file there without specifying a path.

Notes

- The second argument you pass to `WinExec()` would be the same argument with which you would call `ShowWindow()`. That can include `SW_HIDE`, `SW_MAXIMIZE`, `SW_MINIMIZE`, etc. This flag affects the main window of the application you are executing.

- If you are putting a directory path in the command line argument of these API calls, make sure to put double-quotes around it. Why? Because the following is treated as three separate command line arguments because of the spaces.

```
c:\Program Files\Microsoft Office\prog.exe
```

With quotes, it becomes one argument again.

```
"c:\Program Files\Microsoft Office\prog.exe"
```

- When executing a batch file, you are, in fact, executing `CMD.EXE` and passing it the name of the batch file as a command line argument. That's why the batch file can't be the first argument when using `::CreateProcess()`.

- In this example, we use `TerminateProcess()` to kill the executed application. However, this can be bad—the terminated application isn't allowed to clean up before it terminates. A cleaner way to do the same thing would be to send this application a message to terminate. Since MFC doesn't provide one, you will need to create one yourself (see Example 58). Of course, if the application doesn't respond by killing itself, you can still use `TerminateProcess()`.

- Rather than executing another application, you might instead simply create another processing thread within your current application. An application is itself just a thread that can create other threads. Since Windows is a true multitasking operating system, each of these spawned threads can be running concurrently with your application and inform your application when they are done. Since a thread runs in the same program space as your application, transferring data between thread and application isn't the issue it is between applications. For an example of creating a thread, rather than executing another application, please see Example 57.

III

14

- To communicate with this newly created application, please see Example 49. To share data with it, please see Example 50.

CD Notes

- When executing the project on the accompanying CD, set a breakpoint on `OnWzd1Test()` or `OnWzd2Test()` in `Mainfrm.cpp`. Then, click on either "Wzd1" or "Wzd2" in the "Test" menu and watch as the function executes the `wzd.bat` batch file.

Example 55   Changing Your Priority

Objective

You would like to raise or lower your application or thread's priority to either speed up your application or let it slowly execute in the background.

Strategy

To change our application's priority, we will use the `::SetPriorityClass()` API directly. To change a thread's priority, we will use `CWinThread`'s `SetThreadPriority()` function. We will be looking at threads in the next two examples.

Steps

Change Your Application's Priority

1. To change your application's priority, use

```
::SetPriorityClass(
    ::GetCurrentProcess(),          // process handle
    // REALTIME_PRIORITY_CLASS      // highest: thread must run
                                    //     immediately before any
                                    //     other system task
    // HIGH_PRIORITY_CLASS          // high: time-critical threads
    // NORMAL_PRIORITY_CLASS        // normal: thread with equal
                                    //     importance to other
```

Example 55   Changing Your Priority   **529**

```
                                    //       system applications
        IDLE_PRIORITY_CLASS         // low: threads that can run in
                                    //       the background of the
                                    //       entire system

        );
```

2. To get the current priority, use

```
DWORD priority =
    ::GetPriorityClass(
        ::GetCurrentProcess()    // process handle
        );
```

Change Your Thread's Priority

1. To change your thread's priority, use the following from your thread.

```
SetThreadPriority(
    THREAD_PRIORITY_TIME_CRITICAL    // highest priority
    THREAD_PRIORITY_HIGHEST          // next highest
    THREAD_PRIORITY_ABOVE_NORMAL     // etc....
    THREAD_PRIORITY_NORMAL
    THREAD_PRIORITY_BELOW_NORMAL
    THREAD_PRIORITY_LOWEST
    THREAD_PRIORITY_IDLE             // lowest priority
    );
```

Notes

- Setting priorities is not always conducive to multitasking. Counter to common sense, you should give a CPU-intensive application or thread a *lower* priority, never a higher one. Only applications that require little processing should ever be given a high priority. Use a lower priority to allow a CPU-intensive application to run in the background without having to cut it into CPU pieces yourself. Use a higher priority to handle real-time applications that require immediate processing of incoming messages, such as keyboard clicks or messages from a serial device.

III

14

CD Notes

- Modify the `SetPriorityClass()` function in `Wzd.cpp` to have either a lower or higher priority. Make the priority high, create the application, and run it. Use the File/Open menu commands to open a dialog box. Then, start another process on your system, such as using Windows Explorer to search for a file. While that process is on-going, notice that by dragging the edge of this application's dialog box you can temporarily pause the other application.

Example 56   Multitasking Within Your Application—Worker Threads

Objective

You would like to create a program thread to perform math or other CPU-intensive functions separately and concurrently with your application.

Strategy

We will be using MFC's `AfxBeginThread()` function to create our thread. To tell the main application when the thread is done, we will be creating our own windows message, which the thread will send to our application when completed.

Steps

Set Up a Worker Thread

1. Define a data structure that will be used to transfer data to and from the thread. An example structure might look like this.

```
typedef struct t_THREADDATA
{
    HWND hDoneWnd;    // window handle of main app to send messages
    int nData;        // data to process
} THREADDATA;
```

Example 56   Multitasking Within Your Application—Worker Threads   **531**

2. Write a worker thread function using the following syntax. The `pParam` argument is a pointer to the data structure we defined previously. Please refer to the Worker Thread Process for Example 56 (page 534) for a complete listing.

```
UINT WzdThread( LPVOID pParam )
{
    // get data from thread creator
    THREADDATA *pData = ( THREADDATA * )pParam;

    // do calulations
    for ( int i = pData -> nData;i < 1000000;i++ )
    {
    }

    // save data back to thread creator
    pData -> nData = i;

    // tell creator we're done
    ::SendMessage( pData -> hDoneWnd, WM_DONE, 0, 0 );

    return 0;
}
```

The returned value is up to you to define. The parent application can retrieve this code using `GetExitCodeThread()`.

You can also terminate a worker thread with `AfxEndThread(arg)`, where the `arg` value is again defined by you and which is returned by the thread as before.

Create a Worker Thread

To create this thread, start by initializing the data structure with the appropriate data. Then, we call `AfxBeginThread()`, specifying the thread function

III

14

we created previously. Make sure to embed your data structure in the calling class so that it hangs around after you create the thread.

```
THREADDATA m_ThreadData;
:       :       :
ThreadData.nData = 123;
ThreadData.hDoneWnd = m_hWnd;
AfxBeginThread(
    WzdThread,        // static thread process declare
    &m_ThreadData     // data to send to thread
    );
```

Tell the Application When the Thread Is Done

1. To tell the main application when the thread is done, we need to create our own windows message.

```
#define WM_DONE WM_USER + 1
```

2. As seen previously, we then send this message when the thread is done.

```
// tell creator we're done
::SendMessage( pData -> hDoneWnd, WM_DONE, 0, 0 );
```

3. For your application to receive this message, you need to manually add a message handler for it. Start by manually adding an ON_MESSAGE_VOID() message macro to your receiving window's message map. Make sure to include this below the {{}} brackets so that the ClassWizard can't touch it.

```
BEGIN_MESSAGE_MAP( CWzdView, CView )
    // {{AFX_MSG_MAP( CWzdView )
    // }}AFX_MSG_MAP
    ON_MESSAGE_VOID( WM_DONE,OnDone )
END_MESSAGE_MAP()
```

4. Next, add the message handler.

```
void CWzdView::OnDone()
{
}
```

Example 56   Multitasking Within Your Application—Worker Threads   **533**

Make sure to also define this handler in your class's .h file. We're using the ON_MESSAGE_VOID() macro, rather than the ON_MESSAGE() macro, to avoid having to deal with the extra arguments the latter macro requires.

Notes

- You can pass pointers to your thread through the data structure, since both thread and application occupy the same address space. However, you can't pass the pointer to a CWnd structure. Your application keeps track of every instance of CWnd it creates and the window handle to which it refers so that it can instantly translate CWnd pointers into window handles. Each thread also keeps track of the windows it creates. If you were to pass the pointer of a CWnd object to a thread, that thread would have no record of its window handle and would be unable to make that translation for that window. What you can pass instead is the window handle itself. Then, if you would like, you can still wrap that window handle in a CWnd object within the thread itself.

- To prevent two threads from writing to the same area of data at the same time, please see Example 59.

- The AfxBeginThread() function is simply a helper function that creates an instance of the CWinThread class and calls the function you supply from within that thread. When your process returns, or when you call AfxEndThread(), the helper function calls ::PostQuitMessage() to terminate the thread.

- Typical uses for a worker thread include math functions on the columns of a spread sheet or background spell checking in a document application. Please refer to Examples 51 and 52 for a use that involves synchronous socket and port communication.

- If you have several threads running at once, you might also include a thread ID in the "all done" message to tell your application which thread completed. To do this, you will need to use the ON_MESSAGE() message macro with a message handler that looks like this.

```
LRESULT CMainFrame::OnDone( WPARAM wParam,LPARAM lParam )
{
    return 0;
}
```

III

14

CD Notes

- When executing the project on the accompanying CD, set a breakpoint on OnTestWzd() and OnDone() in WzdView.cpp. Then, click on the "Test" and then "Wzd" menu commands and watch as a thread is created and then reports back that it's finished.

Worker Thread Process for Example 56

```
#ifndef WZDTHREAD_H
#define WZDTHREAD_H

#define WM_DONE WM_USER + 1

typedef struct t_THREADDATA
{
    HWND hDoneWnd;
    int nData;
} THREADDATA;

UINT WzdThread( LPVOID pParam );

#endif
// WzdThread.cpp : thread process
//

#include "stdafx.h"
#include "WzdThread.h"

/////////////////////////////////////////////////////////////////////////////
// CWzdThread

UINT WzdThread( LPVOID pParam )
{
    // get data from thread creator
    THREADDATA *pData = ( THREADDATA * )pParam;

    // do calulations
    for ( int i = pData -> nData;i < 1000000;i++ )
    {
```

Example 57   Multitasking Within Your App— User Interface Threads   **535**

```
    }

    // save data back to thread creator
    pData -> nData = i;

    // tell creator we're done
    ::SendMessage( pData -> hDoneWnd, WM_DONE, 0, 0 );

    return 0;
// return value up to you--parent can retrieve with GetExitCodeThread();
// can also call AfxEndThread(0) where the meaning of the argument is up to you
    }
```

Example 57   Multitasking Within Your App—
User Interface Threads

Objective

You would like to create a program thread to perform functions separately and concurrently with your application, but which also requires its own user interface, such as the search and replace function of a document application.

Strategy

As with creating a worker thread, we will use the framework's AfxBeginThread() function to create our user interface thread. However, this time we will have complete control over our thread, rather than just the fragment we were allowed with a worker thread.

Steps

Create a New Thread Class

1. Follow the steps in User Interface Thread Class for Example 57 (page 537) to create a thread class that opens a modeless dialog box.

III

14

Create a User Interface Thread

1. To start this thread, we use

```
CWinThread *pThread = AfxBeginThread( RUNTIME_CLASS( CWzdThread ) );
```

2. Your thread should call `::PostQuitMessage(arg)` to terminate, where `arg` is of your own definition. For your application to get the value of `arg`, it can call

```
int arg = pThread -> GetExitCodeThread();
```

Note: Your application should not directly terminate a thread. A thread must exit itself to allow it to clean up. What you should do, instead, is create your own windows message telling the thread to terminate. The thread can then process that message by calling `::PostQuitMessage()`. Please refer to the next example for how to send a message to your thread.

Notes

- Unlike a worker thread, which is intended more for grunt processing, a user interface thread can have it's own interface and is, in fact, similar to executing another application. The advantage of creating a thread, rather than another application however, is that your thread shares program space with your application, thus simplifying its ability to share data with your application.

- Typically, a user interface thread can be used to do search and replace functions or any service that you don't want tying up your main application, but which still requires an interface. You can leave the interface off entirely and use this type of thread as a way for a window message server to deliver its messages without tying the main server up.

- In a time-critical application where you don't want to have to wait for a worker thread to start, you can build your worker thread logic into a user interface thread and then create the thread ahead of time. Then, when you need something processed, you can send a message to your user interface thread, which is already running and waiting for orders.

Example 57   Multitasking Within Your App— User Interface Threads   **537**

CD Notes

- When executing the project on the accompanying CD, click on the "Test" and then "Wzd" menu commands to create a new thread that contains a modeless dialog box.

User Interface Thread Class for Example 57

```
#if !defined( AFX_WZDTHREAD_H__411AE4C2_E515_11D1_9B80_00AA003D8695__INCLUDED_ )
#define AFX_WZDTHREAD_H__411AE4C2_E515_11D1_9B80_00AA003D8695__INCLUDED_

#if _MSC_VER >= 1000
#pragma once
#endif    // _MSC_VER >= 1000
// WzdThread.h : header file
//

#include "WzdDialog.h"


/////////////////////////////////////////////////////////////////////////
// CWzdThread thread

class CWzdThread : public CWinThread
{
    DECLARE_DYNCREATE( CWzdThread )
protected:
    CWzdThread();     // protected constructor used by dynamic creation

// Attributes
public:

// Operations
public:

// Overrides
    // ClassWizard generated virtual function overrides
    // {{AFX_VIRTUAL( CWzdThread )
public:
    virtual BOOL InitInstance();
    virtual int ExitInstance();
```

III

14

```
    // }}AFX_VIRTUAL


// Implementation
protected:
    virtual ~CWzdThread();

    // Generated message map functions
    // {{AFX_MSG( CWzdThread )
    // NOTE - the ClassWizard will add and remove member functions here.
    // }}AFX_MSG

    DECLARE_MESSAGE_MAP()
private:
    CWzdDialog m_dlg;
};


/////////////////////////////////////////////////////////////////////////////


// {{AFX_INSERT_LOCATION}}
// Microsoft Developer Studio will insert additional declarations immediately
//      before the previous line.

#endif
    // !defined( AFX_WZDTHREAD_H__411AE4C2_E515_11D1_9B80_00AA003D8695__INCLUDED_ )
// WzdThread.cpp : implementation file
//

#include "stdafx.h"
#include "wzd.h"
#include "WzdThread.h"

#ifdef _DEBUG
#define new DEBUG_NEW
#undef THIS_FILE
static char THIS_FILE[] = __FILE__;
#endif


/////////////////////////////////////////////////////////////////////////////
// CWzdThread
```

Example 58   Sending a Message to a User Interface Thread   **539**

```
IMPLEMENT_DYNCREATE( CWzdThread, CWinThread )

CWzdThread::CWzdThread()
{
}

CWzdThread::~CWzdThread()
{
}

BOOL CWzdThread::InitInstance()
{
    m_dlg.Create( IDD_WZD_DIALOG );
    m_dlg.ShowWindow( SW_SHOW );
    m_pMainWnd = &m_dlg;

    return TRUE;    // can end thread by returning FALSE here
}

int CWzdThread::ExitInstance()
{
    m_dlg.DestroyWindow();
    return CWinThread::ExitInstance();
}

BEGIN_MESSAGE_MAP( CWzdThread, CWinThread )
    // {{AFX_MSG_MAP( CWzdThread )
    // NOTE - the ClassWizard will add and remove mapping macros here.
    // }}AFX_MSG_MAP
END_MESSAGE_MAP()
```

III

14

Example 58   Sending a Message to a User Interface Thread

Objective

You would like to send a message to one of your user interface threads.

Strategy

Up until now we have only used `CWnd::SendMessage()` to send a message. However, you can only use `SendMessage()` to send messages to a window, not a thread. Therefore, we will be using another function called `CWin-Thread::PostThreadMessage()` to send our message. The `PostThreadMessage()` function allows you to directly access the message map in the thread class itself. In this example, we will create a custom window message that will cause our user interface thread to terminate.

Note: You can still use `SendMessage()` to send a message to a thread, but only to a window that your thread has created.

Steps

Send a Message to a Thread

1. Define a user defined window message.

```
#define WM_WZDKILLTHREAD WM_USER + 1
```

2. Start the user interface thread as usual, but retain a pointer to the `CWin-Thread` object it creates.

```
m_pThread = AfxBeginThread( RUNTIME_CLASS( CWzdThread ) );
```

3. Now, to send a message to this thread, use `PostThreadMessage()`.

```
m_pThread -> PostThreadMessage( WM_WZDKILLTHREAD,0,0 );
```

4. So that your thread can intercept this message, manually add the following message map macro to your thread class's message map. Make sure to add it outside of the {{}} brackets so that the ClassWizard won't mess with it.

```
BEGIN_MESSAGE_MAP( CWzdThread, CWinThread )
    // {{AFX_MSG_MAP( CWzdThread )
    // }}AFX_MSG_MAP
    ON_THREAD_MESSAGE( WM_WZDKILLTHREAD,OnKillThread )
END_MESSAGE_MAP()
```

Example 58   Sending a Message to a User Interface Thread   **541**

5. Use the following syntax for your message handler. In this example, we also call `::PostQuitMessage()` to terminate the thread.

```
LRESULT CWzdThread::OnKillThread( WPARAM wParam,LPARAM lParam )
{
    ::PostQuitMessage( 0 );
    return 0;                     // returned to PostThreadMessage()
}
```

Also, make sure to define this handler in your thread class's definition file (`.h`).

Notes

- You can't use `PostThreadMessage()` to send a message to worker thread, since you don't have access to its message map. However, you can talk to your worker thread more informally through a data element of the structure you passed to it. For example, you might include a `bKill` flag in this structure that, when set by your application, causes your thread to prematurely abort.

- Your user interface thread can also optionally intercept registered messages (please see Example 49 for more on registered messages). To process a registered message, use the following message macro.

```
ON_REGISTER_THREAD_MESSAGE( registered_message_id,process )
```

Since threads are considered to be part of the application that created them, using registered messages to communicate with them may be a little over-kill considering that each registered message eats up a resource.

CD Notes

- When executing the project on the accompanying CD, click on the "Test" then "Start" menu commands to create a new thread with a modeless dialog box. Then click on "End" to send a message to the thread to terminate.

III

14

Example 59   Sharing Data with Your Threads

Objective

You would like to share your application's data with your threads without running into conflicts caused by two threads writing to the same data address at the same time.

Strategy

We will be using up to three MFC classes: CMutex, CSingleLock, and CMultiLock, to synchronize simultaneous access to a data class by one or more threads.

Steps

Fireproofing Your Data Objects

1. Determine which data classes will be shared among your threads. Embed a CMutex variable in each class definition.

```
class CWzdData : public CObject
{
    :    :    :
    // synchronization protection
    CMutex m_mutex;
    :    :    :
};
```

2. If this data class does not yet have member functions that access its data, add them now. These functions might look something like this.

```
void CWzdData::GetData( int *pInt,float *pFloat,DWORD *pWord )
{
    *pInt = m_nInt;
    *pFloat = m_fFloat;
    *pWord = m_dwWord;
}


void CWzdData::SetData( int nInt,float fFloat,DWORD dwWord )
```

Example 59   Sharing Data with Your Threads   **543**

```
{
    m_nInt = nInt;
    m_fFloat = fFloat;
    m_dwWord = dwWord;
}
```

3. Create an instance of a CSingleLock class on the stack of your SetData() function referencing the embedded CMutex variable. You can then use CSingleLock's Lock() function to prevent multiple access to the data in this function.

```
BOOL CWzdData::SetData( int nInt,float fFloat,DWORD dwWord )
{
    CSingleLock slock( &m_mutex );

    if ( slock.Lock( 1000 ) )      // timeout in milliseconds,
                                   //     default = INFINITE

    {
        // set values--can also be lists and arrays
        m_nInt = nInt;
        m_fFloat = fFloat;
        m_dwWord = dwWord;
        return TRUE;
    }
    return FALSE;                  // timed out!

    // unlocks on return or you can call slock.Unlock();
}
```

If no other thread is currently accessing this data, Lock() returns immediately. Otherwise, Lock() waits the number of milliseconds you specify until it times out and returns FALSE.

4. If the data stored in this class is interrelated with the data stored in another class, you can embed a CMutex variable in both classes and use CMultiLock to wait on both.

```
CMutex mutex[2];
mutex[0] = &mutex1;
mutex[1] = &mutex2;
```

III

14

```
CMultiLock mlock( mutex,2 );     // where 2 is the number of mutexes
if ( mlock.Lock( 1000 ) )
{

}
```

5. For an example data class with these modifications, please refer to the Data Class for Example 59 (page 544).

Notes

- For more on CMutex and related classes, please see Chapter 1. CMutex uses the ::CreateMutex() Windows API that we used in Example 3 to prevent more than one instance of our application from running at one time. The ::CreateMutex() function was not meant to keep track of the instances of an application. We simply use it as a kludge, as you can see in that example.

- See Example 51 for a practical use of mutexes. There, we make a CObList safe from multiple accesses by window socket threads.

CD Notes

- When executing the project on the accompanying CD, set a breakpoint on the GetData() function in WzdData.cpp. Then, click the "Test" and "Wzd" menu commands and watch as three threads try to access the same data.

Data Class for Example 59

```
#ifndef WZDDATA_H
#define WZDDATA_H

#include <afxmt.h>

class CWzdData : public CObject
{
public:
    DECLARE_SERIAL( CWzdData )

    CWzdData();
```

Example 59   Sharing Data with Your Threads   **545**

```
    BOOL GetData( int *pInt,float *pFloat,DWORD *pWord );
    BOOL SetData( int nInt,float fFloat,DWORD dwWord );

    // synchronization protection
    CMutex m_mutex;

    // result data
    int m_nInt;
    float m_fFloat;
    DWORD m_dwWord;

};
#endif
// WzdData.cpp : implementation of the CWzdData class
//

#include "stdafx.h"
#include "WzdData.h"

/////////////////////////////////////////////////////////////////////////////
// CWzdData

IMPLEMENT_SERIAL( CWzdData, CObject, 0 )


CWzdData::CWzdData()
{
    m_nInt = 0;
    m_fFloat = 0.0f;
    m_dwWord = 0;
}

BOOL CWzdData::GetData( int *pInt,float *pFloat,DWORD *pWord )
{
    // we lock here too so that we'll never read half written data
    CSingleLock slock( &m_mutex );

    if ( slock.Lock( 1000 ) )    // timeout in milliseconds, default = INFINITE
    {
        // get values--can also be lists and arrays
```

III

14

```
            *pInt = m_nInt;
            *pFloat = m_fFloat;
            *pWord = m_dwWord;
            return TRUE;
        }
    return FALSE;    // timed out!

    // unlocks on return or you can call slock.Unlock();
}


BOOL CWzdData::SetData( int nInt,float fFloat,DWORD dwWord )
{
    CSingleLock slock(&m_mutex);    // or with CMultiLock can specify several
                                    //    m_mutex's for waiting on several
                                    //    data items

    if ( slock.Lock( 1000 ) )       // timeout in milliseconds,
                                    //    default = INFINITE
    {
        // set values--can also be lists and arrays
        m_nInt = nInt;
        m_fFloat = fFloat;
        m_dwWord = dwWord;
        return TRUE;
    }
    return FALSE;    // timed out!

    // unlocks on return or you can call slock.Unlock();
}
```

15

Chapter 15

Potpourri

This is the "everything-else" chapter of useful examples, ranging from using a timer to flashing a window.

Example 60   Creating Timers   We will show how to create timed events using MFC classes.

Example 61   Creating Sounds   Our application will play wave files.

Example 62   Creating VC++ Macros   We will review the macro syntax supported by MFC as an alternative to writing small functions or inline functions.

Example 63   Using Function Addresses   We will show how to pass the address of the member function of a class so that it can be indirectly called.

Example 64   Manipulating Binary Strings   We will extend the MFC's `CString` class so that it can also handle binary strings.

Example 65   Rebooting Your Machine   We will look at how to programmatically reboot a system.

Example 66   Getting Free Disk Space   We will find out how to determine the amount of space is left on a disk drive.

Example 67   Flashing a Window or Text   We will look at how to flash a text message or the caption bar of a window.

Example 60   Creating Timers

Objective

You would like to know when x seconds have elapsed.

Strategy

We will be using the CWnd::SetTimer() function to create our timer. The Set-Timer() function has two modes. The first allows us to specify a static function that will be called when the timer times out. The second mode sends a WM_TIMER window message to the window that calls SetTimer(), which can be handled like any other windows message.

Steps

Create a Timer That Calls a Static Function When Elapsed

1. Define the timer callback function as a static function of your class by using the following syntax.

```
private:
static void CALLBACK EXPORT TimerProc( HWND hWnd, UINT nMsg,
    UINT nIDEvent, DWORD dwTime );
```

2. Create the timer with that function name, using the following.

```
SetTimer(
    3,                      // event ID, passed to time out process
    1000,                   // time out time in milliseconds
    CWzdView::TimerProc     // timeout process called
    );
```

In this example, the CWzdView::TimerProc() function is called after one (1) second at which point an event ID of three (3) is sent to TimerProc().

Example 60   Creating Timers   **549**

3. Implement the timer function, as follows.

```
void CALLBACK EXPORT CWzdView::TimerProc( HWND hWnd, UINT nMsg,
    UINT nIDEvent, DWORD dwTime )
{
    // hWnd will be this window
    // nMsg will be WM_TIMER
    // nIDEvent will be the event id specified in SetTimer
    // dwTime will be the current system time
}
```

Create a Timer That Will Send a WM_TIMER Message When Elapsed

1. Use the following to create a timer that will generate a WM_TIMER window message when it has expired.

```
SetTimer(
    4,       // event ID
    2000,    // time in milliseconds
    NULL     // causes a WM_TIMER message to be sent to this window
    );
```

In this example, a WM_TIMER message is sent to this window after two (2) seconds with an event ID of four (4).

2. You can use the ClassWizard to add a WM_TIMER message to your window class. The handler it creates will have the following syntax.

```
void CWzdView::OnTimer( UINT nIDEvent )
{
    // nIDEvent will be the event id specified in SetTimer

    CView::OnTimer( nIDEvent );
}
```

III

15

Kill a Timer

1. To kill a timer before it has expired, you can use CWnd::KillTimer().

```
UINT timerID = SetTimer( 3,3000,NULL );
KillTimer(
    timerID    // timer id returned from SetTimer()
    );
```

Note that you identify the timer you want to kill by using the ID value returned when you called SetTimer().

Notes

- There is no way to easily create a repeating timer. You must, instead, call CWnd::SetTimer() each time a timer times out.

- Timers eat up system resources. If your application will require dozens of timers, you might be better served to multiplex a single system timer. Start by determining the smallest unit of time you want to track. Then, when some part of your application wants to create a timer, have it create an entry in a linked list instead. That entry should specify what multiple of the system timer it represents and where it wants to be informed when a time out occurs (e.g., a window handle for a WM_TIMER message or a function address to call). Then, create one system timer set to expire at the smallest unit of time. Upon time out, it should reset its timer, but also scroll through the linked list, incrementing a counter in each entry until they have timed out. When an entry times out, either send a WM_TIMER message to the window pointer specified in the entry or call the specified function address.

CD Notes

- When executing the project on the accompanying CD, set a breakpoint on the OnTestWzd() function. Then click the "Test" and then "Wzd" menu commands and watch as two timers are created and timeout.

Example 61   Creating Sounds   **551**

Example 61   Creating Sounds

Objective

You would like your application to create sounds.

Strategy

We will be using two Windows API calls to create sounds with our application. The ::Beep() API creates a single tone, for which we can set the frequency and duration. The ::sndPlaySound() API will allow our application to play a .wav file.

Steps

Generate a Single Tone

1. To create a simple tone from your application, you can use the following.

```
::Beep(
    1000,    // sound frequency in hertz
    1000     // sound duration in milliseconds
    );
```

The first argument is the tone in hertz. The higher this value, the higher the tone's pitch will be. (The human ear has a frequency range of 20 to 20,000hz.) The second argument is the duration of the tone in milliseconds.

A .wav file is a digitized audio file. To allow your application to play a .wav file, the system on which it runs must have a sound card. You will also need to manually include MFC's multimedia support in your application.

III

Play a .wav File

15

1. To include MFC's support for .wav files, you must include the following in every module that will be calling this API.

```
#include <mmsystem.h>
```

2. You must also add winmm.lib to the link settings for your application.

3. To play a .wav file directly from a disk file, you can use

```
::sndPlaySound(
    "sound.wav",      // file to play
    SND_SYNC |        // or SND_ASYNC to play in another thread
    SND_LOOP |        // play continually (SND_ASYNC must be set too)
    SND_NODEFAULT     // play nothing if error in .wav file
    );
```

4. To stop a .wav file from playing, you can call

```
::sndPlaySound( "",0 );
```

Include a .wav **File in Your Application's Resources**

1. If you would like to include the .wav file in your application's resources, start by clicking on the Developer Studio's "Insert" and "Resource" menu commands to open the "Insert Resource" dialog box. Click on "Import" and then locate and select the .wav file you wish to include. A "WAVE" folder will automatically be created for you in your resources and the .wav file will be added and copied into your project's \res directory.

2. To play a .wav file that's defined in your application's resources, you can use the following. In this example, the .wav file is identified with an ID of IDR_SOUND.

```
// find .wav file name in resources and play
HRSRC hRsrc = FindResource( AfxGetResourceHandle(),
    MAKEINTRESOURCE( IDR_SOUND ), "WAV" );
HGLOBAL hglb = LoadResource( AfxGetResourceHandle(), hRsrc );
::sndPlaySound( ( LPCTSTR )::LockResource( hglb ),
    SND_MEMORY|SND_SYNC|SND_NODEFAULT );
FreeResource( hglb );
```

Notes

- You will need your own sound card, microphone, and multimedia software to create your own .wav file.

Example 62   Creating VC++ Macros   **553**

CD Notes

- When executing the project on the accompanying CD, set a breakpoint on the OnTestWzd() function. Then, click the "Test" and then "Wzd" menu commands and watch as three sounds are generated.

Example 62   Creating VC++ Macros

Objective

You would like to use the full capabilities of VC++ to create a macro.

Strategy

We will simply review MFC's macro support. C++ purists try to use inline code, instead of macros. However, there are still too many things you can do with macros that you can't do with inline code.

Steps

Create Simple Macro Definitions

1. To define a macro, use one of the following.

```
#define WZD1 7    // WZD1 is substituted for 7 when used in code
#define WZD2      // WZD2 is simply defined so that #ifdef WZD2 is TRUE
```

Create Conditional Macros

1. To conditionally compile lines of code, use one of the following.

```
#ifdef WZD1      // if defined, process next
#undef WZD1      // undefine WZD1
#endif
#ifndef WZD1     // if WZD1 not defined, process next
#endif

#if WZD2 == 5    // if WZD2 equals 5, process next
    //
```

III

15

```
#elif WZD2 > 6     // else if WZD6 is greater then 6
   //
#else              // else...
   //
#endif
```

Create Macros with Arguments

1. The syntax for a simple macro that takes arguments is

```
#define WZD3( arg1,arg2 ) \
    arg1 + arg2;    // WZD3( 2,5 ) is substituted for 2+5 in the code
```

When using the continuation character (\ ) there must not be anything on the line after it—*not even a space.*

2. To make a macro argument appear with double quotes in the substitution, precede the argument with a pound sign (#) in the macro.

```
#define WZD4( arg1,arg2 ) \
    arg1 + #arg2;    // WZD4( 1,test ) becomes 1 + "test" in the code
```

3. To make a macro argument appear with single quotes in the substitution, precede the argument with a pound and ampersign sign (#@) in the macro.

```
#define WZD5( arg1,arg2 ) \
    arg1 + #@arg2;    // WZD5( 1,t ) becomes 1 + 't'
```

4. To append a macro argument to a token in your macro, precede the token with a double pound sign (##).

```
#define WZD6( arg1,arg2 ) \
    arg1 + arg2##3;    // WZD6( 1,2 ) becomes 1 + 23
```

Use the Predefined Macros

1. The VC++ compiler predefines four macros that allow you to insert information about your source file into your code. The __FILE__ macro expands when processed to a text string that contains the name of the source file in which it's located (there are two underline characters before and after FILE). For example, the following line in WzdView.cpp

```
Print( __FILE__ );
```

Example 63   Using Function Addresses   **555**

expands when compiled to

```
Print( "WzdView.cpp" );
```

The __LINE__ expands to the current line in the source file. The __DATE__ and __TIME__ macros expand to the date and time the source file was compiled.

Notes

- The __FILE__ and __LINE__ macros can be used to help you more accurately identify an error message. For example, rather than simply displaying the error in your application

```
"File not found"
```

you can instead include the exact spot in your code where this error occurred. The following line

```
"File not found ( __FILE__ @ __LINE__ )."
```

might expand when compiled to

```
"File not found ( WzdView.cpp @ 1232 )"
```

CD Notes

- There is no accompanying project on the CD for this example.

Example 63   Using Function Addresses

Objective

You would like to pass the address of a member function of a class to another function for execution or storage.

Strategy

ANSI C allows us to treat the address of a function as if it were like any other variable. But the functions in a C++ object aren't static, as they are in C. Thus, not only do we have to contend with a function address, but we also have to contend with the address of the object in which that function sits.

III

15

In this example, we will be passing the address of the following member function of the `CWzdView` class to another function. That other function will then call this function indirectly.

```
int CWzdView::ViewFunc( int i,LPCSTR s,BOOL b )
{
    i;s;b;        // process values
    return 0;     // result is returned
}
```

As seen here, the `ViewFunc()` function takes three arguments and returns an integer value.

Steps

1. Define a function pointer type for this member function, as follows.

```
typedef int ( CWzdView::*PMFUNC)(int,LPCSTR,BOOL );
```

In this example, `PMFUNC` becomes the function pointer type. Put this definition where it can be easily included in your code, such as in a file that every code module includes.

2. When passing a pointer to this function, you need to also include a pointer to the instance of the class in which it lives.

```
void Perform( CWzdView *pClass, PMFUNC pFunc );
```

In this example, `PMFUNC` lives in the `CWzdView` class. You can, therefore, call `Perform()` using the following arguments.

```
wzdUtil.Perform(
    xxx,      // a pointer to the object that ViewFunc() sits in
    pFunc     // the pointer to ViewFunc()
    );
```

3. To now call this member function indirectly, you need to combine the object and member function pointer.

```
// pClass is the object pointer;
pFunc is the function pointer i =
    ( pClass -> *func )( 1,"test",TRUE );
```

Example 64   Manipulating Binary Strings   **557**

Notes

- You could also encapsulate this pointer pair (i.e., `Object::Function`) in a C++ class.

```
class CFunc
{
    CFunc( CWzdView *pClass, PMFUNC func )
        {m_pClass = pClass;m_pFunc = pFunc;};
    classPtr *m_pClass;
    funcPtr *m_pFunc;
    Perform( int i,LPCSTR s,BOOL b ){( PClass -> *func )( i,s,b );};
}
```

The advantage to this approach is that a member function address can be passed and stored as a single argument. The drawback is that you would need an encapsulating class for every function type. You would also need to worry about one more class instances to delete. Sometimes, hiding the inner workings of an indirect call can cause more confusion than simplification.

- This is obviously more of an exercise in C++ than it is an MFC. However, since this is such an obscure part of C++, I thought it merited a place here.

CD Notes

- When executing the project on the accompanying CD, set a breakpoint on the `OnTestWzd()` function. Then click the "Options" and then "Wzd" menu commands and watch as a function of a class instance is called indirectly.

Example 64   Manipulating Binary Strings

Objective

You would like to use the MFC string class to manipulate binary strings, too.

Strategy

MFC's CString class allows us to manipulate only a null terminated text string. To make it also let us use its wonderful functionality on a binary string, where nulls might be strewn throughout, we will convert and store that binary string internally as its hex text equivalent. In other words, the bytes 0x23, 0x43, and 0xa4 will be stored in CString as the text string "2343a4".

In this example we will be deriving our own class from the CString class, to which we will add two new member functions that will convert and restore binary strings to text strings.

Steps

1. Manually create a class derived from CString, as seen in the String Class for Example 64 (page 560). The ClassWizard can't create a class derived from CString.

2. Use the Text Editor to add a PutBinary() function to this class. Since the CString class can internally store only string text, we need to convert any binary string into a text representation here. To do this, we will use CString's Format() function to convert each binary byte into a two character hex value.

```
void CWzdString::PutBinary( LPBYTE pByte,int len )
{
    Empty();
    CString hex;
    for ( int i = 0;i < len;i++ )
    {
        hex.Format( "%02X",pByte[i] );
        *this += hex;
    }
}
```

Example 64   Manipulating Binary Strings   **559**

3. Add another function to this class called `GetBinary()`. Regenerating a binary string from a `CString` class then becomes a matter of converting its hexadecimal string back to binary bytes, using `sscanf()`.

```
// returned value is actual binary length
int CWzdString::GetBinary( LPBYTE pByte,int maxlen )
{
    // make sure the string contains only valid hex characters
    if ( SpanIncluding( "0123456789aAbBcCdDeEfF" ) != *this )
    {
        return 0;
    }

    // make sure less then max bytes
    int len = GetLength();
    if ( len > maxlen*2 )
    {
        len = maxlen*2;
    }

    // pad HEX to even number
    CString hex = *this;
    if ( ( len % 2 ) != 0 )
    {
        len++;
        hex += "0";
    }

    // convert to binary
    len/ = 2;
    for ( int i = 0; i < len; i++ )
    {
        int b;
        sscanf( hex.Mid( ( i * 2 ), 2 ), "%02X", &b );
        pByte[i] = ( BYTE )b;
```

III

15

```
      }
    return( len );

}
```

Notes

- Be careful with the `sscanf()` function. In the previous example, we're scanning into an integer variable, even though all we expect is a byte., because `sscanf()` returns an integer no matter what. If you were to have scanned into a byte, the release version of your application would have occasionally and mysteriously crashed as `sscanf()` trashed your stack.

CD Notes

- When executing the project on the accompanying CD, set a breakpoint on the `OnWzdType()` function in `WzdView.cpp`. Then click the "Options" and then "Wzd" menu commands and watch as a binary string is manipulated.

String Class for Example 64

```
// CWzdString.h
//
///////////////////////////////////////////////////////////////////////////////

#if !defined WZDSTRING_H
#define WZDSTRING_H

class CWzdString : public CString
{
public:
    CWzdString::CWzdString() : CString(){;};
    CWzdString::CWzdString( const CString& str ) : CString( str ){;};
    CWzdString::CWzdString( LPCTSTR lpsz ) : CString( lpsz ){;};
    CWzdString::CWzdString( LPCWSTR lpsz ) : CString( lpsz ){;};

// Attributes
public:
```

Example 64   Manipulating Binary Strings   **561**

```
// Operations
public:
    void PutBinary( LPBYTE pByte,int len );
    int GetBinary( LPBYTE pByte,int maxlen );


};


/////////////////////////////////////////////////////////////////////////////

#endif
// WzdString.cpp
//

#include "stdafx.h"
#include "WzdString.h"

#ifdef _DEBUG
#define new DEBUG_NEW
#undef THIS_FILE
static char THIS_FILE[] = __FILE__;
#endif


/////////////////////////////////////////////////////////////////////////////
// CWzdString

// extracts characters from beg to end
CString CWzdString::Extract( int beg,int end)
{
    return Mid( beg - 1,end - beg + 1 );
}

// stores binary as HEX ascii string
void CWzdString::PutBinary( LPBYTE pByte,int len )
{
    Empty();
    CString hex;
    for ( int i = 0;i < len;i++ )
    {
        hex.Format( "%02X",pByte[i] );
        *this+ = hex;
```

III

15

```
        }
    }


// retrieves HEX ascii string as binary
// returned value is actual binary length
int CWzdString::GetBinary( LPBYTE pByte,int maxlen )
{
    // make sure the string contains only valid hex characters
    if ( SpanIncluding( "0123456789aAbBcCdDeEfF" ) != *this )
    {
        return 0;
    }

    // make sure less then max bytes
    int len = GetLength();
    if ( len > maxlen*2 )
    {
        len = maxlen*2;
    }

    // pad HEX to even number
    CString hex = *this;
    if ( ( len % 2 ) != 0 )
    {
        len++;
        hex += "0";
    }

    // convert to binary
    len/ = 2;
    for ( int i = 0; i < len; i++ )
    {
        int b;
        sscanf( hex.Mid( ( i * 2 ), 2 ), "%02X", &b );
        pByte[i] = ( BYTE )b;
    }
    return( len );

}
```

Example 65   Rebooting Your Machine   **563**

Example 65   Rebooting Your Machine

Objective

You would like to programmatically reboot a system, either your own or another system remotely.

Strategy

To shut down our own system, we will use the `::ExitWindowsEx()` API. To shut down some other system, we will use the `::InitiateSystemShutDown()` API. Actually, this last API will also shut down our own system and even allow us to set a delay time in which we can terminate the shut down. One other consideration for shut down is that an NT machine requires an `SE_SHUTDOWN_PRIVILEDGE`. We will get that privilege using `::AdjustTokenPriviledge()`.

Steps

Change Your NT Privilege

1. To reboot an NT machine, you must give your application a shutdown privilege.

```
static HANDLE hToken;
static TOKEN_PRIVILEGES tp;
static LUID luid;
if ( ::OpenProcessToken( GetCurrentProcess(),
    TOKEN_ADJUST_PRIVILEGES | TOKEN_QUERY, &hToken ) )
{
    ::LookupPrivilegeValue( NULL, SE_SHUTDOWN_NAME, &luid );
    tp.PrivilegeCount            = 1;
    tp.Privileges[0].Luid        = luid;
    tp.Privileges[0].Attributes = SE_PRIVILEGE_ENABLED;
    ::AdjustTokenPrivileges( hToken, FALSE, &tp,
            sizeof( TOKEN_PRIVILEGES ), NULL, NULL );
}
```

III

15

Note: Your password account must also allow you to have this privilege.

Note: Add this even if your application will also be running on other systems, such as Windows 98. (It won't hurt.)

Reboot the Local Machine

1. To reboot the local machine, use

```
::ExitWindowsEx(
    // EWX_LOGOFF        // logs user off
    // EWX_SHUTDOWN      // shuts down system
    EWX_REBOOT           // reboots system
    // EWX_POWEROFF      // shuts down and turns system off
                         //     (if system has power-off feature)
    // |EWX_FORCE,       // forces processes to terminate--
                         //     use only in emergency
    0                    // reserved
    );
```

Reboot the Local or Remote Machine

1. To reboot a local or remote machine, use

```
::InitiateSystemShutdown(
    "\\\\OtherMachine",   // computer to shut down--
                          //     NULL == local system
    "Goodbye!",           // a message to user
    60,                   // time to display message (in seconds)
    FALSE,                // TRUE == force processes to terminate
    TRUE                  // TRUE == reboot after shutdown
    );
```

Example 66   Getting Free Disk Space   **565**

2. To cancel a reboot started by `::InitiateSystemShutdown()`, use

```
::AbortSystemShutdown(
    "\\\\OtherMachine"      // computer NOT to shut down--
                            //     NULL == local system
    );
```

Notes

- The remote API allows you to reboot a remote server, in the event it's hung or to install new software.

CD Notes

- When executing the project on the accompanying CD, click on the "Test" menu item. The first menu command will immediately reboot your machine. The second command will reboot your machine after one minute and the last menu command will cancel this second command's reboot.

Example 66   Getting Free Disk Space

Objective

You would like to find out how much space is left on a disk drive.

Strategy

We will be using the `::GetDiskFreeSpaceEx()` API.

Steps

Get Free Disk Space

1. To get the free space on your disk, use the following call.

```
ULARGE_INTEGER freeToCaller;
ULARGE_INTEGER diskCapacity;
ULARGE_INTEGER freeSpace;

::GetDiskFreeSpaceEx(
```

III

15

```
"c:\\",          // pointer to the directory name
&freeToCaller,   // bytes available to caller
&diskCapacity,   // total bytes on disk
&freeSpace       // free bytes on disk
);
```

Notes

- For Windows 95 systems, you need OEM 2 or better to use this function. If your application won't have this function available, you can still use the lesser ::GetDiskFreeSpace() function. However, you are then forced to calculate free space from the returned values.

CD Notes

- When executing the project on the accompanying CD, click on the "Test" and "Wzd" menu items to open a dialog. Press the button to get the current free space on your c: drive.

Example 67   Flashing a Window or Text

Objective

You would like to flash a text message in a dialog box or flash the caption bar of a dialog to get the user's attention.

Strategy

There are two approaches to flashing, but both involve using a timer to intermittently modify a window. In the case of text, we will just hide or show the window in which it's sitting. In the case of a popup or overlapped window, we will activate or inactive it so that its title bar appears to flash. This latter solution is used when you want to get your user's attention due to an error or message.

Example 67   Flashing a Window or Text   **567**

Steps

Create Flashing Text

1. Start a timer (see Example 60 for more on timers). If in a dialog class, start the timer in the `OnInitDialog` message handler.

```
SetTimer(
    1,      // event ID
    250,    // time in milliseconds
    NULL    // causes a WM_TIMER message to be sent to this window
    );
```

2. Add a `WM_TIMER` message handler to alternatively hide and show the text in question.

```
void CWzdDialog::OnTimer( UINT nIDEvent )
{
    if ( nIDEvent == 1 )
    {
        static BOOL t = TRUE;
        m_ctrlStatic.ShowWindow( ( t = !t )?SW_SHOW:SW_HIDE );
    }

    CDialog::OnTimer( nIDEvent );
}
```

Notice we can still use this timer handler for other uses in this dialog class by using the `nIDEvent` ID.

3. When turning off the timer, make sure the text is visible.

Create a Flashing Window

1. Start a timer, as seen previously.

III

15

2. Add a `WM_TIMER` message handler that uses `FlashWindow()` to toggle the caption bar of that window from active to inactive.

```
void CWzdDialog::OnTimer( UINT nIDEvent )
{
    FlashWindow(
        TRUE    // if FALSE, forces caption bar to be active--
                //      useful when flashing stops and
                //      you want bar to be active
        );

    CDialog::OnTimer( nIDEvent );
}
```

3. Make sure to call `FlashWindow(FALSE)` when stopping the flashing to ensure that the window caption bar is active (bright).

Notes

- You can use this solution to flash your main window if your application has a problem in the background while your user is working on something else. Even if your main window isn't visible on the desktop, your task bar entry will flash.

CD Notes

- When executing the project on the accompanying CD, click on the "Test" and "Wzd" menu items to open a dialog. Press the text flashing button to get the text to flash or the window flash button to get the window caption bar to flash.

Appendix A

Message and Override Sequences

In Chapter 1, we reviewed the nature of MFC and how it wraps the Windows API. In particular, we looked at how a window is created, how messages are sent between windows, what classes make up an MFC application, and how to use the ClassWizard to override and enhance this process for our own application. What we didn't cover, and what is probably one of the biggest sources of frustrations for even a proficient MFC programmer, is overriding this process at the wrong point. In other words, adding a window message handler for the wrong message or overriding the wrong MFC class function. What's needed is documentation on what window message is sent or class function is called and in what order for even the simplest of activities.

- Creating, moving, showing, and closing a window.
- Creating and closing a modal or modeless dialog box.
- Creating and closing a Dialog Application.
- Creating and closing an SDI Application, as well as loading and saving SDI documents.

- Creating and closing an MDI Application, as well as loading and saving MDI documents.

This appendix will attempt to document the messages and overrides involved with each of these activities. If none of the examples below applies to your application, you can create your own sequence table just by using the ClassWizard to add every possible message handler and override it has in its list to the affected classes and set a breakpoint on each. You'll be surprised sometimes by what you'll find.

Note: Perhaps one reason these sequences of events is rarely—if ever—documented is that they are vulnerable to rapid change between versions of MFC. Generally speaking, however, I have found these particular sequences to seldom and—then only slightly—change.

Windows

In this section, we will list the messages and functions called to create, move, show, and close any type of window, including overlapped, popup and child.

Note: If this window is being created by a dialog box, such as a control, the messages start with `WM_CTLCOLOR`.

Creating a Window

Once you call `CWnd`'s `CreateEx()` function, you set into motion several messages and function calls. As you can see, before the Windows API is even called, you are given one last chance to change the creation process with `PreCreateWindow()`.

Table A.1 Creating a Window

Handler/Function	Type	Description	Notes
PreCreateWindow()	Ovr	Allows you to change creation arguments before window is created.	1
PreSubclassWindow()	Ovr	Allows you to be the first to subclass this window.	2
OnGetMinMaxInfo()	Msg	Allows you to set limits on how big or small window can become.	3
OnNcCreate()	Msg	Sent to inform you that the nonclient area of window is about to be created.	4
OnNcCalcSize()	Msg	Allows you to change how large the client area of a window will be.	5
OnCreate()	Msg	Sent to inform you the window has been created.	6
OnSize()	Msg	Sent to inform you the window size is changing.	7
OnMove()	Msg	Sent to inform you that the window is moving.	7
OnChildNotify()	Ovr	Called as part of message reflection to tell you that the parent is about to be told that the window was just created. Only called for Child windows!	8

That's it. The window has been created. Notice that at no time was the window drawn on the screen. The system sends the WM_PAINT and WM_NCPAINT messages later.

Notes

1. You should override the CWnd::PreCreateWindow() member function to change the creation arguments used to create a window. Why not just call CreateEx() with the correct arguments? Because sometimes you don't have access to them at that point, as is the case when your application opens the main window. You application starts by picking a default window style and Window Class and then gives you a chance to change them by calling this function. Because there aren't that many cases in which your application automatically opens a window for you, this member function is usually overridden in the CMainFrame class and nowhere else.

A

2. You should override the `CWnd::PreSubclassWindow()` member function to be the first to subclass a window. This can be important because whatever Window Procedure is the last to subclass a window may not receive all the messages it needs. For example, MFC overrides this function for the `CDragListBox` class to be the first to subclass the control before you get a chance.

3. The `WM_GETMINMAXINFO` message is sent by the system to allow you to keep a window from getting too big or too small. This can happen when the user has grabbed the corner of a resizable window and drags.

4. The `WM_NCCREATE` message is sent by the system to inform you that the nonclient area of the window is about to be created. Don't bother to monkey with the arguments passed to you here with the hope of affecting the default Window Procedure. MFC saves the original message and uses that when calling the default Window Procedure. You can, however, call the default Window Procedure yourself and bypass MFC's attempt at keeping you safe.

5. This `WM_NCCALCSIZE` message is sent by the system to allow you to change how big the client area of a window will be. The client area is usually the space left over after the border and scroll bars, etc. of the nonclient area have been drawn.

6. The `WM_CREATE` message is sent by the system *after* it has created the window. This is a good point to create any windows that this window owns, such as control windows.

7. The `WM_SIZE` and `WM_MOVE` messages are sent to position and size the window according to the size and position specified by the `::CreateWindowEx()` call. This is a good spot to resize the windows that belong to this window, especially if their size depends on this window's size. Overlapped windows do not get these messages, except when you use `ShowWindow()` to show the window.

8. This `WM_PARENTNOTIFY` message has been reflected by the MFC class that controls the parent window of this window. A `WM_PARENTNOTIFY` message is sent to the parent at this point to notify it that its child window has been created. Since this window found out about it first with the `WM_CREATE` message, there really isn't much use for this message. Only child windows get this message.

Moving a Window

The following sequence occurs when you move or resize a window using `MoveWindow()`.

Table A.2 Moving a Window

Handler/Function	Type	Description	Notes
OnWindowPosChanging()	Msg	Informs you that a window's x, y, or even z position is about to change.	9
OnGetMinMaxInfo()	Msg	Allows you to set limits on how big or small window can become.	3
OnNcCalcSize()	Msg	Allows you to change how large the client area of a window will be.	5
OnWindowPosChanged()	Msg	Informs you that a hidden window was shown and that WM_SIZE and WM_MOVE messages are about to be sent.	10
OnSize()	Msg	Sent to inform you the window size is changing.	7
OnMove()	Msg	Sent to inform you that the window is moving.	7

Notes

9. The WM_WINDOWPOSCHANGING message is sent by the system in response to a function changing a window's position, either in the x-y axis or its z-order, to indicate the window is about to move.

10. The WM_WINDOWPOSCHANGED message is sent by the system just after it shows a hidden window but before it sends the WM_SIZE and WM_MOVE messages for an x-y move.

Showing a Child Window

The following sequence occurs when you show a Child window using the `ShowWindow()` function.

A

Table A.3 Showing a Child Window

Handler/Function	Type	Description	Notes
OnShowWindow()	Msg	Sent by ShowWindow() function.	11
OnWindowPosChanging()	Msg	Informs you that a window's x, y, or even z position is about to change.	9
OnWindowPosChanged()	Msg	Informs you that a hidden window was shown and that WM_SIZE and WM_MOVE messages are about to be sent.	10

Notes

11. The WM_SHOWWINDOW message is sent by the ShowWindow() function. Anything you can do here would be better done in OnCreate() or OnSize().

Showing an Overlapped or Popup Window

For overlapped and popup windows, the sequence starts the same.

Table A.4 Showing an Overlapped or Popup Window

Handler/Function	Type	Description	Notes
OnShowWindow()	Msg	Sent by ShowWindow() function.	11
OnWindowPosChanging()	Msg	Informs you that a window's x, y, or even z position is about to change.	9
OnQueryNewPalette()	Msg	Allows you to change the system palette.	12
OnWindowPosChanging()	Msg	Informs you that a window's x, y, or even z position is about to change.	9
OnNcPaint()	Msg	Allows you to paint the nonclient area of the window yourself.	13
OnEraseBkgnd()	Msg	Allows you to pick the color to paint background of window with.	14
OnWindowPosChanged()	Msg	Informs you that a hidden window was shown and that WM_SIZE and WM_MOVE messages are about to be sent.	10
OnNcActivate()	Msg	Tells you to draw the nonclient area in its active color.	15
OnActivate()	Msg	Tells window it is about to become active.	16
OnSetFocus()	Msg	Tells window it is about to receive keyboard input focus.	17

Handler/Function	Type	Description	Notes
OnNcPaint()	Msg	Allows you to paint the nonclient area of the window yourself.	13
OnWindowPosChanged()	Msg	Informs you that a hidden window was shown and that WM_SIZE and WM_MOVE messages are about to be sent.	10
OnSize()	Msg	Sent to inform you the window size is changing. Sent here by Overlapped window only.	7
OnMove()	Msg	Sent to inform you that the window is moving. Sent here by Overlapped window only.	7

Notes

12. The WM_QUERYNEWPALETTE message is sent by the system to tell a window it's about to become the active window and it gets a chance to take over the system palette. For systems without enough video memory to allow True Color, the system palette arbitrates what colors are displayed on the screen where the window with input focus gets the top priority.

13. The WM_NCPAINT message is sent by the system to allow you to draw (paint) the nonclient area of the window yourself.

14. The WM_ERASEBKGND message is sent by the system if you allow it to draw the nonclient area to allow you to set the background color of the window.

15. The WM_NCACTIVATE message is sent by the system to tell a window to draw it's nonclient area in its active color.

16. The WM_ACTIVATE message is sent to a window to indicate it is about to become active to allow it to perform any color changes in the client area.

17. The WM_SETFOCUS message is sent to a window to tell it that it now has input focus, which means keyboard input is now sent to this window.

Closing a Window

The following sequence occurs when you close any type of window.

A

Table A.5 Closing a Window

Handler/Function	Type	Description	Notes
OnClose()	Msg	Sent when the close button has been clicked.	18
OnDestroy()	Msg	Sent when this window is about to be destroyed.	19
OnNcDestroy()	Msg	Sent after the window has been destroyed.	20
PostNcDestroy()	Ovr	Called by CWnd as final act of processing OnNcDestroy().	21

Notes

18. The WM_CLOSE message is sent when the close button in the nonclient area of a window has been pressed. Or, if the window has no close button, you can be the one to send this message. This would be the time to ask the user if they really want to close the window. If they don't, just don't call the base class member function, which sends a WM_DESTROY message to the window.

19. The WM_DESTROY message is sent by anyone wanting to destroy this window. This is a good spot to send your own window destroy messages you might have to send, but remember that the system will automatically destroy your child windows for you.

20. The WM_NCDESTROY message is sent by the system after your window has been destroyed. It performs some MFC cleanup.

21. This CWnd member function is called by CWnd's default implementation of the WM_NCDESTROY message. This is your absolute last chance to destroy everything associated with this window. If the CWnd object controlling this window has no more use and will not be destroyed otherwise, this would be a good spot to deconstruct it with

```
delete this;
```

Dialog Boxes

In this section, we will list the messages and functions called to create and close a dialog box. A dialog box can be created in two modes: modeless and modal.

- A modal dialog box will not relinquish control until it is closed. It has its own message pump to allow it to perform specialized processing of its child control windows (e.g., tabbing between edit boxes).
- A modeless dialog box uses the application's message pump, which allows any window to take input focus, but you lose the functionality of the modal message pump.

To add your own functionality to a dialog box, you must first derive a class from CDialog and use the ClassWizard to include one of the following functions/handlers.

Creating a Modal Dialog box

The following sequence occurs when you create a modal dialog box using

```
CMyDialog dlg;
dlg.DoModal();
```

Table A.6 Opening a Modal Dialog Box

Handler/Function	Type	Description	Notes
DoModal()	Ovr	Overrides the DoModal() member function.	
PreSubclassWindow()	Ovr	Allows you to be the first to subclass this window.	2
OnCreate()	Msg	Sent to inform you the window has been created.	6,27
OnSize()	Msg	Sent to inform you the window size is changing.	7
OnMove()	Msg	Sent to inform you that the window is moving.	7
OnSetFont()	Msg	Sent to allow you to change the font used with the controls in this dialog box.	22
OnInitDialog()	Msg	Sent to allow you to initialize the controls in your dialog box or create new ones.	23
OnMove()	Msg	Sent to inform you that the window is moving.	7
OnShowWindow()	Msg	Sent by ShowWindow() function.	11
OnCtlColor()	Msg	Sent by parent to allow you to change this dialog box's color.	24
OnChildNotify()	Ovr	Sent as a result of a WM_CTLCOLOR message.	8,25

A

Notes

22. The WM_SETFONT message is sent by the system when creating dialog box to allow you to override the font specified in the resource file for a dialog box. The font is used when the dialog box draws text in one of its controls.

23. The WM_INITDIALOG message is sent by the system to allow you to initialize or create any additional control windows.

24. The WM_CTLCOLOR message is sent by the system to allow you to determine the color with which to draw the dialog box.

25. This member function is called by the default message handler processing the WM_CTLCOLOR message and seems to be included so that you can put all of your child notification processing in OnChildNotify.

Closing a Modal Dialog box

The sequence to close a dialog box is very similar to closing a window.

Table A.7 Closing a Modal Dialog Box

Handler/Function	Type	Description	Notes
OnClose	Msg	Sent when the close button has been clicked.	18
OnKillFocus()	Msg	Sent just before window loses keyboard input focus.	26
OnDestroy()	Msg	Sent when this window is to be destroyed.	19
OnNcDestroy()	Msg	Sent after the window has been destroyed.	20
PostNcDestroy()	Ovr	Called by CWnd as final act of processing OnNcDestroy().	21

Notes

26. This tells your window it will no longer receive key strokes from the keyboard.

Creating a Modeless Dialog box

The following sequence occurs when you create a modeless dialog box using

```
CMyDialog dlg;
dlg.Create(…);
```

Table A.8 Opening a Modeless Dialog Box

Handler/Function	Type	Description	Notes
PreSubclassWindow()	Ovr	Allows you to be the first to subclass this window.	2
OnCreate()	Msg	Sent to inform you the window has been created.	6,27
OnSize()	Msg	Sent to inform you the window size is changing.	7
OnMove()	Msg	Sent to inform you that the window is moving.	7
OnSetFont()	Msg	Sent to allow you to change the font used with the controls in this dialog box.	22

That's the end. As mentioned previously, a modeless dialog box does not use the special DoModalLoop() message pump, meaning some things you get for free in a modal dialog box you have to do manually here. Of course, the advantage of a modeless dialog box is that the rest of your application isn't frozen out—other windows can still be used when the dialog box is up.

Notes

27.The WM_CREATE message is sent after the dialog window is created but before any control windows have been created. Since a modeless dialog doesn't send a WM_INITDIALOG message, you will need to wait until the Create() function has returned before calling your own InitDialog() function.

Dialog Applications

In this section, we will list the messages and functions called to create and close a Dialog application. If you recall from Chapter 1, a Dialog application simply puts up a modal dialog box for its user interface. Dialog applications have no document or view.

Creating a Dialog Application

Other than the extra step needed to create an application, the sequence needed to create a dialog application is identical to the previous sequence for creating a modal dialog box.

A

Table A.9 Creating a Dialog Application

Handler/Function	Type	Description	Notes
InitInstance()	Ovr	Mandatory—this initializes and determines what type of application this will be.	28
See Table A.6 Opening a Modal Dialog Box			29
DoModalLoop()	Ovr	Performs special processing of control windows in a dialog box.	

Notes

28. This member function is called by CWinApp as it starts up the application. In fact, a Dialog Application spends all of its time in your derivation of this function. This is the best spot to initialize your application (e.g., open databases, register Window Classes, etc.). Then, to create a Dialog Application, the AppWizard puts the following three lines

```
CXxxDlg dlg;
m_pMainWnd = &d
int nResponse = dlg.DoModal();
```

which creates the modal dialog box. (Xxx is the name of your project.)

29. At this point, the sequence becomes identical to creating a modal dialog box, except for two differences.

   • The AppWizard automatically adds a message handler for WM_INTDIALOG, which adds an "About…" menu item to the system menu. The system menu is the one that appears when you click on the icon in the top-left corner of an overlapped window. OnInitDialog also now sets this icon. SDI and MDI applications set the icon automatically when they load a frame window, as we will discuss in the following text.

   • The AppWizard also adds some functionality here to draw this application's icon in the task bar when it is minimized. Again, this is done invisibly for you in an SDI or MDI application.

Closing a Dialog Application

The sequence to close a Dialog application is identical to closing a window or a modal dialog box, with one additional step to close the application itself.

Table A.10 Closing a Dialog Application

Handler/Function	Type	Description	Notes
OnClose()	Msg	Sent when the close button has been clicked.	18,30
OnDestroy()	Msg	Sent when this window is to be destroyed.	19
OnNcDestroy()	Msg	Sent after the window has been destroyed.	20
PostNcDestroy()	Ovr	Called by CWnd as final act of processing OnNcDestroy().	21
ExitInstance()	Ovr	Called to terminate your application.	31

Notes

30. This is the only spot to ask the user if they are sure they want to terminate. If they don't, just don't call the base class OnClose() function, which sends a WM_DESTROY message to the dialog box.

31. This is called by CWinApp to terminate your application. This is a good spot to close databases and deallocate any application resources.

SDI Applications

In this section, we will list the sequence of messages and functions used to create and close an SDI application. We will also look at the sequences that a SDI application uses to:

- create a new document,
- open an existing document, and
- save a document.

Creating an SDI Application

The following sequence occurs when you create a SDI application.

Table A.11 Opening an SDI Application

Handler/Function	Type	Description	Notes
InitInstance()	Ovr	Mandatory—this initializes and determines the type of application this will be.	32
New Document Created or Existing Document Loaded—See Tables A.12 and A.13.			

A

Handler/Function	Type	Description	Notes
Run()	Ovr	CWinApp's background processing function.	33
OnIdle()	Ovr	Allows you to do your own background processing.	34

Notes

32. This member function is called by CWinApp as it starts up the application and allows you to customize the initialization of your application (e.g., open databases, register Window Classes, etc.).

When the AppWizard creates your project, it overrides the InitInstance() function for you and fills it with some initializations of its own. For an SDI application, it also adds the following three steps.

- Creation of a document template.
- Parsing and processing of the command line.
- Displaying the Main Window.

The document template is created with the following lines of code.

```
CSingleDocTemplate* pDocTemplate;
pDocTemplate = new CSingleDocTemplate(
    IDR_MAINFRAME,
    RUNTIME_CLASS( CExamplesDoc ),
    RUNTIME_CLASS( CMainFrame ),          // main SDI frame window
    RUNTIME_CLASS( CExamplesView )
    );
AddDocTemplate( pDocTemplate );
```

A document template associates the Frame, Document ,and View classes to use when loading a particular document. There can be more than one document template per application allowing your application to process more than one type of document, although this is rare and usually reserved for large MDI applications, such as the Developer Studio.

SDI Applications with Multiple Document Templates

Note these two changes to user-application interaction if you add other document templates to your SDI application.

- Whenever a user creates or opens a document using your application, MFC prompts them with a dialog box to ask which template to use.

MFC fills this dialog box with the template names it finds in the resource file under the resource ID, which in the previous case is IDR_MAINFRAME.

- If the user creates or loads a document type that has already been loaded, MFC does not create another Frame, Document, and View class object. Instead, it uses the existing objects. However, if they open a new document type, a second set of objects will be created.

The command line is parsed and processed with these lines of code.

```
CCommandLineInfo cmdInfo;
ParseCommandLine( cmdInfo );
if ( !ProcessShellCommand( cmdInfo ) )
    return FALSE;
```

The command line is parsed in ParseCommandLine() and processed in ProcessShellCommand(). An empty command line causes ProcessShellCommand() to create a new document. A file name on the command line causes ProcessShellCommand() to attempt to open it as a document.

The ProcessShellCommand() function creates or opens a document by calling the OnCmdMsg() member function of CWinApp with one of two predefined MFC commands: ID_FILE_NEW to create a new document or ID_FILE_OPEN to open an existing document.

The final lines of code in InitInstance() involve showing and painting the Main Window.

```
m_pMainWnd->ShowWindow(SW_SHOW);
m_pMainWnd->UpdateWindow();
```

where m_pMainWnd contains the pointer to the Main Window created by ProcessShellCommand().

33. This CWinApp member function is called after the application is safely up and running. This function spends the rest of its life doing background processing in an idle loop and looking for posted messages in the application's message queue—especially the WM_CLOSE message, which causes it to terminate the loop and the application.

34. This CWinApp member function allows you to override the Run() function's idle loop so that you, too, can perform background processing in your application.

A

Creating a New SDI Document

The following sequence creates a new SDI document. If the Document, Frame, and View class objects do not yet exist, such as at program initialization, they are also created. An SDI document is created with the following steps.

- If no document class object exists, one is created.
- If no frame class object and window exits, one is created.
- If a frame class object is created, a view class object and window are also created.
- The document is initialized.
- The frame and view are activated.

Table A.12 Creating a New SDI Document

Handler/Function	Type	Description	Notes
CWinApp::OnFileNew()	Msg	Sent when "New" has been clicked in the main menu.	35
CSingleDocTemplate:: OpenDocumentFile()	Ovr	Called by default implementation of OnFileNew(). Creates an empty document if file name argument is NULL. Calls the following functions to create Document, Frame, and View Class objects, if needed.	36

Create document class object if none exists...

CDocTemplate:: CreateNewDocument()	Ovr	Creates a new document class object using the CRuntimeClass from the document template and CreateObject().	37

Create frame class object and frame window if they do not exist...

CDocTemplate:: CreateNewFrame()	Ovr	Creates a new frame class object using the CRuntimeClass from the document template and CreateObject().	
CMainFrame:: LoadFrame()	Ovr	Causes the frame window to be created and loaded with resources.	
CFrameWnd::Create()	Ovr	Creates the frame window.	
CMainFrame:: PreCreateWindow()	Ovr	Allows you to override style, class of frame window before it is created from template.	1,38

CMainFrame:: PreCreateWindow()	Ovr	Not a misprint—called this second time for a reason. (See note).	1,38
CMainFrame:: PreSubclassWindow()	Ovr	Allows frame window to be presub-classed.	2
CMainFrame:: OnGetMinMaxInfo()	Msg	Allows frame window to have maximum or minimum size.	3
CMainFrame:: OnCreate()	Msg	Informs that frame window has been created.	6,39
CMainFrame:: OnCreateClient()	Ovr	Called by LoadFrame() to created view class object and window. OnCreateClient() calls the following functions.	

If creating a new frame, create a new View Class object and view window...

CFrameWnd:: CreateView()	Ovr	Creates a new view class object using the CRuntimeClass from the document template and CreateObject().	
CView::Create()	Ovr	Creates the view window.	
CView:: PreCreateWindow()	Ovr	Allows you to override the style and class of the view window before it is created from the template.	1
CView:: PreSubclassWindow()	Ovr	Allows you to presubclass the view window.	2
CView::OnCreate()	Msg	Informs you that the view window has been created.	6,40
CDocument:: OnChangedViewList()	Ovr	Called to tell document class that a view has been added to it.	
CView::OnSize()	Msg	Size the view window.	7
CView::OnMove	Msg	Move the view window.	7
CView:: OnChildNotify()	Ovr	Informed that the view window has just been created.	8
CView::OnShowWindow()	Msg	Show the view window.	11

If frame and view created, size them to each other....

CMainFrame::OnMove()	Msg	Move the frame window around the view window.	7
CMainFrame::OnSize()	Msg	Size the frame window around the view window.	7
CMainFrame:: RecalcLayout()	Ovr	Called to reposition view and control bars within frame window.	

A

CView:: CalcWindowRect()	Ovr	Called to determine the size of the entire window based on the size you want the client area to be.	
CView::OnMove()	Msg	Move view within frame window.	7
CView::OnSize()	Msg	Size view within frame window.	7
CMainFrame:: RecalcLayout()	Ovr	Once more...	
CView:: CalcWindowRect()	Ovr	...with feeling.	

Finish Creating Document....

CSingleDocTemplate:: SetDefaultTitle()	Ovr	Set the default title of this document.	
CDocument::SetTitle()	Ovr	Set the real title.	
CDocument:: OnNewDocument()	Ovr	Do any initializations—usually left for the next function.	41
CDocument:: DeleteContents()	Ovr	Initializes member variables of CDocument—used when an existing document is loaded.	41

Update View for the first time....

CDocTemplate:: InitialUpdateFrame()	Ovr	Tell views to initialize for the first time.	
CMainFrame:: InitialUpdateFrame()	Ovr	Tell views to initialize for the first time.	
CView:: OnInitialUpdate()	Ovr	Tell views to initialize for the first time.	42
CView::OnUpdate()	Ovr	Called anytime the document has changed to update the view.	43

Activate frame window....

CMainFrame:: ActivateFrame()	Ovr	Called to activate frame.	
CMainFrame:: OnQueryNewPalette()	Msg	Allows you to load up the system palette with your own colors.	12
CMainFrame:: OnActivateApp()	Msg	Informs you that application is being activated.	
CView:: OnActivateView()	Msg	Informs you that view is being activated.	
CMainFrame:: OnActivate()	Msg	Informs you that frame is being activated.	

CMainFrame:: OnShowWindow()	Msg	Called to show frame.	11
CMainFrame:: OnEraseBkgnd()	Msg	Called to allow frame background color to be changed.	
CMainFrame::OnPaint()	Msg	Called to draw client area of frame window.	
CView::OnPaint()	Msg	Called to draw view, but use OnDraw() instead.	
CView::OnDraw()	Ovr	Called by CView processing of OnPaint().	44

35. The ID_FILE_NEW Command Message indicates the user wants to load an empty document. This message is sent either by the ProcessShellCommand() function at application initialization or when the user clicks on "New" in the "File" menu. The ID_FILE_NEW message itself is one of several predefined MFC messages. You can process this message yourself to create a custom document or you can let MFC handle it by creating a new MFC-styled document. The sequence that follows is how MFC creates a new document.

36. The OpenDocumentFile() function is a member function of the Document Template class, CDocTemplate. You can call OpenDocumentFile() anywhere in your application to create a new document. Just pick the Document Template you want to create and call

```
pTemplate -> OpenDocumentFile( NULL );
```

37. The CreateNewDocument() function is another member function of the Document Template class, CDocTemplate, and creates a document object from the template you defined in the InitInstance() function using CreateObject(). However, you don't have to use this function to create a document object. Instead, you can just use the old standby

```
Doc = new CDocument;
```

38. The first time PreCreateWindow() is called, MFC is trying to determine with which Window Class to create the frame window. If you don't specify a Window Class here, MFC creates one automatically using the AfxFrameOrView as a template and AfxRegisterWndClass() to actually create it and name it. Note that even the frame window uses AfxWndProc as its window procedure. The second time this function is called is when the actual frame window is being created.

39. This is the usual spot in CMainFrame to create any control bars (e.g., dialog bars, status bar, toolbars, etc.).

A

40. This is a good spot to create any controls in the view.

41. You can override this function to initialize your document's member variables in some particular way. However, if you use the default implementation of this function and override DeleteContents() instead to initialize variables, you won't have to do it again for opening an existing file. DeleteContents() is called by both.

42. This is a good spot to initialize any controls in your view.

43. This is a good spot to update any controls or graphics in your view when the document changes.

44. This function is actually called by OnPaint(). Override this function instead of OnPaint() so that the functionality to print the screen to the printer is built-in if you're using the device context to draw to the screen. Any controls in the view will not be automatically printed.

Opening an Existing SDI Document

The following sequence loads an existing SDI document from a file. If the Document, Frame, and View Class objects do not yet exist, such as at program initialization, they are also created. However, since we have already reviewed that case, we will only examine the case when these objects already exist.

Table A.13 Opening an Existing SDI Document

Handler/Function	Type	Description	Notes
CWinApp::OnFileOpen()	Msg	Sent when "Open" has been clicked in the main menu.	45
CSingleDocTemplate:: OpenDocumentFile()	Ovr	Called by default implementation of OnFileOpen(). Loads a specified file into the document. Optionally creates the Document, Frame, and View Class objects, if needed.	
CDocument:: SaveModified()	Ovr	Called to make sure current document is not modified—if it is, user is given chance to save it.	
CDocument:: OnOpenDocument()	Ovr	Default implementation calls the following functions to load an existing file.	
CDocument:: DeleteContents()	Ovr	Initializes CDocument's member variables.	41

Handler/Function	Type	Description	Notes
CDocument:: Serialize()	Ovr	Uses Serialize feature of MFC to load file into member variables.	
CDocument:: SetPathName()	Ovr	Sets path name of file just loaded.	
CDocument::SetTitle()	Ovr	Sets title of document.	
CWinApp:: AddToRecentFileList()	Ovr	Adds file's path name to list of recently opened files in the "File" menu.	
CView:: OnInitialUpdate()	Ovr	Tell views to initialize for the first time.	42
CView::OnUpdate()	Ovr	Called anytime the document has changed to update the view.	43

Notes

45. The OnFileOpen() handles another predefined Command Message, ID_FILE_OPEN.

Saving an SDI Document

The following sequence saves a modified SDI document to a file.

Table A.14 Saving an SDI Document

Handler/Function	Type	Description	Notes
CDocument:: OnFileSave()	Msg	Sent when "Save" has been clicked in the main menu.	46
CDocument:: OnSaveDocument()	Ovr	Default implementation calls the following functions to save a document to a file.	47
CDocument:: Serialize()	Ovr	Uses Serialize feature of MFC to serialize member variables out to a file.	48
CDocument:: SetPathName()	Ovr	Sets path name to file just saved.	49
CDocument:: SetTitle()	Ovr	Sets the document name to the name of document just saved.	49
CWinApp:: AddToRecentFileList()	Ovr	Adds file's path name to list of recently opened files in the "File" menu.	49

A

Notes

46. The `OnFileSave()` handles yet another predefined Command Message, `ID_FILE_SAVE`. Another way of getting here is with `ID_FILE_SAVE_AS`, which is handled by `CDocument::OnFileSaveAs()`.

47. This function opens the file name passed to it, then opens an archive with that file name and calls `Serialize()` to save the file to disk.

48. `Serialize()` allows you to automatically save your document to disk. If you are storing to the system registry or a database instead, you would override `OnFileSave()` and store your data there.

49. These functions are only called if the path name changed during a "Save As" command.

Closing an SDI Application

The following sequence closes an SDI. This sequence assumes the document has not been modified. Refer to Table A.14 for the sequence used to save a modified document.

Table A.15 Closing an SDI Document

Handler/Function	Type	Description	Notes
`CMainFrame::OnClose()`	Msg	Sent when someone clicks on the close button or the "Exit" item under the "File" menu.	50
`CDocument::CanCloseFrame()`	Ovr	Called by `OnClose`; the default implementation calls the next function to see if document is modified and if the user wants to save.	
`CDocument::SaveModified()`	Ovr	Asks user to save a modified document.	
`CMainFrame::OnShowWindow()`	Msg	Hides main frame window.	11
`CView::OnActivateView()`	Msg	Deactivates view.	
`CMainFrame::OnActivate()`	Msg	Deactivates frame.	
`CMainFrame::OnActivateApp()`	Msg	Deactivates application.	

Handler/Function	Type	Description	Notes
CView::OnKillFocus()	Msg	Removes keyboard input focus from this application.	26
CDocument:: OnCloseDocument()	Ovr	Gives the document a chance to close. Default implementation calls next function.	
CDocument:: DeleteContents()	Ovr	Initializes document class by deallocating all memory.	41
CMainFrame:: DestroyWindow()	Ovr	Destroys main frame window.	
CMainFrame:: OnDestroy()	Msg	Sent to destroy main frame window.	19
CView::OnDestroy()	Msg	Sent to destroy view window.	19
CView::OnActivateView	Msg	Deactivates view.	
CView:: PostNcDestroy()	Ovr	Called after view window has been destroyed.	21
CDocument:: OnChangedViewList()	Ovr	Tells document that it has just lost a view.	
CMainFrame:: PostNcDestroy()	Ovr	Called after the frame window has been destroyed.	21
CWinApp:: ExitInstance()	Ovr	Allows you to deallocate anything allocated for this application.	30

Notes

50. This is where you want to put an "Are you sure?" message. If they answer "no", don't call the base class function that sends a WM_DESTROY to the main frame window. You can just wait for the frame to check to see if the document has been modified with the SaveModified() function.

MDI Applications

Creating an MDI Application

In this section, we will list the messages and functions called to create and close an MDI application. An MDI application is similar to an SDI application, except that an MDI application has one Main Frame Window filled with one or more child frame windows.

We will also look at the sequences that an MDI application uses to

- create a new document,

A

- open an existing document,
- save an MDI document, and
- close an MDI child frame.

Table A.16 Creating an MDI Application

Handler/Function	Type	Description	Notes
InitInstance()	Ovr	Mandatory—this initializes and determines the type of application this will be.	51
Main Window Created—See Table A.17			
New Document Created or Existing Document Loaded—See Tables A.18 and A.19.			
Run()	Ovr	CWinApp's background processing function.	32
OnIdle()	Ovr	Allows you to do your own background processing.	33

51. This member function is called by CWinApp as it starts up the application and allows you to customize the initialization of your application (e.g., open databases, register Window Classes, etc.).

When the AppWizard creates your project, it overrides the InitInstance() function for you and fills it with some initializations of its own. For an MDI application, it also adds the following four steps.

- Creates a document template.
- Creates the Main Frame Window.
- Parses and processes the command line.
- Displays the Main Window.

The document template is created with the following lines of code.

```
CMultiDocTemplate* pDocTemplate;
pDocTemplate = new CMultiDocTemplate(
    IDR_EXAMPLTYPE,
    RUNTIME_CLASS( CExamplemdiDoc ),
    RUNTIME_CLASS( CChildFrame ),          // custom MDI child frame
    RUNTIME_CLASS( CExamplemdiView )
    );
AddDocTemplate( pDocTemplate );
```

A document template associates the Frame, Document, and View Classes to use when loading a particular document. There can be more than one document template per application allowing your application to process more than one type of document, although this is rare and usually reserved for large MDI applications, such as the Developer Studio.

Notice that the Frame Class used here is a child frame class. Now, whenever a new or existing document is loaded, a new child frame, view, and document class object are also created and reside in the Main Frame Window. When is the main frame window created?

The main frame window is created next with the following code lines.

```
// create main MDI Frame window
CMainFrame* pMainFrame = new CMainFrame;
if ( !pMainFrame -> LoadFrame( IDR_MAINFRAME ) )
    return FALSE;
m_pMainWnd = pMainFrame;
```

The LoadFrame() function is used again just as in an SDI application with the resulting sequence of function calls and messages:

Table A.17 Creating Main Frame

Handler/Function	Type	Description	Notes
CMainFrame:: LoadFrame()	Ovr	Calls the following functions to load the frame.	
CMainFrame:: PreCreateWindow()	Ovr	Called in case you want to specify a Window Class for the frame window.	1,38
CMainFrame:: PreCreateWindow()	Ovr	Called in case you want to change the window style or size of the frame window.	1,38
CMainFrame:: PreSubclassWindow()	Ovr	Allows you to presubclass the window.	2
CMainFrame:: OnGetMinMaxInfo()	Msg	Allows you to to put limits on the size of the Main Frame Window.	3
CMainFrame:: OnCreate()	Msg	The Main Frame Window has been created—time to create toolbars, dialog bars, and status bars.	6
CMainFrame:: OnCreateClient()	Ovr	Time to fill client area of the MDI application with the MDICLIENT Window Class window.	

A

Handler/Function	Type	Description	Notes
CMainFrame:: RecalcLayout()	Ovr	Called to reposition view and control bars within frame window.	

After the main frame window is created, the command line is parsed and processed with these lines of code.

```
CCommandLineInfo cmdInfo;
ParseCommandLine( cmdInfo );
if ( !ProcessShellCommand( cmdInfo ) )
   return FALSE;
```

The command line is parsed in ParseCommandLine() and processed in ProcessShellCommand(). An empty command line causes ProcessShellCommand() to create a new document. A file name on the command line causes ProcessShellCommand() to attempt to open it as a document.

The ProcessShellCommand() function creates or opens a document by calling the OnCmdMsg() member function of CWinApp with one of two predefined MFC commands: ID_FILE_NEW to create a new document or ID_FILE_OPEN to open an existing document just as with an SDI application.

The final lines of code in InitInstance() involve showing and painting the Main Window

```
m_pMainWnd -> ShowWindow( SW_SHOW );
m_pMainWnd -> UpdateWindow();
```

where m_pMainWnd contains the pointer to the Main Window created previously.

Creating a New MDI Document

The following sequence creates a new MDI document. The document, child frame, and View Class objects are always created when a new MDI document is opened. Other than that, the sequence is almost identical to creating a new SDI document, seen Previously.

Table A.18 Creating a New MDI Document

Handler/Function	Type	Description	Notes
CWinApp::OnFileNew()	Msg	Sent when "New" has been clicked in the main menu.	35
CMultiDocTemplate:: OpenDocumentFile()	Ovr	Called by default implementation of OnFileNew(). Creates an empty document if file name argument is NULL. Calls the following functions to create document, child frame and view class objects.	36

Create document class object...

CDocTemplate:: CreateNewDocument()	Ovr	Creates a new document class object using the CRuntimeClass from the document template and CreateObject().	37

Create child frame class object and child frame window...

CDocTemplate:: CreateNewFrame()	Ovr	Creates a new frame class object.	
CChildFrame:: LoadFrame()	Ovr	Causes the child frame window to be created and loaded with resources.	
CChildFrame:: PreCreateWindow()	Ovr	Allows you to override the style and class of teh frame window before it is created from the template.	1,38
CChildFrame::Create()	Ovr	Creates the child frame window.	
CMainFrame:: RecalcLayout()	Ovr	Called to reposition view and control bars within frame window.	
CChildFrame:: PreCreateWindow()	Ovr	Not a misprint—called this second time for a reason. See note.	1,38
CChildFrame:: PreSubclassWindow()	Ovr	Allows frame window to be presubclassed.	2
CChildFrame:: OnGetMinMaxInfo()	Msg	Allows frame window to have maximum or minimum size.	3
CChildFrame:: PreCreateWindow()	Ovr	Allows you to override the style or class of the frame window before it is created from the template.	1,38
CChildFrame:: OnCreate()	Msg	Informs that the frame window has been created.	6,39
CChildFrame:: OnCreateClient()	Ovr	Called by LoadFrame() to created the view class object and window. LoadFrame() calls the next functions.	

A

Create a new view class object and view window...

CFrameWnd:: CreateView()	Ovr	Creates a new view class object using the CRuntimeClass from the document template and CreateObject().	
CView::Create()	Ovr	Creates the view window.	
CView::PreCreateWindow()	Ovr	Allows you to override the style and class of the view window before it is created from the template.	1
CView:: PreSubclassWindow()	Ovr	Allows you to presubclass the view window.	2
CView::OnCreate()	Msg	Informs you that the view window has been created.	6,40
CDocument:: OnChangedViewList()	Ovr	Called to tell document class that a view has been added to it.	
CView::OnSize()	Msg	Size the view window.	7
CView::OnMove	Msg	Move the view window.	7
CView:: OnChildNotify()	Ovr	Informed that the view window has just been created.	8
CView::OnShowWindow()	Msg	Show the view window.	11

If child frame and view created, size them to each other....

CMainFrame::OnMove()	Msg	Move the frame window around the view window.	7
CMainFrame::OnSize()	Msg	Size the frame window around the view window.	7
CMainFrame:: RecalcLayout()	Ovr	Called to reposition view and control bars within the frame window.	
CView:: CalcWindowRect()	Ovr	Called to determine the size of the entire window based on the size you want the client area to be.	
CView::OnMove()	Msg	Move the view within the frame window.	7
CView::OnSize()	Msg	Size the view within the frame window.	7
CMainFrame:: RecalcLayout()	Ovr	Once more...	
CView:: CalcWindowRect()	Ovr	...with feeling.	
CChildFrame:: OnMDIActivate()	Ovr	Activate the child frame.	

Finish Creating Document....

CDocument::SetTitle()	Ovr	Set the real title.	
CDocument:: OnNewDocument()	Ovr	Do any initializations—usually left for the next function.	41
CDocument:: DeleteContents()	Ovr	Initializes member variables of CDocument—used when an existing document is loaded.	41

Update View for the first time....

CDocTemplate:: InitialUpdateFrame()	Ovr	Tell views to initialize for the first time.	
CMainFrame:: InitialUpdateFrame()	Ovr	Tell views to initialize for the first time.	
CView:: OnInitialUpdate()	Ovr	Tell views to initialize for the first time.	42
CView::OnUpdate()	Ovr	Called anytime the document has changed to update the view.	43

Activate main and child frame windows....

CChildFrame:: ActivateFrame()	Ovr	Called to activate the Main Frame Window.	
CChildFrame:: OnMDIActivate()	Ovr	Called to activate the child frame window.	
CChildFrame:: OnShowWindow()	Msg	Sent to show the child frame window.	
CView::OnActivate-View()	Msg	Sent to activate this view.	
CMainFrame:: OnShowWindow()	Msg	Sent to show the Main Frame Window.	11
CMainFrame:: OnQueryNewPalette()	Msg	Sent to allow the Main Frame Window to fill the system palette with its own colors.	12
CMainFrame:: OnEraseBkgnd()	Msg	Sent to allow the Main Frame Window to change the color used to draw the background.	
CMainFrame:: OnActivateApp()	Msg	Sent to activate this application.	
CView::OnActivate-View()	Msg	Sent to activate this view.	
CMainFrame::OnActivate()	Msg	Sent to activate this frame window.	

A

CView::OnPaint()	Msg	Called to draw the view, but use OnDraw() instead.	
CView::OnDraw()	Ovr	Called by CView processing of OnPaint().	44

Opening an Existing MDI Document

The following sequence occurs when you open an existing MDI document. Again, this is very similar to an SDI document, except that the Document, Frame, and View Class objects are always created.

Table A.19 Open an Existing MDI Document

Handler/Function	Type	Description	Notes
CWinApp::OnFileOpen()	Msg	Sent when "Open" has been clicked in the main menu.	35
CMultiDocTemplate:: OpenDocumentFile()	Ovr	Called by the default implementation of OnFileNew(). Creates an empty document if the file name argument is NULL. Calls the following functions to create the document, child frame, and View Class objects.	36

Create document class object ...

CDocTemplate:: CreateNewDocument()	Ovr	Creates a new document class object.	37

Create child frame class object and window...

CChildFrame:: LoadFrame()	Ovr	Calls the following functions to load the child frame.	
CChildFrame:: PreCreateWindow()	Ovr	Allows you to use your own Window Class to create the child frame window.	1,38
CMainFrame:: RecalcLayout()	Ovr	Called to reposition teh view and control bars within the frame window.	
CChildFrame:: PreCreateWindow()	Ovr	Allows you to change other properties of the child frame window.	1,38
CChildFrame:: PreSubclassWindow()	Ovr	Allows you to presubclass a window.	2
CChildFrame:: OnGetMinMaxInfo()	Ovr	Allows you to set limits on the size of the child frame window.	3

CChildFrame::OnCreate()	Msg	Informs that the child frame window has been created—time to create toolbars, dialog bars, and status bar for this child frame.	6
CChildFrame::OnCreateClient()	Ovr	Time to create the view for this child frame using the functions that follow.	

Create view class object and window...

CView::PreCreateWindow()	Ovr	Allows you to change the properties of the created view window.	1
CView::PreSubclassWindow()	Ovr	Allows you to presubclass the view window.	2
CView::OnCreate()	Ovr	Informs that the view window has been created—time to create any controls for this view window.	6
CDocument::OnChangedViewList()	Ovr	Tells document that a view has been added.	
CView::OnSize()	Msg	Sizes view—a good time to size any child windows of this view.	7
CView::OnMove()	Msg	Moves view window.	7
CView::OnChildNotify()	Ovr	Tells you that the view window was just created.	8
CView::OnShowWindow()	Msg	Shows view window.	11

Size view to child frame...

CChildFrame::OnMove()	Msg	Moves the child frame window.	7
CChildFrame::OnSize()	Msg	Sizes the child frame window.	7
CChildFrame::RecalcLayout()	Ovr	Called to reposition the view and control bars within the frame window	
CView::CalcWindowRect()	Ovr	Called to determine the size of the entire window based on the size you want the client area to be.	

Finish opening document...

CDocument::OnOpenDocument()	Ovr	Default implementation calls the following functions to load an existing file.	
CDocument::DeleteContents()	Ovr	Initializes CDocument's member variables.	41
CDocument::Serialize()	Ovr	Uses the Serialize feature of MFC to load the file into member variables.	

A

CDocument::SetPathName()	Ovr	Sets the path name of the file just loaded.	
CDocument::SetTitle()	Ovr	Sets the title of the document.	
CWinApp::AddToRecentFileList()	Ovr	Adds fthe ile's path name to the list of recently opened files in the "File" menu.	
CView::OnInitialUpdate()	Ovr	Tell views to initialize for the first time.	42
CView::OnUpdate()	Ovr	Called anytime the document has changed to update the view.	43

Activate child frame and view...

CChildFrame::ActivateFrame()	Ovr	Called to activate the Main Frame Window.	
CChildFrame::OnMDIActivate()	Ovr	Called to activate the child frame window.	
CView::OnActivateView()	Msg	Sent to activate this view.	
CView::OnKillFocus()	Msg	Removes keyboard focus from the view...	26
CChildFrame::OnSetFocus()	Msg	...and adds it to the child frame window.	
CChildFrame::OnShowWindow()	Msg	Shows the child frame window.	11
CView::OnActivateView()	Msg	Sent to activate this view.	

Saving an MDI Document

The following sequence occurs when you save a modified MDI document.

Table A.20 Saving an MDI Document

Handler/Function	Type	Description	Notes
CDocument::OnFileSave()	Msg	Sent when "Save" has been clicked in the main menu.	46
CDocument::OnSaveDocument()	Ovr	Default implementation calls the following functions to save a document to a file.	47
CDocument::Serialize()	Ovr	Uses the Serialize feature of MFC to serialize member variables out to a file.	48
CDocument::SetPathName()	Ovr	Sets the path name to the file just saved.	49

Handler/Function	Type	Description	Notes
CDocument::SetTitle()	Ovr	Sets the document name to the name of document just saved.	49
CWinApp:: AddToRecentFileList()	Ovr	Adds the file's pathname to the list of recently opened files in the "File" menu.	49

Closing an MDI Child Frame

The following sequence occurs when you close an MDI child frame. This sequence assumes this child frame does not have a modified document. See Table A.19 for the sequence used to save an MDI document.

Table A.21 Closing an MDI Child Frame

Handler/Function	Type	Description	Notes
CChildFrame:: OnClose()	Msg	Sent when someone clicks the close button on the child frame window.	50
CDocument:: CanCloseFrame()	Ovr	Calls the next function to determine if it can close the child frame.	
CDocument:: SaveModified()	Ovr	Checks if document is modified and, if so, offers to save it.	52
CDocument:: OnCloseDocument()	Ovr	Default implementation calls the next function.	
CDocument:: DeleteContents()	Ovr	Deallocates document memory.	41
CChildFrame:: DestroyWindow()	Ovr	Called to destroy the child frame window.	
CChildFrame:: OnShow-Window()	Msg	Hides the child frame window.	11
CChildFrame:: OnMDI-Activate()	Msg	Deactivates the child frame window.	
CView:: OnActivateView()	Msg	Deactivates the view in the child frame.	
CView::OnKillFocus()	Msg	Removes keyboard input focus from the view.	26
CChildFrame:: OnDestroy()	Msg	Sent to destroy the child frame window.	19
CView::OnDestroy()	Msg	Sent to destroy the view window.	19
CView:: PostNcDestroy()	Ovr	Called after the view window has been destroyed.	21

A

Handler/Function	Type	Description	Notes
CView:: OnChangedViewList()	Ovr	Tells document that one view has been removed.	
CChildFrame:: PostNcDestroy()	Ovr	Called after the child frame has been destroyed.	21

52. The default implementation opens a file dialog box and asks you to name the location in which to save the file. If you aren't saving to a file, but are saving to a database or system registry, you must override this function and not use the default.

Closing an MDI Application

The following sequence occurs when you close an MDI application. This sequence assumes that all MDI Child frames have already been closed. See Table A.20 for the sequence when a Child Frame is closed.

Table A.22 Closing an MDI Application

Handler/Function	Type	Description	Notes
CMainFrame::OnClose()	Msg	Sent when the user clicks on the close button in the Main Frame Window.	50
CWinApp:: SaveAllModified()	Ovr	The SaveModified() function is called for every open document.	
CMainFrame:: OnShowWindow()	Msg	Main frame window is hidden.	11
CMainFrame:: OnActivate()	Msg	Main frame window is deactivated.	
CMainFrame:: OnActivateApp()	Msg	Application is deactivated.	
CMainFrame:: DestroyWindow()	Ovr	Main frame window is told to destruct.	
CMainFrame:: OnDestroy()	Msg	Sent to destroy the Main Frame Window.	19
CMainFrame:: PostNcDestroy()	Ovr	Called after the Main Frame Window is destroyed.	21
CWinApp:: ExitInstance()	Ovr	Called to allow you to deallocate anything allocated for this application.	

Customizing

The sequences depicted previously are for the three standard applications—Dialog, SDI, and MDI—performing standard document loads and saves using Serialize(). There are obviously many more permutations (e.g., using a database instead of Serialize(), loading dialog bars in the frame window, etc.) but to list them all would fill this book. By listing these, I hoped to give you a flavor for what MFC is doing with the idea that you can not only modify it, but even create your own sequences. The following are, perhaps, the two most useful considerations.

• You can create applications that are hybrids of Dialog, SDI, and MDI applications.

• You can use Document Templates to create documents whenever you want—not just as a result of clicking on "New" or "Open".

Hybrid Applications

A lot of fuss has been paid to what type of application you want to create: Dialog, SDI, or MDI. However, you can create an application that is a combination of all three. For example, what if you wanted your application to come up as a dialog application in one situation, but as an MDI application in another. To do that, you would modify the InitInstance() member function of CWinApp.

For example, if you have an MDI application that you want to occasionally come up as a simple Dialog application to do some automated creation process, you can modify your InitInstance() to do the following.

• If a command line flag is set to -D, execute these code lines to create a Dialog application.

```
CExampledlgDlg dlg;
m_pMainWnd = &dlg;
int nResponse = dlg.DoModal();
return FALSE;;
```

Otherwise, execute these code lines to start an MDI application.

```
CMultiDocTemplate* pDocTemplate;
pDocTemplate = new CMultiDocTemplate(
    IDR_EXAMPLTYPE,
    RUNTIME_CLASS( CExamplemdiDoc ),
    RUNTIME_CLASS( CChildFrame ),          // custom MDI child frame
```

A

```
        RUNTIME_CLASS( CExamplemdiView )
     );
AddDocTemplate( pDocTemplate );
CMainFrame* pMainFrame = new CMainFrame;
if ( !pMainFrame -> LoadFrame( IDR_MAINFRAME ) )
return FALSE;
m_pMainWnd = pMainFrame;
pMainFrame -> ShowWindow( m_nCmdShow );
pMainFrame -> UpdateWindow();
```

Please see Example 4.

Fun with Documents

The MFC classes provide an automatic method to load and save documents using the predefined command messages ID_FILE_NEW, ID_FILE_OPEN, and ID_FILE_SAVE. However, you can load a document-frame-view combination at any time, just by calling the OpenDocumentFile() member function of CDocTemplate.

For example, when you're opening a document in an MDI application and you want another document to open at the same time, just save the document template you want to open later

```
m_pSaveDocTemplate = new CMultiDocTemplate(
    IDR_EXAMPLTYPE,
    RUNTIME_CLASS( CExamplemdiDoc ),
    RUNTIME_CLASS( CChildFrame ),          // custom MDI child frame
    RUNTIME_CLASS( CExamplemdiView)
    );
```

and then override the OpenDocumentFile() member function and add this

```
m_pSaveDocTemplate -> OpenDocumentFile( xxx );
```

B

Appendix B

Drawing Structures

This appendix gives layout information for each of three drawing-related file formats: icon files, bitmap files, and dialog templates. Each description begins with an overview of the file's general structure. Each section is then detailed in byte-ordered layout figures.

Icon and Cursor Files

Icon and cursor files share a common format. A single type code (in the file header) distinguishes the two. Both icon and cursor files may contain information about multiple objects. At the most abstract level, they consist of just two sections.

- File Header Section
- Picture Data Section

Each of these sections, however, contains structures that are repeated once for each of the icons (or cursors) contained in the file.

File Header Section

The first few bytes of the header give the type of the file and the count of objects contained in the file.

0000	0000
0002	0001=icon, 0002=cursor
0004	# of icons/cursors in this file

This "global" information is then followed by a "directory" of all the icons or cursors contained in the file.

Individual Icon and Cursor Headers

Each entry in the "directory" is a more detailed header structure that specifies the dimensions and color format for the individual object. These entries also include a pointer to the actual bitmap data for the object. The following structure is repeated once for each icon or cursor in the file.

0000	width	height
0002	2=B&W, 16=16 colors, 0=256 colors	
0004	x of cursor hot point	
0006	y of cursor hot point	
0008	size of Icon/Cursor Data	
000a	"	
000c	offset into this file of Icon/Cursor Data	
000e	"	

Picture Data Section

This portion of the file contains a separate section of bitmap data for each icon or cursor. Each object's picture data is represented by a three part structure consisting of a header, a color table, and a table of pixel values. This three part structure is repeated for each icon or cursor in the file.

Picture Data Header

This 40-byte header describes the graphical qualities of the object: its size, shape, and number of color planes.

0000	size of header (40 bytes)
0002	
0004	width of icon/cursor
0006	
0008	height * 2
000a	
000c	# of planes
000e	size of pixel pointers in bit data
0010	
0012	
0014	# of bytes in 1st plane
0016	
0018	# of bytes in 2nd plane
001a	
001c	# of bytes in 3rd plane
001e	
0020	# of bytes in 4th plane
0022	
0024	# of bytes in 5th plane
0026	

Color Table

The color table is present only if the object is not monochrome.

DWORD RGB data one entry per color

B

Bit Data

In a color image, each entry in this table indexes an RGB value in the color table. In a monochrome image, each entry is either zero or one (black or white).

pixel pointers into color table
one entry per pixel—
if monochrome, just 0 or 1

Notes

- You can create an icon or cursor file for your application using the Icon or Cursor Editor. When you create a new cursor or icon, these editors create a default 32∞32 pixel 16 color icon or a 32∞32 pixel monochrome cursor. To create other types of icons and cursors, click on the Developer Studio's "Image" and "New Device Image…" menu commands to open the "New Image" dialog. If the type of icon or cursor you want to create isn't listed, you can click the "Custom" button to create your own. You can create a color cursor this way. After you've created your icon or cursor, you can delete the default one by selecting it in the editor's combo box. Then click on the Studio's "Image" and "Delete Device Image" menu commands.

- There is currently no Windows API or MFC class function that will turn the data in this file directly into an icon or cursor. What you can do instead is create a memory file that you can feed to the appropriate icon or cursor loading function.

Bitmap Files

Bitmap (.bmp) files follow a structure similar to that of the bitmap portion in an icon or cursor file. At the most abstract level, each bitmap file consists of these four sections.

- File Header
- Bitmap Header
- Color Table
- Picture Data

File Header

The file header structure (structure type BITMAPFILEHEADER) includes a file type signature ("BM") and file size and layout information.

0000	"BM"
0002	size of file in bytes
0004	"
0006	0000
0008	0000
0010	offset into file for Picture Data
0012	"

Bitmap Header

The file header is followed by a bitmap information header (structure type BITMAPINFOHEADER). This structure specifies the dimensions, compression type and color format format for the image.

0000 0002	size of header in bytes (40)
0004	width of bitmap in pixels
0008	height of bitmap in pixels
0010	# of planes, must be 1
0012	bits per pixel (see notes)
0014	compression type 0 = none
0016	image size if compressed
0018	horizontal pixels/meter
001c	vertical pixels/meter
0020	# of colors used in Color Table (see notes)
0024	# of important colors in Color Table

B

Color Table

The color table varies in size depending upon the number of colors available (see notes). When 24-bit color is used, the color table is omitted because each entry in the picture data encodes the RGB value explicitly

DWORD RGB data one entry per color

Picture Data

The picture data is a (possibly compressed) table of pixel values. Depending upon the color coding scheme, the uncompressed pixel values may be interpreted differently.

- If monochrome, this table is interpreted as a series of bits, one bit per pixel, where 0=black and 1=white.

- If 16 or 256 color, this table is interpreted as a series of 4- or 8-bit pointers into the color table, one pointer per pixel.

- If high or true color, the picture data is interpreted as a series of 24-bit RGB values, each defining a pixel color.

Notes

- The Bitmap Header contains a "bits per pixel" value that determines how big the color table will be and what the Picture Data represents. Its values can be 1, 4, 8, and 24 and they're interpreted as follows.

  - If the bits per pixel is 1, this is a monochrome bitmap. There is no color table and the picture data is simply a series of zeros (0) for black and ones (1) for white.

  - If the bits per pixel is either 4 or 8, this is a 16- or 256-color bitmap. The Color Table is up to 16 double words long for a 16-color bitmap or 256 for a 256-color bitmap. If, however, the value in "# of colors used in Color Table" member of the Bitmap Header is not zero, this value determines how many entries are in the Color Table.

  - If the bits per pixel is 24, this is a True Color bitmap. There is no Color Table because the Picture Data itself contains a series of complete RGB values defining the color for each pixel in the bitmap.

- For more detailed information about bitmaps and other file formats, see http://www.wotsit.org.

Dialog Templates

A dialog template consists of a section of dialog information followed by multiple sections of control information, one per control. The dialog information is coded in a DLGTEMPLATE structure followed by an optional FONTINFO structure. The control information is coded in DLGITEMTEMPLATE structures. Thus, the template has the following general structure.

- Dialog Information
  - Header
  - Menu
  - Class
  - Title
  - Font
- Control Information
  - Controls
  - : : :
  - Control 'n'

Dialog Information

Though the dialog information is mostly contained in a single DLGTEMPLATE structure, it is useful to view the dialog information as consisting of a file Header followed by menu, class, title, and font information.

Header

The dialog's style, position, and size are all coded in what I'll call the header portion of the DLGTEMPLATE structure. This portion of the template also contains a count of the number of controls that belong to the dialog.

0000	style
0002	
0004	extended style
0006	
0008	# of controls

B

0010	x position in dialog units (DU)
0012	y position in DU's
0014	width in DU's
0016	height in DU's

Menu

While the menu information is just a "field" in the DLGTEMPLATE structure, this field is too complex to qualify as part of a true C structure. There are three possible alternatives.

0 if none

or

-1
ordinal of a menu resource (see notes)

or

unicode name of menu resource : : :
0

Class

The class information is recorded using a similar convention. Again, there are three alternatives.

0 if using default

or

-1
ordinal of a class name (see notes)

or

unicode class name : : :
0

Title

If the dialog box is to have it's own title bar, the caption must appear in this field. If there is no title bar, this field contains a null.

0 if none

or

unicode title
: : :
0

Font

Font information is recorded in a separate FONTINFO structure. This structure is present only if DS_FONT has been set in the dialog's style parameter.

point size
unicode font name
: : :
0

Control Information

The control information is a sequence of structures, one for each control. Each of these structures conforms to the following general form.

- Header
- Class
- Title
- Creation Data

The Header, Class, and Title information is contained in a single DLGITEMTEM-PLATE structure. The structure of the Creation Data is unique to each class of control.

B

Header

The first few fields of the DLGITEMTEMPLATE structure parallel the header fields in the DLGTEMPLATE structure. These fields specify the style, position, size, and ID for the control.

0000	style
0002	
0004	extended style
0006	
0008	x position in dialog units (DU)
0010	y position in DU's
0012	width in DU's
0014	height in DU's
0016	control ID

Class

This field specifies the control's class, either as an ordinal or as a variable length string. The alternative formats are

-1
ordinal of a class name

where:
80 = "BUTTON"
81="EDIT"
82="STATIC"
83="LISTBOX"
84="SCROLLBAR"
85="COMBOBOX"

or

unicode class name
: : :
0

Title

The control is identified either by a variable length name or an ordinal. The alternatives are

0 if none

-1
ordinal of a resource

can be a string, icon, or bitmap

or

unicode title
: : :
0

Creation Data

0 for none or # of bytes of data
peculiar to each control, passed in CreateEx() when control is created.

Notes

- Ordinals are an internal pointer system used by the resource compiler to access resources (e.g., strings, bitmaps, icons, etc.) with a minimum of space. Since these pointers are assigned internally, you're in trouble if you need to decipher them yourself.

- Using the previous structures, you can dynamically create, edit, or combine dialog templates. To load an existing template for editing, use

```
HRSRC hResource = ::FindResource( AfxGetInstanceHandle(),
    lpszTemplateName, RT_DIALOG );
HGLOBAL hTemplate = LoadResource( AfxGetInstanceHandle(),
    hResource );
DLGTEMPLATE *pTemplate = ( DLGTEMPLATE* )LockResource( hTemplate );
```

B

The `pTemplate` pointer now points to your dialog template and can be manipulated with `memcpy()`, etc. You can then create a dialog box out of the modified template with

```
CreateDlgIndirect( pTemplate, pWndOwner, AfxGetInstanceHandle() );
// also run extra creation data
ExecuteDlgInit( MAKEINTRESOURCE( IDD ) );
```

Make sure to clean up with

```
UnlockResource( hTemplate );
FreeResource( hTemplate );
```

Appendix C

Using RoboHELP®

Earlier in this book, Examples 35 and 36 showed how to add on-line help to your application. We assumed that in most companies, the actual help files themselves (.hlp and .cnt) are supplied by the company's tech writers. This simplified the creation of the application. However, if this isn't true in your company or you happen to be a tech writer, this appendix is for you.

Most developers don't use the help editing systems provided with MFC. Judging from the systems' complexity of use, it's easy to see that Microsoft doesn't want to be in the help-authoring field. Perhaps this is partly because of the wide variety of help-authoring tools already out there. In this appendix, we will use one of the most widely used help-authoring systems to create the .hlp and .cnt files for your application's on-line help, RoboHELP by Blue Sky Software.

Two types of on-line help can be created for an application.

Menu Help can be selected by a user through the "Help" menu and allows the user to read the help like a book or search the help files for a particular subject.

Context Sensitive Help allows the user to automatically search for a subject based on the control or dialog box they have open.

We will start by creating Menu Help.

Example C68   Creating Menu Help Files with RoboHELP

Objective

You would like to create the `.hlp` and `.cnt` on-line help files required by Example 35 using RoboHELP.

Strategy

We will start by organizing our help into topics and books. Then, we'll write each topic using Microsoft Word and intersperse our text with optional controls, such as buttons, that will allow our user to jump to related topics. At this point, we'll also determine which words in each topic to include in our help index. Finally, we will use RoboHELP to organize these topics by book and create the `.hlp` and `.cnt` files.

Steps

Plan Your Menu Help

1. Decide what you want to include in your menu help.
2. Break subjects up into topics, which should have the same size and scope of a section in a paper manual.
3. Organize similar topics together. These collections of topics are called books.

Create a New Help Project

1. Upon starting RoboHELP, you will notice the RoboHELP Explorer opening on the left side of the screen and an "Open a Help Project" dialog in the middle. Select "Create a New Help Project" and click "OK".
2. A sort of Help Project Wizard will then appear. Pick a help project type of "Application Help", a Visual C++ Help Template, and make your primary target WinHelp 4.

3. The last step is to enter a title and a name for this help project. The name is used when creating the .hlp and .cnt help files, as well as the directory in which this project is stored. The title appears in WinHelp when using these help files.

4. RoboHELP will now start Microsoft Word and position it to fill the right side of the screen. Both of these applications will now work together to create your help files.

Create New Topics

1. Use the File/New/Topic menu commands or toolbar button to create a new topic. Enter a topic title and click "OK". The topic title will appear over your topic, so make it descriptive.

Note: The Comment edit box is for your own internal comment—don't enter the text for your topic here.

---

2. Control will now shift to Microsoft Word. Type in your topic using Word. Keep it simple for now. The next few steps will show you how to dress up your presentation.

3. Use Word's character formatting commands to highlight your text with different fonts, colors, point sizes, and styles.

4. Use RoboHELP's feature commands to add hotspots to your text to allow your user to jump to related topics. These commands are available from a new toolbar in Word that appears when running in conjunction with RoboHELP. Table C.1 shows the types of hotspots you can add.

C

Table C.1 Help Features

Name	Typical Appearance	Typical Use
Popup	Green text with a dotted underline	Opens a small descriptive window to elaborate on a word or concept.
Jump	Green text with a solid underline	Allows the user to jump to another topic in this help file, a topic in another help file, or even to a website.
Authorable Button	A button with text that you specify	Functions in the same way as a jump.
Mini Button	A small blank button	Functions in the same way as a jump.
Graphic Button	A button with a graphic image that you specify	Functions in the same way as a jump.
Short Cut Button	A small button with an up and left arrow	Opens a graphic image.

5. Use RoboHELP's "New Help Image" toolbar button to add any graphics to your topic. Not only can a graphic illustrate your help, but you can also use RoboHELP to define hotspots over the image that, when clicked, can open popups or jump to other topics. A graphic file with hotspots is called a SHED file (from Microsoft's Segmented Hypergraphic Editor, also know as HotSpot Editor [shed.exe]). To turn your image into a SHED file, just select it using "New Help Image" and click on the "Shed" button to define the hotspots.

6. Use RoboHELP's "New Multimedia" toolbar button to add an .avi or .wav file to your help. You can provide these .avi and .wav files yourself or, if you have RoboHELP's Office product, you can create an .avi file using the "Software Video Camera" utility. This utility repeatedly captures your screen as you demonstrate your product and turns those captures into an .avi file.

7. In one final pass through this topic, use the "Add to Index" toolbar button to add any keywords you would like to appear in your help system's index. Remember that a vast majority of your users will be using these keywords to locate a topic.

Create and Organize New Books

1. Maximize the RoboHELP Explorer window to reveal a list of all topics in your help project in a right-hand window. Then, open your help system's table of contents view by clicking on the "TOC" tab (located at the bottom) of the left-hand view.

2. Create your books using the File/New/Book menu commands or toolbar button. Note that this command is available only when viewing your help's table of contents.

3. Now, drag and drop your topics into their appropriate books. You can also use drag and drop to change the order of any book or topic. If a topic has no related topics, simply add it alone.

Note: The first topic that appears in the table of contents is what appears when your user clicks on the "Index" command in your application's "Help" menu. This topic should, therefore, be somewhat of an index on the rest of your help topics with plenty of jumps.

Create and Test the Help Files

1. Compile your help project using the File/Compile menu commands.

2. You can test your help project from within RoboHELP with the File/Run menu commands or you can copy the .hlp and .cnt files from your help project's directory to the location at which your application expects them. This is typically in the same subdirectory as your application's executable. However, you can change this location programmatically by refering to Example 36.

Notes

- As I mentioned in Example 36, the AppWizard will add some of this functionality itself when you select "Context help" in Step 4. However, the addition of these extra help files can make an already sprawling application even more complicated. Since you can just as easily start from scratch with RoboHELP, I have found it best to keep the help project separate from the application project, combining them only in the final product. Creating them is usually the job of a tech writer and product manager, anyway.

C

Example C69 Creating Context Help Files with RoboHELP

Objective

You would like to create the .hlp and .cnt Context Help files required by Example 36 using RoboHELP.

Strategy

As mentioned in the previous example, Menu Help is composed of topics that are organized by books. Context Help is nothing more than individual topics that aren't organized by book, but which can be accessed by the help ID your application provides WinHelp—an ID that we saw how to provide for WinHelp in Example 36. In this example, we will load these IDs into RoboHELP and create new topics one-by-one, referencing those IDs.

Steps

Load Help IDs into RoboHELP

1. First, follow the last example to create the Menu Help for your project.

2. Load any control help IDs into RoboHELP by using the File/Import/Map File... menu commands. These help IDs will be in the resource.hm file in your application's directory, which was created and filled with IDs every time you clicked on "Help ID" in the Dialog Editor for a control.

3. Next, load any menu or dialog help IDs. You must create these IDs first by using the makehm.exe utility that comes with Visual C++, using the following commands.

```
makehm ID_,HID_,0x10000 IDM_,HIDM_,0x10000 resource.h >>"myhelp.hm"
makehm IDP_,HIDP_,0x30000 resource.h >>" myhelp.hm"
makehm IDR_,HIDR_,0x20000 resource.h >>" myhelp.hm"
makehm IDD_,HIDD_,0x20000 resource.h >>" myhelp.hm"
makehm IDW_,HIDW_,0x50000 resource.h >>" myhelp.hm"
```

You then load the myhelp.hm file, as before.

4. The menu help IDs created previously only included the new commands you added to the standard menu. You must also load the help IDs for the

standard menu commands (e.g., File/New, File/Open, etc.), which can be found in the `Afxhelp.hm` file in your MFC's `include\` directory.

Create Context Help Topics

1. Now, use the File/New/Topic menu commands or the associated toolbar button to open the "New Topic" dialog box.

2. Assign a help ID to this topic by adding it to the "Help ID" edit box in this dialog box. The help ID is nothing more than the command or control ID with an `H` prefix. For example, the help ID for `ID_FILE_NEW` is `HID_FILE_NEW`. Rather than type in each ID, you can also click on the browse button to the left of this edit box, which will allow you to select from all of the help IDs you loaded in the last section.

3. Upon clicking "OK", this new topic will be added to your help project. You can then use Microsoft Word to add an entry for it.

4. Repeat this process for each context help ID you want to support. You don't have to support all of them, particularly all of the IDs added by `Afxhelp.hm`.

5. Don't add this topic to your table of contents, as you did with your menu help topics. You can, however, use jumps in this topic to transfer your user to the other topics in your help system. Please refer to Appendix C for the types of jumps you can add using RoboHELP.

Create and Test the Help Files

1. Compile your help project as before using the File/Compile menu commands. However, testing Context Sensitive Help has to be done with the real application.

Notes

- Another way to think of Context Sensitive Help is as a simple way for you to automatically refer your user to a particular page in your on-line help. If you don't need all that horse power to help your user, you might instead consider using the Bubble Help created in Example 37, which displays a short help message when ever the user allows the mouse cursor to rest over a control or an area of a window.

C

Index

M

What's on the Downloadable Files?

VC++ MFC Extensions by Example is accompanied by downloadable files, which includes all of the examples described in the book. Each example is contained in a Visual C++ v5.0 project subdirectory with a name that corresponds to the example in the book. Each project subdirectory also includes a Wizard subdirectory that contains just the source necessary to add that example to your application. Also included is the author's SampleWizard™, a utility that uses these Wizard subdirectories to add the source from an example directly to your application. (The project for the SampleWizardTM utility is included on the CD.) You can even add your own examples to the SampleWizard's collection.

Using SampleWizard™

- Pick the section, chapter, and example in the book that contains the type of example you're looking for.

- Determine how the example source will be inserted into your application.

- View the example and insert by one of the following four methods: (a) simply read the file, (b) copy and paste sections of the file into your source, (c) copy the entire displayed file tothe clipboard, or (d) copy any of the example files to your application's target directory.

    For additional information on using SampleWizard™, see the ReadMe Downloadable file.

The files that were originally on a CD can be downloaded from: ftp://ftp.cmpbooks.com/pub/VC_MFC_Extensions.zip. References to CD files apply also to the downloadable files.

T - #0019 - 311024 - C0 - 235/191/37 [39] - CB - 9781138412408 - Gloss Lamination